Lattice-Based Cryptosystems

Jiang Zhang · Zhenfeng Zhang

Lattice-Based Cryptosystems

A Design Perspective

Springer

Jiang Zhang
State Key Laboratory of Cryptology
Beijing, China

Zhenfeng Zhang
Institute of Software
Chinese Academy of Sciences
Beijing, China

ISBN 978-981-15-8429-9 ISBN 978-981-15-8427-5 (eBook)
https://doi.org/10.1007/978-981-15-8427-5

This Springer imprint is published by the registered company Springer Nature Singapore Pte Ltd.
The registered company address is: 152 Beach Road, #21-01/04 Gateway East, Singapore 189721, Singapore

Preface

As we enter into the post-quantum era, traditional public-key cryptosystems face great challenges and will fail to protect our private information over the Internet. Due to several potential benefits, e.g., worst-case hardness assumptions and security against quantum computers, cryptosystem based on the hardness of lattice problems is a very promising candidate for providing security against quantum adversaries, and a lot of effort has been devoted to constructing lattice-based cryptosystems. In the past years, we were asked many times for providing a list of books on lattice-based cryptography for graduated students and young researchers, but to the best of our knowledge, there is actual no book on the design of lattice-based cryptosystems. Thus, we think it is very necessary to write a book on lattice-based cryptography such that it is easier for more people to access and enter into this research area. The proposed book mainly focuses on the construction of lattice-based cryptosystems, which can serve as a textbook for graduate-level courses. The book will first cover some basic mathematics on lattices, and then show how to use those math tools in constructing lattice-based cryptosystems, including public-key encryption, identity-based encryption, attribute-based encryption, key exchange and digital signatures.

Beijing, China
July 2020

Jiang Zhang
Zhenfeng Zhang

Preface

Acknowledgements

We would like to thank Yu Chen, Jintai Ding, Özgür Dagdelen, Shuqin Fan, Aijun Ge, Phong Q. Nguyen, Michael Snook and Yu Yu for contributing to our coauthored papers that provide the main material of this book. We would also like to thank Yanfei Guo, Shuai Han, Eike Kiltz, Shengli Liu, Wenhao Wang, Xiang Xie, Kang Yang, Rui Zhang, Xusheng Zhang and the anonymous reviewers for helpful discussions and suggestions on the preliminary version of those papers.

Jiang Zhang is supported in part by the National Key Research and Development Program of China (Grant Nos. 2018YFB0804105, 2017YFB0802005), the National Natural Science Foundation of China (Grant Nos. 62022018, 61602046), and the Young Elite Scientists Sponsorship Program by CAST (Grant No. 2016QNRC001). Zhenfeng Zhang is supported in part by the National Key Research and Development Program of China (Grant No. 2017YFB0802005) and the National Natural Science Foundation of China (Grant No. U1536205).

Contents

Acronyms

ABE	Attribute-based Encryption
AKE	Authenticated-Key Exchange
CVP	Closest Vector Problem
FE	Functional Encryption
GS	Group Signature
IBE	Identity-based Encryption
KE	Key Exchange
LWE	Learning with Errors
PKE	Public Key Encryption
SVP	Shortest Vector Problem
SIS	Short Integer Solution
SIVP	Shortest Independent Vector Problem

Chapter 1
Introduction

Abstract This book mainly focuses on constructing provably secure cryptosystems from lattices. In this chapter, we first give an introduction to cryptography and provable security. Then, we give a brief overview on post-quantum cryptography. Finally, we provide some milestones on lattice-based cryptography, and outline the contents of this book.

1.1 Cryptography and Provable Security

In history, cryptography was only referred as an art to encrypt and decrypt messages. This conception radically changed after an explosion of research in cryptography since the late 1980s. Nowadays, cryptography has been widely accepted as a science to study the techniques for securing digital information, transactions and distributed computations [16]. One of the main features of modern cryptography is provable security which originated in the seminal work of Goldwasser and Micali [14] in 1984. After decades of development, provable security is not only considered as one of the most productive and valuable contributions by theorists in cryptography [7], but also viewed as an attribute of a proposed cryptosystem by standards bodies and implementers [2]. One import thing to mention is that provable security does not actually prove security in an absolute sense, and is just a relative notion depending on the hardness assumption, the security model and the security reduction:

- *Hardness assumption*: typically consisting of mathematical problems in number theory (e.g., discrete logarithm problems) or some basic cryptographic primitives (e.g., pseudorandom functions).
- *Security model*: consisting of *attack model* and *security goal*:
  - *Attack model*: specifying the knowledge that the adversary has the ability of the adversary to interact with honest parties, and the restrictions that are put on the adversary. For instance, when considering the chosen-ciphertext security for public-key encryption, the adversary has the challenge public key and can access a decryption oracle to decrypt any ciphertext of his choice (modeling that he can interact with the honest user who can decrypt the ciphertexts under the

J. Zhang and Z. Zhang, *Lattice-Based Cryptosystems*,
https://doi.org/10.1007/978-981-15-8427-5_1

challenge public key), but cannot send the challenge ciphertext to the decryption oracle (after seeing the challenge ciphertext).
  - *Security goal*: specifying the security features that have to be satisfied by a cryptosystem, which is usually abstracted from the application and functionality of the cryptosystem. For example, when considering public-key encryption, the standard security goal is ciphertext indistinguishability, which says that it should be computationally infeasible for an adversary knowing the public key to distinguish the encryptions of two different messages (with equal length).

- *Security Reduction*: specifying how to transform an adversary that satisfies the attack model and breaks the security goal of a cryptoystem to an algorithm for breaking the underlying hardness assumptions.

A cryptosystem is said to be provable secure only if precise descriptions of the hardness assumption, the security model, and the security reduction is given. Clearly, the real meaning of provable security for a particular cryptosystem depends on the complexity of breaking the underlying hardness assumption, the accuracy of modeling the abilities of the real world adversary and the tightness/efficiency of the security reduction. Since the researchers may find better algorithm to solve previously known hard problems and the adversary may employ advanced technology to obtain extra information, the implication of provable security may also change as the development of the information and computation techniques.

1.2 Post-quantum Cryptography

In 1997, Shor [30] proposed the first quantum polynomial time algorithms for solving the problems of factorizations and discrete logarithms, which are known to be hard for classical algorithms and have been used to construct many widely used cryptosystems such as the OAEP encryption [3] and the DSA signature [9]. As the rapid development of quantum technology in recent years, classical cryptography based on traditional number theory problems faces great challenges and would be susceptible to large-scale quantum computers in the near future. For this reason, the governments and organizations around the world have launched different projects one after the other to develop new cryptographic technologies for the quantum era. In August 2015, the American National Security Agency (NSA) released a major policy statement on the transitioning from traditional cryptographic algorithms to quantum resistant algorithms [24]. Later, the American National Institute of Standards and Technology (NIST) initiated a global process to solicit and standardize one or more quantum-resistant public-key cryptography algorithms [23]. In May 2018, the China Association for Science and Technology (CAST) has released a report on the 60 major science and technology problems in twelve research fields, which considers the design of quantum-resistant cryptographic algorithms as one of the six major problems in the field of information technology.

The goal of post-quantum cryptography is to develop cryptosystems which are secure against adversaries armed with large-scale quantum and classical computers and can interoperate with existing cryptographic applications. Since the best known quantum search algorithm—Grover's algorithm—only providing a quadratic speed-up in comparison with classical search algorithms, and an exponential speed-up for search problems is impossible [4], the impact of quantum algorithms on symmetric-key encryption will not be drastic. In particular, doubling the key size will be sufficient for symmetric-key algorithms to preserve security [22]. For this reason, the focus of post-quantum cryptography is public-key cryptography. Currently, there are mainly four directions to design post-quantum public-key cryptography, i.e., lattice-based cryptography, code-based cryptography, multivariate polynomial cryptography and hash-based signatures. A brief overview of these four directions is given below.

Lattice-based cryptography. Lattice-based cryptography dates back to the seminal work of Ajtai [1] in 1996, who showed how to construct provable secure hash functions based on hard lattice problems. In recent years, lattice-based cryptography has attracted significant interest and has been very successful not only in constructing cryptosystems with typical functionalities such as public-key encryption [28] and digital signatures [12], but also in realizing many powerful cryptosystems such as fully homomorphic encryption [11] and functional encryption [13]. Due to the specific algebraic structures, most lattice-based cryptosystems are relatively simple and highly parallelizable, and some lattice-based encryptions, signatures and key exchanges already have comparable, even better computational efficiency in comparison with existing counterparts based on traditional number theory hard problems. In addition to the above benefits, lattice-based cryptosystems are typically provably secure under worst-case hardness assumptions, i.e., the average-case security of lattice-based cryptosystems are based on the worst-case hardness of certain lattice problems, which is a unique advantage known only for lattice-based cryptography.

Code-based cryptography. The first code-based cryptosystem is McEliece, which was proposed by Robert McEliece in 1978 [18] and has not been broken during the last forty years. The McEliece cryptosystem is quite fast but has much larger key size than that of the RSA encryption [29] and the Elgamal encryption [10] proposed almost at the same period, which prevented it from being widely used in applications. The security of the McEliece cryptosystem is very sensitive to the use of the binary Goppa code, many attempts have been made to reduce the key sizes by replacing the binary Goppa code with other error-correcting codes, but failed in preserving the security of the cryptosystem. Although the security of code-based cryptography is related to the fact from the complexity theory that syndrome decoding in an arbitrary linear code is difficult [5], most known code-based cryptosystems typically use codes with special algebraic structures that allow efficient syndrome decoding, and the designers mainly focus on finding appropriate tricks (usually without theoretical guarantees) to hide the structures of those codes. Code-based cryptography seems more successful in designing encryption schemes.

Multivariate polynomial cryptography. In 1983, Ong and Schnorr made the first attempt to construct multivariate signature [25]. Although this signature scheme was found insecure [27], it seemed to initiate the study of multivariate polynomial cryptography [6]. The security of multivariate polynomial cryptography is related to the problem of solving nonlinear multivariate equations over finite fields (MQ), which is known to be NP-hard. However, there are no multivariate polynomial cryptosystems whose security is guaranteed by the NP-hardness of the MQ problem. The past few decades have witnessed several multivariate cryptosystems, but most of them have been broken. The reason is that the MQ problems underlying most multivariate cryptosystems can be efficiently solved given some trapdoors, and the designers usually failed to hide those trapdoors in their multivariate cryptographic constructions from the adversary. Multivariate polynomial cryptography seems more successful in designing digital signatures.

Hash-based signatures. Just like the name suggests, this direction only focuses on constructing digital signatures from hash functions. The concept of digital signatures was introduced by Diffie and Hellman [8], and the first realization was due to Rivest, Shamir and Adleman from integer factorization [29]. Digital signatures are the only known public-key cryptosystem whose existence is implied by the primitives such as hash functions in the so-called world of "minicrypt" [15]. The research on designing digital signatures from hash functions began with the seminal work of Lamport [17], who showed how to construct digital signatures with one-time security from any one-way functions. By relying on the Merkle-hash tree, one can construct digital signatures with multiple security solely from hash functions [19]. Hash-based signatures can be made very efficient if one allows the signer to keep a state of previously signed messages. There are also stateless hash-based signatures with worse efficiency. For now, the community still does not know how to construct other public-key cryptosystems beyond signatures solely from hash functions.

1.3 Lattice-Based Cryptography

Lattice-based cryptography began with the seminal work of Ajtai [1], who gave the first collision-resistant hash function on random lattices in 1996. Since then, lattice-based cryptography has gained great success. We illustrate a few milestones in the history of lattice-based cryptography. Following the work of Ajtai [1], Micciancio and Regev [21] formally introduced the small integer solutions (SIS) problem, and proved that it is as hard as some lattice problems such as the shortest vector problem (SVP), and the shortest independent vectors problem (SIVP) in the worst-case by using Gaussian measures on lattices. Since then, Gaussian distribution plays an important role in the research of lattice-based cryptography. In 2005, Regev [28] introduced another important problem called learning with errors (LWE), and proved that solving the LWE problem in the average-case is not easier than solving some lattice problems such as SIVP in the worst-case. The introduction of SIS and LWE greatly promoted the development of lattice-based cryptography, and made the design of lattice-based cryptosystems more accessible to non-experts on lattices. Actually, most lattice-based cryptosystems are directly based on the SIS and/or LWE problems and can be understood by people with little knowledge on lattices. In 2008, Gentry et al. [12] showed how to use lattice trapdoors in constructing signatures and identity-based encryptions, which initiated the study of other advanced cryptosystems such attribute-based encryptions and functional encryptions on lattices.

This book focuses on constructing provably secure lattice-based cryptosystems and will cover several main topics in this area, including the constructions of public-key encryption, identity-based encryption, attribute-based encryption, key change and digital signatures. We will try not to involve too much mathematical background on lattices so that anyone with basic knowledge on linear algebra and cryptography can access the book. We refer the readers to [20] for more information on the complexity of lattice problems, to [26] for an overview of lattice-based cryptography in the past decades, and to [6] for an introduction of post-quantum cryptography.

This book is structured as follows. Chapter 2 recalls some necessary mathematical background on lattices at a level of simply presenting the basic definitions and tools. Chapter 3 mainly shows how to use a new message encoding in constructing an efficient chosen-ciphertext secure PKE from LWE in the standard model. Chapter 4 mainly focuses on how to construct lattice-based IBEs with shorter key sizes by using a primitive called lattice-based programmable hash functions. Chapter 5 presents two constructions of ciphertext-policy attribute-based encryption schemes supporting different access policies from lattices. Chapter 6 mainly focuses on how to construct authenticated key exchange (AKE) from lattices in both the public key infrastructure setting and the password-only setting. Chapter 7 mainly shows how to construct short signature from lattices in the standard model and group signature in the random oracle model.

References

1. Ajtai, M.: Generating hard instances of lattice problems (extended abstract). In: Proceedings of the Twenty-Eighth Annual ACM Symposium on Theory of Computing, STOC '96, pp. 99–108. ACM (1996)
2. Bellare, M.: Practice-oriented provable-security. In: Damgård, I. (ed.) Lectures on Data Security. Lecture Notes in Computer Science, vol. 1561, pp. 1–15. Springer, Heidelberg (1999)
3. Bellare, M., Rogaway, P.: Optimal asymmetric encryption. In: De Santis, A. (ed.) Advances in Cryptology - EUROCRYPT '94. Lecture Notes in Computer Science, vol. 950, pp. 92–111. Springer, Berlin/Heidelberg (1995)
4. Bennett, C.H., Bernstein, E., Brassard, G., Vazirani, U.: Strengths and weaknesses of quantum computing. SIAM J. Comput. **26**(5), 1510–1523 (1997)
5. Berlekamp, E., McEliece, R., van Tilborg, H.: On the inherent intractability of certain coding problems. IEEE Trans. Inf. Theory **24**(3), 384–386 (1978)
6. Bernstein, D.J., Buchmann, J., Dahmen, E.: Post-quantum Cryptography. Springer, Heidelberg (2009)

7. Damgård, I.: A "proof-reading" of some issues in cryptography. In: Arge, L., Cachin, C., Jurdziński, T., Tarlecki, A. (eds.) Automata, Languages and Programming. Lecture Notes in Computer Science, vol. 4596, pp. 2–11. Springer, Heidelberg (2007)
8. Diffie, W., Hellman, M.: New directions in cryptography. IEEE Trans. Inf. Theory **22**(6), 644–654 (1976)
9. Digital Signature Standard (DSS): National institute of standards and technology (nist), fips pub 186-4, http://csrc.nist.gov/publications/fips.html
10. ElGamal, T.: A public key cryptosystem and a signature scheme based on discrete logarithms. In: Blakley, G., Chaum, D. (eds.) Advances in Cryptology. Lecture Notes in Computer Science, vol. 196, pp. 10–18. Springer, Heidelberg (1984)
11. Gentry, C.: Fully homomorphic encryption using ideal lattices. In: Proceedings of the 41st Annual ACM Symposium on Theory of Computing, STOC '09, pp. 169–178. ACM (2009)
12. Gentry, C., Peikert, C., Vaikuntanathan, V.: Trapdoors for hard lattices and new cryptographic constructions. In: Proceedings of the 40th Annual ACM Symposium on Theory of Computing, STOC '08, pp. 197–206. ACM (2008)
13. Goldwasser, S., Kalai, Y., Popa, R.A., Vaikuntanathan, V., Zeldovich, N.: Reusable garbled circuits and succinct functional encryption. In: Proceedings of the Forty-Fifth Annual ACM Symposium on Theory of Computing, STOC '13, pp. 555–564. ACM (2013)
14. Goldwasser, S., Micali, S.: Probabilistic encryption. J. Comput. Syst. Sci. **28**(2), 270–299 (1984)
15. Impagliazzo, R.: A personal view of average-case complexity. Structure in Complexity Theory Conference (1995)
16. Katz, J., Lindell, Y.: Introduction to Modern Cryptography. CRC Press (2007)
17. Lamport, L.: Constructing digital signatures from a onw-way function. Technical Report SRI-CSL-98, SRI Intl. Computer Science Laboratory (1979)
18. McEliece, R.: A public-key cryptosystem based on algebraic coding theory. The Deep Space Network Progress Report, DSN PR 42–44, pp. 114–116 (1978)
19. Merkle, R.C.: A certified digital signature. In: Gilles, B. (ed.) Advances in Cryptology - CRYPTO '89. Lecture Notes in Computer Science, vol. 435, pp. 218–238. Springer, Berlin/Heidelberg (1990)
20. Micciancio, D., Goldwasser, S.: Complexity of Lattice Problems: A Cryptographic Perspective, vol. 671. Springer, Netherlands (2002)
21. Micciancio, D., Regev, O.: Worst-case to average-case reductions based on gaussian measures. SIAM J. Comput. **37**, 267–302 (2007)
22. NIST: Report on Post-Quantum Cryptography. http://csrc.nist.gov/publications/detail/nistir/8105/final.html
23. NIST: Post-quantum cryptography standardization (2016). http://csrc.nist.gov/Projects/Post-Quantum-Cryptography
24. NSA: National security agency. Cryptography today (August 2015). https://www.nsa.gov/ia/programs/suiteb_cryptography/
25. Ong, H., Schnorr, C.P.: Signatures through approximate representations by quadratic forms. In: David, C. (ed.) Advances in Cryptology - CRYPTO '83. Lecture Notes in Computer Science, pp. 117–131. Springer, Heidelberg (1984)
26. Peikert, C.: A decade of lattice-based cryptography. Cryptology ePrint Archive, Report 2015/939 (2015)
27. Pollard, J., Schnorr, C.: An efficient solution of the congruence $x^2 + ky^2 = m(\text{mod } n)$. IEEE Trans. Inf. Theory **33**(5), 702–709 (1987)
28. Regev, O.: On lattices, learning with errors, random linear codes, and cryptography. In: Proceedings of the Thirty-Seventh Annual ACM Symposium on Theory of Computing, STOC '05, pp. 84–93. ACM (2005)
29. Rivest, R.L., Shamir, A., Adleman, L.: A method for obtaining digital signatures and public-key cryptosystems. Commun. ACM **21**(2), 120–126 (1978)
30. Shor, P.: Polynomial-time algorithms for prime factorization and discrete logarithms on a quantum computer. SIAM J. Comput. **26**(5), 1484–1509 (1997)

Chapter 2
Lattices

Abstract Lattice problems have been investigated since 1800s by many giant mathematicians such as Lagrange, Gauss, Hermite and Minkowski, but they became a hot topic in the area of complexity theory only after the remarkable invention of the LLL algorithm in 1982 and were used in constructing cryptosystem by Ajtai in 1996. In this chapter, we will give some background on lattices, including the definitions of lattices and hard problems, discrete Gaussian distributions, Small Integer Solutions, Learning with Errors as well as the notion of lattice trapdoors.

2.1 Definition

2.1.1 Notation

The set of real numbers (resp., integers) is denoted by $\mathbb{R}$ (resp., $\mathbb{Z}$). The function log denotes the natural logarithm. Vectors are in column form and denoted by bold lower case letters (e.g., $\mathbf{x}$). We view a matrix simply as the set of its column vectors and denoted by bold capital letters (e.g., $\mathbf{X}$).

Denote ℓ_2 and ℓ_∞ norm by $\|\cdot\|$ and $\|\cdot\|_\infty$, respectively. Define the norm of a matrix $\mathbf{X}$ as the norm of its longest column (i.e., $\|\mathbf{X}\| = \max_i \|\mathbf{x}_i\|$). If the columns of $\mathbf{X} = \{\mathbf{x}_1, \ldots, \mathbf{x}_k\}$ are linearly independent, let $\widetilde{\mathbf{X}} = \{\widetilde{\mathbf{x}}_1, \ldots, \widetilde{\mathbf{x}}_k\}$ denote the Gram-Schmidt orthogonalization of vectors $\mathbf{x}_1, \ldots, \mathbf{x}_k$ taken in that order. For $\mathbf{X} \in \mathbb{R}^{n\times m}$ and $\mathbf{Y} \in \mathbb{R}^{n\times m'}$, $[\mathbf{X}\|\mathbf{Y}] \in \mathbb{R}^{n\times(m+m')}$ denotes the concatenation of the columns of $\mathbf{X}$ followed by the columns of $\mathbf{Y}$. And for $\mathbf{X} \in \mathbb{R}^{n\times m}$ and $\mathbf{Y} \in \mathbb{R}^{n'\times m}$, $[\mathbf{X}; \mathbf{Y}] \in \mathbb{R}^{(n+n')\times m}$ is the concatenation of the rows of $\mathbf{X}$ followed by the rows of $\mathbf{Y}$.

The natural security parameter throughout this book is κ, and all other quantities are implicitly functions of κ. Let $\mathsf{poly}(n)$ denote an unspecified function $f(n) = O(n^c)$ for some constant c. We use standard notation O, ω to classify the growth of functions. If $f(n) = O(g(n) \cdot \log^c n)$, we denote $f(n) = \tilde{O}(g(n))$. We say a function $f(n)$ is negligible if for every $c > 0$, there exists a N such that $f(n) < 1/n^c$ for all $n > N$. We use $\mathsf{negl}(n)$ to denote a negligible function of n, and

J. Zhang and Z. Zhang, *Lattice-Based Cryptosystems*,
https://doi.org/10.1007/978-981-15-8427-5_2

we say a probability is overwhelming if it is $1 - \mathsf{negl}(n)$. For any element $0 \le v \le q$, we denote $\mathsf{BitDecomp}_q(v) \in \{0, 1\}^k$ as the k-dimensional bit-decomposition of v, where $k = \lceil \log_2 q \rceil$. The distance of two discrete random variables X and Y over a (countable) set A as

$$\Delta(X, Y) := \frac{1}{2} \sum_{a \in A} |\Pr[X = a] - \Pr[Y = a]| .$$

2.1.2 Lattices

Let $\mathbb{R}^n$ be the n-dimensional Euclidean space. A lattice in $\mathbb{R}^n$ is the set

$$\mathcal{L}(\mathbf{b}_i, \ldots, \mathbf{b}_m) = \left\{ \sum_{i=1}^{m} x_i \mathbf{b}_i : x_i \in \mathbb{Z} \right\}$$

of all integral combinations of m linearly independent vectors $\mathbf{b}_1, \ldots, \mathbf{b}_m \in \mathbb{R}^n$. The integers m and n are called the rank and dimension of the lattice, respectively. The sequence of vectors $\mathbf{b}_1, \ldots, \mathbf{b}_m$ is called a lattice basis and it is conveniently represented as a matrix

$$\mathbf{B} = [\mathbf{b}_1, \ldots, \mathbf{b}_m] \in \mathbb{R}^{n \times m}.$$

The dual lattice of Λ, denoted Λ^*, is defined to be

$$\Lambda^* = \left\{ \mathbf{x} \in \mathbb{R}^n : \forall \, \mathbf{v} \in \Lambda, \; \langle \mathbf{x}, \mathbf{v} \rangle \in \mathbb{Z} \right\}.$$

The fundamental parallelepiped $\mathcal{P}(\mathbf{B})$ of $\Lambda = \mathcal{L}(\mathbf{B})$ is defined as

$$\mathcal{P}(\mathbf{B}) = \left\{ \mathbf{B}\mathbf{x} \in \mathbb{R}^n : \mathbf{x} = (x_1, \ldots, x_n)^T \in \mathbb{R}^m, \; 0 \le x_i < 1 \right\}.$$

The determinant (or volume) of a lattice Λ is defined as the volume of $\mathcal{P}(\mathbf{B})$, which is also equal to $\sqrt{\det(\mathbf{B}^T\mathbf{B})}$. Let $\mathcal{B}_m(\mathbf{0}, r) = \{\mathbf{x} \in \mathbb{R}^m : \|\mathbf{x}\| < r\}$ be the m-dimensional open ball of radius r centered at $\mathbf{0}$. For any m-dimensional lattice Λ, the i-th minimum $\lambda_i(\Lambda)$ is the shortest radius r such that $\mathcal{B}_m(\mathbf{0}, r)$ contains i linearly independent lattice vectors. Formally,

$$\lambda_i(\Lambda) = \inf\{r : \dim(\mathsf{span}(\Lambda \cap \mathcal{B}_m(\mathbf{0}, r))) \ge i\}.$$

For any rank n lattice Λ, $\lambda_1(\Lambda), \ldots, \lambda_n(\Lambda)$ are constants, and $\lambda_1(\Lambda)$ is the length of the shortest vector in Λ.

We now recall several major lattice problems.

Definition 2.1 (*Approximate Shortest Vector Problem*, SVP_γ) Given a basis $\mathbf{B}$ of an n-dimensional lattice $\Lambda = \mathcal{L}(\mathbf{B})$, the goal of the SVP_γ problem is to find a lattice vector $\mathbf{v} \in \Lambda$, such that $\|\mathbf{v}\| \leq \gamma(n) \cdot \lambda_1(\Lambda)$, where the approximation factor $\gamma = \gamma(n)$ is a function of the dimension n.

Definition 2.2 (*Approximate Shortest Independent Vector Problem,* SIVP_γ) Given a basis $\mathbf{B}$ of an n-dimensional lattice $\Lambda = \mathcal{L}(\mathbf{B})$, the goal of the SIVP_γ problem is to find a set of n linearly independent lattice vectors $\mathbf{V} = \{\mathbf{v}_1, \ldots, \mathbf{v}_n\} \subset \Lambda$, such that $\|\mathbf{V}\| \leq \gamma(n) \cdot \lambda_n(\Lambda)$, where the approximation factor $\gamma = \gamma(n)$ is a function of the dimension n.

Definition 2.3 (*Approximate Closest Vector Problem,* CVP_γ) Given a basis $\mathbf{B}$ of an n-dimensional lattice $\Lambda = \mathcal{L}(\mathbf{B})$ and a point $\mathbf{t}$ in the space spanned by $\mathbf{B}$. The goal of the CVP_γ problem is to find a set of n linearly independent lattice vectors $\mathbf{S} = \{\mathbf{s}_1, \ldots, \mathbf{s}_n\} \subset \Lambda$, such that $\|\mathbf{S}\| \leq \gamma(n) \cdot \mathsf{dist}(\mathbf{t}, \Lambda)$, where $\mathsf{dist}(\mathbf{t}, \Lambda)$ denotes the distance between $\mathbf{t}$ and the lattice vector closest to $\mathbf{t}$ (i.e., $\mathsf{dist}(\mathbf{t}, \Lambda) = \min_{\mathbf{v} \in \Lambda} \|\mathbf{v} - \mathbf{t}\|$), and the approximation factor $\gamma = \gamma(n)$ is a function of the dimension n.

For $\gamma(n) \leq \sqrt{2}$, the above problems are provably NP-hard (note that the SVP_γ is only provably NP-hard under randomized reductions) [20]. As the approximate factor $\gamma(n)$ goes larger, these problems become easier. In particular, if $\gamma(n) = 2^{n \log \log n / \log n}$, the above problems can be solved in polynomial time by using the LLL algorithm [16]. For now, almost all lattice-based cryptosystems can only be proven secure under lattice problems with approximate factors $\gamma(n) \geq \mathsf{poly}(n)$, which are unlikely to be NP-hard (unless NP=P). Fortunately, the currently best known algorithms, even for quantum algorithms, still require exponential time (and space) to solve the above lattice problems with polynomial approximate factors.

2.1.3 q-Ary Lattices

Lattice-based cryptosystems usually consider q-ary lattices defined by a matrix over $\mathbb{Z}_q$. Formally, let $\mathbf{A} \in \mathbb{Z}_q^{n \times m}$ be a full-rank matrix for some positive integers n, m, q, we consider two subsets of $\mathbb{Z}^m$ defined by $\mathbf{A}$:

$$\Lambda_q^{\perp}(\mathbf{A}) = \left\{ \mathbf{e} \in \mathbb{Z}^m \ \ s.t.\ \mathbf{A}\mathbf{e} = \mathbf{0} \ (\mathrm{mod}\ q\ ) \right\}.$$

$$\Lambda_q(\mathbf{A}) = \left\{ \mathbf{y} \in \mathbb{Z}^m \ \ s.t.\ \exists \mathbf{s} \in \mathbb{Z}^n,\ \mathbf{A}^{\mathbf{T}}\mathbf{s} = \mathbf{y} \ (\mathrm{mod}\ q\ ) \right\}.$$

By definition, we have $\Lambda_q^{\perp}(\mathbf{A}) = \Lambda_q^{\perp}(\mathbf{CA})$ for any invertible $\mathbf{C} \in \mathbb{Z}_q^{n \times n}$. One can easily check that the two sets defined above are lattices with full-rank m, i.e., they are discrete subgroups of $\mathbb{Z}^m$. Moreover, the two lattices are dual to each other when properly scaled, as $\Lambda_q^{\perp}(\mathbf{A}) = q\Lambda_q(\mathbf{A})^*$ and $\Lambda_q(\mathbf{A}) = q\Lambda_q^{\perp}(\mathbf{A})^*$.

For any fixed $\mathbf{u}$, we can define a coset of $\Lambda_q^{\perp}(\mathbf{A})$ as

$$\Lambda_q^{\mathbf{u}}(\mathbf{A}) = \Big\{ \mathbf{e} \in \mathbb{Z}^m \;\; s.t. \; \mathbf{Ae} = \mathbf{u} \; (\text{mod } q \,) \Big\}.$$

Let $\mathsf{dist}(\mathbf{z}, \Lambda_q(\mathbf{A}))$ be the distance of the vector $\mathbf{z}$ from the lattice $\Lambda_q(\mathbf{A})$. For any $\mathbf{A} \in \mathbb{Z}_q^{n\times m}$, define $Y_{\mathbf{A}} = \{\tilde{\mathbf{y}} \in \mathbb{Z}_q^m : \forall a \in \mathbb{Z}_q \backslash \{0\}, \mathsf{dist}(a\tilde{\mathbf{y}}, \Lambda_q(\mathbf{A})) \geq \sqrt{q}/4\}$. The following lemma is implicit in [14]:

Lemma 2.1 ([14]) *Let κ be the security parameter. Let integers n_1, n_2, m and prime q satisfy $m \geq (n_1 + n_2 + 1)\log q$ and $n_1 = 2(n_2 + 1) + \omega(\log \kappa)$. Then, for all but a negligible fraction of $\mathbf{B} \in \mathbb{Z}_q^{m\times n_1}$, the probability that there exist numbers $a, a' \in \mathbb{Z}_q\backslash\{0\}$, vectors $\mathbf{w} \neq \mathbf{w}' \in \mathbb{Z}_q^{n_2}$, and a vector $\mathbf{c} \in \mathbb{Z}_q^m$, s.t.*

$$\mathsf{dist}(a\mathbf{y}, \Lambda_q(\mathbf{B}^t)) \leq \sqrt{q}/4 \; and \; \mathsf{dist}(a'\mathbf{y}', \Lambda_q(\mathbf{B}^t)) \leq \sqrt{q}/4$$

is negligible in κ over the uniformly random choice of $\mathbf{U} \xleftarrow{\$} \mathbb{Z}_q^{m\times(n_2+1)}$, where $\mathbf{y} = \mathbf{c} - \mathbf{U}\begin{pmatrix} 1 \\ \mathbf{w} \end{pmatrix}$ and $\mathbf{y}' = \mathbf{c} - \mathbf{U}\begin{pmatrix} 1 \\ \mathbf{w}' \end{pmatrix}$.

The following lemma is an application of the generalized leftover hash lemma [10]:

Lemma 2.2 ([1]) *Let q be a prime, $m > (n + 1)\log q + \omega(\log n)$ and $k = poly(n)$. Let the matrices $\mathbf{A}, \mathbf{B}, \mathbf{R}$ be uniformly and randomly chosen from $\mathbb{Z}_q^{n\times m}$, $\mathbb{Z}_q^{n\times k}$, and $\{-1, 1\}^{m\times k}$, respectively. Then for all vectors $\mathbf{w} \in \mathbb{Z}_q^m$, the distribution of $(\mathbf{A}, \mathbf{AR}, \mathbf{R}^T\mathbf{w})$ is statistically close to the distribution of $(\mathbf{A}, \mathbf{B}, \mathbf{R}^T\mathbf{w})$.*

2.1.4 Ideal Lattices

Ideals lattices are a special class of lattices with additional algebraic structures. Formally, let K be finite field of degree n, and let R be a ring of integers whose additive group is isomorphic to $\mathbb{Z}^n$, i.e., R is a set of all $\mathbb{Z}$-linear combinations of some basis $\{b_1, \ldots, b_n\} \subset R$. An (integral) ideal $\mathcal{I}$ of R is a non-trivial (i.e., $\mathcal{I} \neq \emptyset$ and $\mathcal{I} \neq \{0\}$) additive subgroup that is closed under multiplication by R, i.e., for any $r \in R$ and $x \in \mathcal{I}$ we have $r \cdot x \in \mathcal{I}$. A fraction ideal $\mathcal{I} \in R$ is a subset of K such that $d\mathcal{I} \subseteq R$ is an integral ideal for some $d \in R$. A useful fact is that a (fractional) ideal $\mathcal{I}$ can also be generated as the set of all $\mathbb{Z}$-linear combinations of some basis $\{u_1, \ldots, u_n\} \in R$. Let σ be an additive isomorphism mapping R to some lattice $\sigma(R) \in \mathbb{R}^n$. The family of ideal lattices for the ring R and embedding σ is the set of all lattices $\sigma(\mathcal{I})$ for ideals $\mathcal{I}$ in R. Since a fractional ideal can be generated by a $\mathbb{Z}$-basis $\{u_1, \ldots, u_n\} \in R$, its corresponding ideal lattice has a basis $\{\sigma(u_1), \ldots, \sigma(u_n)\} \subset \mathbb{R}^n$. As a concrete example, one can imagine that $K = \mathbb{R}[X]/(X^n + 1)$ and $R = \mathbb{Z}[X]/(X^n + 1)$ for some $n = 2^k$, and σ is the coefficient embedding which maps a polynomial in R into its coefficient vector in $\mathbb{Z}^n$.

By the above definitions, given an ideal $\mathcal{I}$ of R one can define lattice problems such as Ideal-SVP by treating its ideal lattice as a normal lattice. We note that the family of ideal lattices for a ring can be defined by using different embeddings. Since any two embeddings are related to each other simply by a fixed linear transformation on $\mathbb{R}^n$ by definition, the lattice problems are actually equivalent under any two embeddings up to factor introduced by the corresponding linear transformation [19]. In addition to the coefficient embedding, the canonical embedding derived by the ring embeddings is also typical used in lattice-based cryptography [19].

2.2 Discrete Gaussians

For any real $s > 0$ and vector $\mathbf{c} \in \mathbb{R}^n$, define the Gaussian function on $\Lambda \subset \mathbb{Z}^n$ centered at $\mathbf{c}$ with parameter s:

$$\forall \mathbf{x} \in \Lambda,\ \rho_{s,\mathbf{c}}(\mathbf{x}) = \exp\Big(-\pi \frac{\|\mathbf{x}-\mathbf{c}\|^2}{s^2}\Big).$$

Let $\rho_{s,\mathbf{c}}(\Lambda) = \sum_{\mathbf{x}\in\Lambda} \rho_{s,\mathbf{c}}(\mathbf{x})$. Define the discrete Gaussian distribution over Λ with center $\mathbf{c}$, and parameter s as

$$\forall \mathbf{y} \in \Lambda,\ D_{\Lambda,s,c}(\mathbf{y}) = \frac{\rho_{s,\mathbf{c}}(\mathbf{y})}{\rho_{s,\mathbf{c}}(\Lambda)}.$$

The subscripts s and $\mathbf{c}$ are taken to be 1 and $\mathbf{0}$ (respectively) when omitted.

Lemma 2.3 ([6, 17]) *For any real $s, t > 0$, $c \geq 1$, $C = c \cdot \exp(\frac{1-c^2}{2}) < 1$, integer $m > 0$, and any $\mathbf{y} \in \mathbb{R}^m$ we have the followings hold:*

- $\Pr_{\mathbf{x} \stackrel{\$}{\leftarrow} D_{\mathbb{Z}^m,s}}[\|\mathbf{x}\|_\infty > t \cdot s] \leq 2e^{-\pi t^2}$
- $\Pr_{\mathbf{x} \stackrel{\$}{\leftarrow} D_{\mathbb{Z}^m,s}}[\|\mathbf{x}\| > c \cdot \frac{1}{\sqrt{2\pi}} \cdot s\sqrt{m}] \leq C^m$
- $\Pr_{\mathbf{x} \stackrel{\$}{\leftarrow} D_{\mathbb{Z}^m,s}}[|\langle \mathbf{x}, \mathbf{y}\rangle| > t \cdot s\|\mathbf{y}\|] \leq 2e^{-\pi t^2}$

Lemma 2.4 ([21]) *Let integer $n > 0$, and q a power of some prime $p \geq 2$. Let integer $m \geq n \log_2 q + \omega(\log n)$. Then, for any $\ell = \mathsf{poly}(n)$ and real $r \geq \omega(\sqrt{\log n})$, the distribution $(\mathbf{A}, \mathbf{AR})$ is statistically close to uniform over $\mathbb{Z}_q^{n\times m} \times \mathbb{Z}_q^{n\times\ell}$, where $\mathbf{A} \stackrel{\$}{\leftarrow} \mathbb{Z}_q^{n\times m}$ and $\mathbf{R} \stackrel{\$}{\leftarrow} (D_{\mathbb{Z}^m,r})^\ell$.*

The following lemma is implicit in the proof of [21, Thm. 6.3], which can be proven by combining [24, Thm. 3.1] and [25, Corollary 3.10].

Lemma 2.5 ([21]) *Let $r \geq \omega(\sqrt{\log n})$. For any vectors $\mathbf{v} \in \mathbb{Z}^m$, $\mathbf{c} \in \mathbb{R}^m$, randomly choose $\mathbf{r} \stackrel{\$}{\leftarrow} D_{\mathbb{Z}^m,r,\mathbf{c}}$ and $e \stackrel{\$}{\leftarrow} D_{\mathbb{Z},\alpha q r\cdot\sqrt{m}}$, then the distribution $\mathbf{r}^T\mathbf{v} + e$ is statistically close to $D_{\mathbb{Z},s}$, where $s = r \cdot \sqrt{\|\mathbf{v}\|^2 + m(\alpha q)^2}$.*

Following [11, 21], we say that a random variable X over $\mathbb{R}$ is subgaussian with parameter s if for all $t \in \mathbb{R}$, the (scaled) moment-generating function satisfies

$\mathbb{E}(\exp(2\pi t X)) \le \exp(\pi s^2 t^2)$. For any lattice $\Lambda \subset \mathbb{R}^m$ and $s > 0$, $D_{\Lambda,s}$ is subgaussian with parameter s. Besides, any B-bounded symmetric random variable X (i.e., $|X| \le B$) is subgaussian with parameter $B\sqrt{2\pi}$ [21]. For random subgaussian matrix, we have the following result from the non-asymptotic theory of random matrices [26].

Lemma 2.6 *Let $\mathbf{X} \in \mathbb{R}^{n\times m}$ be a random subgaussian matrix with parameter s. There exists a universal constant $C \approx 1/\sqrt{2\pi}$ such that for any $t \ge 0$, we have $s_1(\mathbf{X}) \le C \cdot s \cdot (\sqrt{m} + \sqrt{n} + t)$ except with probability at most $2\exp(-\pi t^2)$.*

We now recall the definition of smoothing parameter introduced by Micciancio and Regev [22].

Definition 2.4 (*Smoothing Parameter* [22]) For any n-dimensional lattice Λ and positive real $\epsilon > 0$, the smoothing parameter $\eta_\epsilon(\Lambda)$ is the smallest real $s > 0$ such that $\rho_{1/s}(\Lambda^* \backslash \{\mathbf{0}\}) \le \epsilon$.

Informally, the smoothing parameter has the following geometric meaning: for any $s \ge \eta_\epsilon(\Lambda)$, adding a randomly and independently choosing vector $\mathbf{e} \xleftarrow{\$} D_{\mathbb{Z}^n,s}$ to each lattice point in $\Lambda = \mathcal{L}(\mathbf{B})$ gives a distribution which is statistically close to the "uniform" distribution over $\mathbb{Z}^n$. Since the uniform distribution over an infinite set is not well-defined, we can state this more formally: the statistical distance between the distribution of $D_{\mathbb{Z}^n,s}$ mod $\mathcal{P}(\mathbf{B})$ and the uniform distribution over $\mathcal{P}(\mathbf{B}) \cap \mathbb{Z}^n$ is at most 2ϵ, i.e., $\Delta(D_{\mathbb{Z}^n,s} \bmod \mathcal{P}(\mathbf{B}), \mathcal{U}(\mathcal{P}(\mathbf{B}) \cap \mathbb{Z}^n)) \le 2\epsilon$. Besides, we have the following lemma from [12]:

Lemma 2.7 ([12]) *Let Λ, Λ' be m-dimensional lattices, with $\Lambda' \subseteq \Lambda$. Then, for any $\epsilon \in (0, 1/2)$, any $s \ge \eta_\epsilon(\Lambda')$, and any $\mathbf{c} \in \mathbb{R}^m$, the distribution of $(D_{\Lambda,s,\mathbf{c}} \bmod \Lambda')$ is within distance at most 2ϵ of uniform over $(\Lambda \bmod \Lambda')$.*

Moreover, we have the following useful facts related to the smoothing parameter and Gaussian distributions from [12, 14, 22].

Lemma 2.8 ([22]) *For any m-dimensional lattice Λ, $\eta_\epsilon(\Lambda) \le \sqrt{m}/\lambda_1(\Lambda^*)$, where $\epsilon = 2^{-m}$, and $\lambda_1(\Lambda^*)$ is the length of the shortest vector in lattice Λ^*.*

Lemma 2.9 *For any positive integer $m \in \mathbb{Z}$, vector $\mathbf{y} \in \mathbb{Z}^m$ and large enough $s \ge \omega(\sqrt{\log m})$, we have that $\Pr_{\mathbf{x} \xleftarrow{\$} D_{\mathbb{Z}^m,s}}[\mathbf{x} = \mathbf{y}] \le 2^{1-m}$.*

Lemma 2.10 ([12]) *Let integers $n, m \in \mathbb{Z}$ and prime q satisfy $m \ge 2n \log q$. Then, for all but an at most $2q^{-n}$ fraction of $\mathbf{A} \in \mathbb{Z}_q^{n\times m}$, we have that (1) the columns of $\mathbf{A}$ generate $\mathbb{Z}_q^n$, (2) $\lambda_1^\infty(\Lambda_q(\mathbf{A})) \ge q/4$, and 3) the smoothing parameter $\eta_\epsilon(\Lambda_q^\perp(\mathbf{A})) \le \omega(\sqrt{\log m})$ for some $\epsilon = \mathsf{negl}(\kappa)$.*

Lemma 2.11 ([12]) *Assume the columns of $\mathbf{A} \in \mathbb{Z}_q^{n\times m}$ generate $\mathbb{Z}_q^n$, and let $\epsilon \in (0, 1/2)$ and $s \ge \eta_\epsilon(\Lambda_q^\perp(\mathbf{A}))$. Then for $\mathbf{e} \sim D_{\mathbb{Z}^m,s}$, the distribution of the syndrome $\mathbf{u} = \mathbf{A}\mathbf{e} \bmod q$ is within statistical distance 2ϵ of uniform over $\mathbb{Z}_q^n$.*

Furthermore, fix $\mathbf{u} \in \mathbb{Z}_q^n$ and let $\mathbf{v} \in \mathbb{Z}^m$ be an arbitrary solution to $\mathbf{A}\mathbf{v} = \mathbf{u} \bmod q$. Then the conditional distribution of $\mathbf{e} \sim D_{\mathbb{Z}^m,s}$ given $\mathbf{A}\mathbf{e} = \mathbf{u} \bmod q$ is exactly $\mathbf{v} + D_{\Lambda_q^\perp(\mathbf{A}),s,-\mathbf{v}}$.

Lemma 2.12 ([12, 14]) *Let integers n, m and prime q satisfy $m \geq 2n \log q$. Let $\gamma \geq \sqrt{q} \cdot \omega(\sqrt{\log n})$. Then, for all but a negligible fraction of $\mathbf{A} \in \mathbb{Z}_q^{n \times m}$, and for any $\mathbf{z} \in Y_{\mathbf{A}} = \{\tilde{\mathbf{y}} \in \mathbb{Z}_q^m : \forall a \in \mathbb{Z}_q \backslash \{0\}, \mathsf{dist}(a\tilde{\mathbf{y}}, \Lambda_q(\mathbf{A})) \geq \sqrt{q}/4\}$, the distribution of $(\mathbf{Ae}, \mathbf{z}^t\mathbf{e})$ is statistically close to uniform over $\mathbb{Z}_q^n \times \mathbb{Z}_q$, where $\mathbf{e} \sim D_{\mathbb{Z}^m,\gamma}$.*

In this following, we show an adaptive version of Lemma 2.12:

Lemma 2.13 *Let positive integers $n, m \in \mathbb{Z}$ and prime q satisfy $m \geq 2n \log q$. Let $\gamma \geq 4\sqrt{mq}$. Then, for all but a negligible fraction of $\mathbf{A} \in \mathbb{Z}_q^{n \times m}$, $Y_{\mathbf{A}} = \{\tilde{\mathbf{y}} \in \mathbb{Z}_q^m : \forall a \in \mathbb{Z}_q \backslash \{0\}, \mathsf{dist}(a\tilde{\mathbf{y}}, \Lambda_q(\mathbf{A})) \geq \sqrt{q}/4\}$, and for any (even unbounded) function $h : \mathbb{Z}_q^n \to Y_{\mathbf{A}}$, the distribution of $(\mathbf{Ae}, \mathbf{z}^t\mathbf{e})$ is statistically close to uniform over $\mathbb{Z}_q^n \times \mathbb{Z}_q$, where $\mathbf{e} \sim D_{\mathbb{Z}^m,\gamma}$ and $\mathbf{z} = h(\mathbf{Ae})$.*

Proof By Lemma 2.10, for all but a negligible fraction of $\mathbf{A} \in \mathbb{Z}_q^{n \times m}$, the columns of $\mathbf{A}$ generate $\mathbb{Z}_q^n$ and the length $\lambda_1(\Lambda_q(\mathbf{A}))$ (in the ℓ_2 norm) of the shortest vector in $\Lambda_q(\mathbf{A})$ is at least $q/4$ (since $\lambda_1(\Lambda_q(\mathbf{A})) \geq \lambda_1^\infty(\Lambda_q(\mathbf{A})) \geq q/4$). Moreover, we have that the smoothing parameter $\eta_\epsilon(\Lambda_q^\perp(\mathbf{A})) \leq \omega(\sqrt{\log m})$ for some negligible ϵ. In the following, we always assume that $\mathbf{A}$ satisfies the above properties. Since $\gamma \geq 4\sqrt{mq} > \eta_\epsilon(\Lambda_q^\perp(\mathbf{A}))$, by Lemma 2.11 the distribution of $\mathbf{Ae}$ mod q is within statistical distance 2ϵ of uniform over $\mathbb{Z}_q^n$, where $\mathbf{e} \sim D_{\mathbb{Z}^m,\gamma}$. Furthermore, fix $\mathbf{u} \in \mathbb{Z}_q^n$ and let $\mathbf{v}$ be an arbitrary solution to $\mathbf{Av} = \mathbf{u}$ mod q, the conditional distribution of $\mathbf{e} \sim D_{\mathbb{Z}^m,\gamma}$ given $\mathbf{Ae} = \mathbf{u} \bmod q$ is exactly $\mathbf{v} + D_{\Lambda_q^\perp(\mathbf{A}),\gamma,-\mathbf{v}}$. Thus, it is enough to show that for arbitrary $\mathbf{v} \in \mathbb{Z}^m$ and $\mathbf{z} = h(\mathbf{Av}) \in Y_{\mathbf{A}}$, the distribution $\mathbf{z}^t\mathbf{e}$ is statistically close to uniform over $\mathbb{Z}_q$, where $\mathbf{e} \sim D_{\Lambda_q^\perp(\mathbf{A}),\gamma,-\mathbf{v}}$.

Now, fix $\mathbf{v} \in \mathbb{Z}^m$ and $\mathbf{z} = h(\mathbf{Av}) \in Y_{\mathbf{A}}$, let $\mathbf{A}' = \begin{pmatrix} \mathbf{A} \\ \mathbf{z}^t \end{pmatrix} \in \mathbb{Z}_q^{(n+1)\times m}$. By the definition $Y_{\mathbf{A}} = \{\tilde{\mathbf{y}} \in \mathbb{Z}_q^m : \forall a \in \mathbb{Z}_q \backslash \{0\}, \mathsf{dist}(a\tilde{\mathbf{y}}, \Lambda_q(\mathbf{A})) \geq \sqrt{q}/4\}$, we have that the rows of $\mathbf{A}'$ are linearly independent over $\mathbb{Z}_q$. In other words, the columns of $\mathbf{A}'$ generate $\mathbb{Z}_q^{n+1}$. Let $\mathbf{x}$ be the shortest vector of $\Lambda_q(\mathbf{A}')$. Note that the lattice $\Lambda_q(\mathbf{A}')$ is obtained by adjoining the vector $\mathbf{z}$ to $\Lambda_q(\mathbf{A})$. Without loss of generality we assume $\mathbf{x} = \mathbf{y} + a\mathbf{z}$ for some $\mathbf{y} \in \Lambda_q(\mathbf{A})$ and $a \in \mathbb{Z}_q$. Then, if $a = 0$, we have $\|\mathbf{x}\| \geq q/4$ by the fact that $\lambda_1(\Lambda_q(\mathbf{A})) \geq q/4$. Otherwise, for any $a \in \mathbb{Z}_q \backslash \{0\}$, we have $\|\mathbf{x}\| \geq \mathsf{dist}(a\mathbf{z}, \Lambda_q(\mathbf{A})) \geq \sqrt{q}/4$. In all, we have that $\lambda_1(\Lambda_q(\mathbf{A}')) = \|\mathbf{x}\| \geq \sqrt{q}/4$. By Lemma 2.8 and the duality $\Lambda_q(\mathbf{A}') = q \cdot (\Lambda_q^\perp(\mathbf{A}'))^*$, we have $\eta_\epsilon(\Lambda_q^\perp(\mathbf{A}')) \leq 4\sqrt{mq} \leq \gamma$ for $\epsilon = 2^{-m}$.[1]

Since the columns of $\mathbf{A}' \in \mathbb{Z}_q^{(n+1)\times m}$ generate $\mathbb{Z}_q^{n+1}$, we have the set of syndromes $\{u = \mathbf{z}^t\mathbf{e} : \mathbf{e} \in \Lambda_q^\perp(\mathbf{A})\} = \mathbb{Z}_q$. By the fact $\Lambda_q^\perp(\mathbf{A}') = \Lambda_q^\perp(\mathbf{A}) \cap \Lambda_q^\perp(\mathbf{z}^t)$, the quotient group $(\Lambda_q^\perp(\mathbf{A})/\Lambda_q^\perp(\mathbf{A}'))$ is isomorphic to the set of syndromes $\mathbb{Z}_q$ via the mapping $\mathbf{e} + \Lambda_q^\perp(\mathbf{A}') \mapsto \mathbf{z}^t\mathbf{e}$ mod q. This means that computing $\mathbf{z}^t\mathbf{e}$ mod q for some $\mathbf{e} \in \Lambda_q^\perp(\mathbf{A})$ is equivalent to reducing $\mathbf{e}$ modulo the lattice $\Lambda_q^\perp(\mathbf{A}')$. By Lemma 2.7, for any $\epsilon = \mathsf{negl}(n)$, any $\gamma \geq \eta_\epsilon(\Lambda_q^\perp(\mathbf{A}'))$ and any $\mathbf{v} \in \mathbb{Z}^m$, the distribution of $D_{\Lambda_q^\perp(\mathbf{A}),\gamma,-\mathbf{v}}$ mod $\Lambda_q^\perp(\mathbf{A}')$ is within statistical distance at most 2ϵ of uniform over

[1] It is possible to set a smaller γ by a more careful analysis with $\epsilon = \mathsf{negl}(n)$.

$(\Lambda_q^{\perp}(\mathbf{A})/\Lambda_q^{\perp}(\mathbf{A}'))$. Thus, the distribution $\mathbf{z}^t\mathbf{e}$ is statistically close to uniform over $\mathbb{Z}_q$, where $\mathbf{e} \sim D_{\Lambda_q^{\perp}(\mathbf{A}),\gamma,-\mathbf{v}}$. This completes the proof. □

Finally, we recall rejection sampling technique in Theorem 2.1 from [18].

Theorem 2.1 (Rejection Sampling [18]) *Let V be a subset of $\mathbb{Z}^m$ in which all the elements have norms less than T, $\alpha = \omega(T\sqrt{\log m})$ be a real, and $\psi : V \to \mathbb{R}$ be a probability distribution. Then there exists a constant $M = O(1)$ such that the distribution of the following algorithm* Samp_1 :

1: $\mathbf{c} \xleftarrow{\$} \psi$.
2: $\mathbf{z} \xleftarrow{\$} D_{\mathbb{Z}^m,\alpha,\mathbf{c}}$.
3: output $(\mathbf{z}, \mathbf{c})$ *with probability* $\min\left(\frac{D_{\mathbb{Z}^m,\alpha}(\mathbf{z})}{M D_{\mathbb{Z}^m,\alpha,\mathbf{c}}(\mathbf{z})}, 1\right)$.

is within statistical distance $\frac{2^{-\omega(\log m)}}{M}$ *from the distribution of the following algorithm* Samp_2 :

1: $\mathbf{c} \xleftarrow{\$} \psi$.
2: $\mathbf{z} \xleftarrow{\$} D_{\mathbb{Z}^m,\alpha}$.
3: output $(\mathbf{z}, \mathbf{c})$ *with probability* $1/M$.

Moreover, the probability that Samp_1 *outputs something is at least* $\frac{1-2^{-\omega(\log m)}}{M}$. *More concretely, if $\alpha = \tau T$ for any positive τ, then $M = e^{12/\tau+1/(2\tau^2)}$ and the output of algorithm* Samp_1 *is within statistical distance* $\frac{2^{-100}}{M}$ *of the output of* Samp_2*, and the probability that $\mathcal{A}$ outputs something is at least* $\frac{1-2^{-100}}{M}$.

2.3 Small Integer Solutions

The small integer solution (SIS) problem was introduced by Ajtai [2], but its name is due to Micciancio and Regev [22], who improved Ajtai's connection between SIS and worst-case lattice problems.

Definition 2.5 (*Small Integer Solution (in the ℓ_2 norm)*) The SIS problem $\mathrm{SIS}_{n,m,q,\beta}$ with parameters (n, m, q, β) is: Given a uniformly random matrix $\mathbf{A} \in \mathbb{Z}_q^{n\times m}$, find a non-zero vector $\mathbf{e} \in \mathbb{Z}^m$ such that $\mathbf{Ae} = \mathbf{0} \bmod q$ and $\|\mathbf{e}\| \le \beta$.

Definition 2.6 (*Inhomogeneous Small Integer Solution (in the ℓ_2 norm)*) The ISIS problem $\mathrm{ISIS}_{n,m,q,\beta}$ with parameters (n, m, q, β) is: Given a uniformly random matrix $\mathbf{A} \in \mathbb{Z}_q^{n\times m}$ and a random syndrome $\mathbf{u} \in \mathbb{Z}_q^n$, find an integer vector $\mathbf{e} \in \mathbb{Z}^m$ such that $\mathbf{Ae} = \mathbf{u} \bmod q$ and $\|\mathbf{e}\| \le \beta$.

The ISIS problem is an inhomogenous variant of SIS. Both problems were shown to be as hard as certain worst-case lattice problems.

Lemma 2.14 ([12]) *For any positive integers $n, m \in \mathbb{Z}$, real $\beta = \mathsf{poly}(n)$ and prime $q \ge \beta \cdot \omega(\sqrt{n\log n})$, the average-case problems $\mathrm{SIS}_{n,m,q,\beta}$ and $\mathrm{ISIS}_{n,m,q,\beta}$ are as hard as the worst-case problem SIVP_γ with $\gamma = \beta \cdot \tilde{O}(\sqrt{n})$.*

2.4 (Ring)-Learning with Errors

In this section, we introduce the Learning with Errors problem and its variants.

2.4.1 *Learning with Errors*

Let $n \in \mathbb{Z}$ and $q = q(n)$ be positive integers, and let $\alpha > 0$ be a real. Let $D_{\mathbb{Z}^n,\alpha}$ be some discrete Gaussian distribution over $\mathbb{Z}_q$, and let $\mathbf{s} \in \mathbb{Z}_q^n$ be some vector. Define $A_{\mathbf{s},\alpha} \subseteq \mathbb{Z}_q^n \times \mathbb{Z}_q$ as the distribution of the variable $(\mathbf{a}, \mathbf{a}^T\mathbf{s} + x)$, where $\mathbf{a} \xleftarrow{\$} \mathbb{Z}_q^n$, $x \xleftarrow{\$} D_{\mathbb{Z}^n,\alpha}$, and all the operations are performed in $\mathbb{Z}_q$. For m independent samples $(\mathbf{a}_1, y_1), \ldots, (\mathbf{a}_m, y_m)$ from $A_{\mathbf{s},\chi_\alpha}$, we denote it in a matrix form $(\mathbf{A}, \mathbf{y}) \in \mathbb{Z}_q^{n\times m} \times \mathbb{Z}_q^m$, where $\mathbf{A} = (\mathbf{a}_1, \ldots, \mathbf{a}_m)$ and $\mathbf{y} = (y_1, \ldots, y_m)^T$.

Definition 2.7 (*Learning with Errors (the search version)*) The LWE problem $\mathrm{LWE}_{n,m,q,\alpha}$ with parameters (n, m, q, α) is: for randomly chosen $\mathbf{s} \xleftarrow{\$} \mathbb{Z}_q^n$, given m samples from $\mathrm{A}_{\mathbf{s},\alpha}$, outputs $\mathbf{s} \in \mathbb{Z}_q^n$.

The decisional variant of $\mathrm{LWE}_{n,m,q,\alpha}$ is that, for a uniformly chosen $\mathbf{s} \xleftarrow{\$} \mathbb{Z}_q^n$, distinguish $A_{\mathbf{s},\chi_\alpha}$ from the uniform distribution over $\mathbb{Z}_q^n \times \mathbb{Z}_q$ with m samples. For certain modulus q, the average-case decisional LWE problem is polynomially equivalent to its worst-case search version [5, 23, 25].

Lemma 2.15 ([13, 25]) *let $q = q(n)$ be a prime, and let $\alpha > 2\sqrt{2n}$. If there exists an efficient (possibly quantum) algorithm that solves* $\mathrm{LWE}_{n,m,q,\alpha}$ *for any polynomial bounded $m = \mathsf{poly}(n)$, then there exists an efficient quantum algorithm for solving the worst-case* SIVP_γ *problem with $\gamma = \tilde{O}(nq/\alpha)$ in the ℓ_2 norm.*

Due to the hardness of SIVP_γ, the LWE problem with $\alpha = 2^{-n^\epsilon}$ is still believed to be hard for some constant $\epsilon < 1/2$. For appropriate choice of parameters, we also have the following useful lemma:

Lemma 2.16 (Unique Witness) *Let $n, k > 0$ be integers. Let $q = p^k$ for some prime $p \geq 2$, and let $m \geq n \log_2 q + \omega(\log n)$. Then, for all but a negligible fraction of $\mathbf{A} \in \mathbb{Z}_q^{n\times m}$, and for any $\mathbf{u} \in \mathbb{Z}_q^m$, there exists at most one pair $(\mathbf{s}, \mathbf{e}) \in \mathbb{Z}_q^n \times \mathbb{Z}^m$ such that $\|\mathbf{e}\|_\infty < q/8$ and $\mathbf{u} = \mathbf{A}^T\mathbf{s} + \mathbf{e}$.*

Proof The proof is adapted from [12, Lemma 5.3]. For any $\mathbf{u} \in \mathbb{Z}_q^m$, we assume that there exist two tuples $(\mathbf{s}, \mathbf{e}) \neq (\mathbf{s}', \mathbf{e}') \in \mathbb{Z}_q^n \times \mathbb{Z}^m$, such that $\|\mathbf{e}\|_\infty, \|\mathbf{e}'\|_\infty < q/8$ and $\mathbf{u} = \mathbf{A}^T\mathbf{s} + \mathbf{e} = \mathbf{A}^T\mathbf{s}' + \mathbf{e}'$. Letting $\tilde{\mathbf{s}} = \mathbf{s} - \mathbf{s}'$ and $\tilde{\mathbf{e}} = \mathbf{e}' - \mathbf{e}$, we have that $\mathbf{A}^T\tilde{\mathbf{s}} = \tilde{\mathbf{e}}$ for some $\tilde{\mathbf{s}} \neq \mathbf{0}$ and $\|\tilde{\mathbf{e}}\|_\infty < q/4$. Now, it suffices to show that for all but an at most $2^{-\omega(\log n)} = \mathsf{negl}(n)$ fraction of $\mathbf{A} \in \mathbb{Z}_q^{n\times m}$, the vector $\mathbf{A}^T\tilde{\mathbf{s}}$ has norm $\|\mathbf{A}^T\tilde{\mathbf{s}}\|_\infty \geq q/4$ for any $\tilde{\mathbf{s}} \in \mathbb{Z}_q^n \backslash \{\mathbf{0}\}$.

Formally, consider the open ℓ_∞ "cube" $\mathcal{V}$ of radius $q/4$ (i.e., each edge has length $q/2$). Denote $(\mathbb{Z}_q^n)^* \subseteq \mathbb{Z}_q^n$ as the set of vectors such that each vector has at least one

coordinate which is invertible in $\mathbb{Z}_q$. For any fixed non-zero $\tilde{\mathbf{s}} \in \mathbb{Z}_q^n$, we can write $\tilde{\mathbf{s}} = p^{k'}\tilde{\mathbf{s}}'$ for some integer $k' \in \{0, \ldots, k-1\}$ and $\tilde{\mathbf{s}}' \in (\mathbb{Z}_q^n)^*$. Then, for a uniformly random choice of $\mathbf{A} \in \mathbb{Z}_q^{n\times m}$, we have that $\mathbf{A}^T\tilde{\mathbf{s}}'$ is uniformly over $\mathbb{Z}_q^m$, and that $\mathbf{A}^T\tilde{\mathbf{s}} = p^{k'}\mathbf{A}^T\tilde{\mathbf{s}}'$ is uniformly over $\mathbb{Z}_q^m \cap p^{k'}\mathbb{Z}^m$. Denote $S_{k'} = \mathcal{V} \cap p^{k'}\mathbb{Z}^m$, which contains at most $(p^{k-k'}/2)^m$ points. Thus, over the uniformly random choice of $\mathbf{A} \in \mathbb{Z}_q^{n\times m}$, the probability that $\mathbf{A}^T\tilde{\mathbf{s}} \in S_{k'}$ is at most $(p^{k-k'}/2)^m/p^{(k-k')m} \le 2^{-m}$. Taking a union bound over all non-zero $\tilde{\mathbf{s}} \in \mathbb{Z}_q^n$, the probability that $\mathbf{A}^T\tilde{\mathbf{s}} \in S_0$ is at most $2^{-\omega(\log n)}$ (note that $S_{k-1} \subset \cdots \subset S_0 = \mathcal{V} \cap \mathbb{Z}^m$ by definition). Since S_0 contains all integer vectors with ℓ_∞ norm $< q/4$, we have that for all but an at most $2^{-\omega(\log n)}$ fraction of $\mathbf{A} \in \mathbb{Z}_q^{n\times m}$, and for any non-zero $\tilde{\mathbf{s}} \in \mathbb{Z}_q^n$, the vector $\mathbf{A}^T\tilde{\mathbf{s}}$ has norm $\|\mathbf{A}^T\tilde{\mathbf{s}}\|_\infty \ge q/4$.
□

2.4.2 Ring-Learning with Errors

Let the integer n be a power of 2, and consider the ring $R = \mathbb{Z}[x]/(x^n + 1)$. For any positive integer q, we define the ring $R_q = \mathbb{Z}_q[x]/(x^n + 1)$ analogously. For any polynomial $a(x)$ in R (or R_q), we identify a with its coefficient vector in $\mathbb{Z}^n$ (or $\mathbb{Z}_q^n$). Then, we define the norm of a polynomial to be the (Euclidean) norm of its coefficient vector.

Lemma 2.17 *For any $s, t \in R$, we have $\|s \cdot t\| \le \sqrt{n} \cdot \|s\| \cdot \|t\|$ and $\|s \cdot t\|_\infty \le n \cdot \|s\|_\infty \cdot \|t\|_\infty$.*

The discrete Gaussian distribution $D_{R,\alpha}$ over the ring R can be naturally defined as the distribution of ring elements whose coefficient vectors are distributed according to the discrete Gaussian distribution $D_{\mathbb{Z}^n,\alpha}$ for some positive real α. Now we come to the statement of the Ring-LWE problem; we will use a special case detailed in [19]. Let R_q be defined as above, and $s \xleftarrow{\$} R_q$. We define $\tilde{A}_{s,\alpha}$ to be the distribution of the pair $(a, as + x) \in R_q \times R_q$, where $a \xleftarrow{\$} R_q$ is uniformly chosen and $x \xleftarrow{\$} D_{R,\alpha}$ is independent of a.

Definition 2.8 (*Ring-LWE*) The RLWE problem $\mathrm{RLWE}_{n,m,q,\alpha}$ with parameters (n, m, q, α) is: for randomly chosen $s \xleftarrow{\$} R_q$, distinguish $\tilde{A}_{s,\chi_\alpha}$ from the uniform distribution over $R_q \times R_q$ with at most m samples.

The following lemma says that the hardness of the RLWE problem can be reduced to some lattice problems such as the shortest independent vectors problem (SIVP) over ideal lattices.

Lemma 2.18 *Let n be a power of 2, α be a real number in $(0, 1)$, and q be a prime such that $q \bmod 2n = 1$ and $\alpha > \omega(\sqrt{\log n})$. Define $R_q = \mathbb{Z}_q[x]/(x^n + 1)$ as above. Then, there exists a polynomial time quantum reduction from $\mathrm{SIVP}_{\tilde{O}(q\sqrt{n}/\alpha)}$ in the worst-case to average-case $\mathrm{RLWE}_{n,m,q,\beta}$, where $\beta = \sqrt{2}\alpha(nm/\log(nm))^{1/4}$.*

2.4.3 Other Variants

It has been proven that the (ring-)LWE problem $\mathrm{LWE}_{n,m,q,\alpha}$ (or $\mathrm{RLWE}_{n,m,q,\alpha}$) is still hard even if the secret is chosen according to the error distribution $D_{\mathbb{Z}^n,\alpha}$ (or $D_{R,\alpha}$) rather than uniformly [5, 19]. This variant is known as the *normal form* and is preferable for controlling the size of the error term [7, 8].

Besides, the (ring-)LWE is still hard when scaling the error by a constant t relatively prime to q [8], i.e., using the pair $(\mathbf{A}, \mathbf{A}\mathbf{s} + t\mathbf{e}) \in \mathbb{Z}_q^{m\times n} \times \mathbb{Z}_q^m$ rather than $(\mathbf{A}, \mathbf{A}\mathbf{s} + \mathbf{e}) \in \mathbb{Z}_q^{m\times n} \times \mathbb{Z}_q^m$ (or using the pair $(a, as + tx) \in R_q \times R_q$ rather than $(a, as + x)$ in the ring setting). Several lattice-based cryptographic schemes have been constructed based on this variant [7, 8].

2.5 Trapdoor Generation

In 1999, Ajtai [3] showed how to sample an essentially uniform matrix $\mathbf{A}$ together with a short basis of $\Lambda_q^\perp(\mathbf{A})$. This basis generation algorithm has been significantly improved in [4].

Lemma 2.19 ([4]) *For any $\delta_0 > 0$, there is a PPT algorithm* $\mathsf{BasisGen}(1^n, 1^m, q)$ *that, on input a security parameter n, an odd prime $q = poly(n)$, and integer $m \geq (5 + 3\delta_0)n \log q$, outputs a statistically $(mq^{-\delta_0 n/2})$-close to uniform matrix $\mathbf{A} \in \mathbb{Z}_q^{n\times m}$ and a basis $\mathbf{T_A} \subset \Lambda_q^\perp(\mathbf{A})$ such that with overwhelming probability $\|\mathbf{T_A}\| \leq O(n \log q)$ and $\|\widetilde{\mathbf{T}}_\mathbf{A}\| \leq O(\sqrt{n \log q})$. In particular, if let $\delta_0 = \frac{1}{3}$, we can choose $m \geq \lceil 6n \log q \rceil$.*

The following lemma is implied by [9, Lem. 3.2 and Lem. 3.3], which shows that there is an efficient algorithm to extract a random basis for $(\mathbf{A}\|\mathbf{B})$ by using a short basis of $\mathbf{A}$.

Lemma 2.20 ([9]) *There is a PPT algorithm* $\mathsf{ExtRndBasis}(\mathbf{A}', \mathbf{T_A}, s)$ *which takes a matrix $\mathbf{A}' = (\mathbf{A}\|\mathbf{B}) \in \mathbb{Z}_q^{n\times(m+m')}$, a basis $\mathbf{T_A} \in \mathbb{Z}_q^{m\times m}$ of $\Lambda_q^\perp(\mathbf{A})$, and a real $s \geq \|\widetilde{\mathbf{T}}_\mathbf{A}\| \cdot \omega(\sqrt{\log m})$ as inputs, outputs a random basis $\mathbf{T}_{\mathbf{A}'}$ of $\Lambda_q^\perp(\mathbf{A}')$ satisfying $\|\mathbf{T}_{\mathbf{A}'}\| \leq s(m + m')$ and $\|\widetilde{\mathbf{T}}_{\mathbf{A}'}\| \leq s\sqrt{m + m'}$, where $\mathbf{B} \in \mathbb{Z}_q^{n\times m'}$.*

Equipped with the above lemma, and the proof technique of [1, Thm. 4], we obtain the following useful lemma:

Lemma 2.21 *Let $q > 2, m > n$, there is a PPT algorithm* `ExtBasisRight` *$(\mathbf{A}', \mathbf{R}, \mathbf{T}_B, s)$ which takes matrix $\mathbf{A}' = (\mathbf{C}\|\mathbf{A}\|\mathbf{A}\mathbf{R} + \mathbf{B}) \in \mathbb{Z}_q^{n\times(2m+m')}$, a uniformly and randomly chosen $\mathbf{R} \in \{-1, 1\}^{m\times m}$, a basis $\mathbf{T_B}$ of $\Lambda_q^\perp(\mathbf{B})$, and a Gaussian parameter $s > \|\widetilde{\mathbf{T}}_\mathbf{B}\| \cdot \sqrt{m} \cdot \omega(\log m)$, outputs a basis $\mathbf{T}_{\mathbf{A}'}$ of $\Lambda_q^\perp(\mathbf{A}')$ satisfying $\|\mathbf{T}_{\mathbf{A}'}\| \leq s(2m + m')$ and $\|\widetilde{\mathbf{T}}_{\mathbf{A}'}\| \leq s\sqrt{2m + m'}$, where $\mathbf{A}, \mathbf{B} \in \mathbb{Z}_q^{n\times m}$ and $\mathbf{C} \in \mathbb{Z}_q^{n\times m'}$.*

Proof As shown in the proof of [1, Thm. 4], we can use $\mathbf{T_B}$ to efficiently sample a basis $\mathbf{T}_{\hat{\mathbf{A}}}$ for the matrix $\hat{\mathbf{A}} = (\mathbf{A}\|\mathbf{AR} + \mathbf{B})$ satisfying $\|\widetilde{\mathbf{T}}_{\hat{\mathbf{A}}}\| \leq \|\widetilde{\mathbf{T}}_{\mathbf{B}}\| \cdot \sqrt{m} \cdot \omega(\sqrt{\log m})$, then we can apply Lemma 2.20 to obtain a basis $\mathbf{T}_{\mathbf{A}'}$ for $\mathbf{A}' = (\mathbf{C}\|\hat{\mathbf{A}})$ satisfying $\|\widetilde{\mathbf{T}}_{\mathbf{A}'}\| \leq s\sqrt{2m + m'}$. Besides, by the property of the ExtRndBasis algorithm, the claim still holds no matter how the (columns of) matrix $\mathbf{C}$ appears in $\mathbf{A}'$. □

The following `SuperSamp` algorithm allows us to sample a random matrix $\mathbf{B}$ together with a short basis $\mathbf{T_B}$ such that the columns of $\mathbf{B}$ lie in a prescribed affine subspace of $\mathbb{Z}_q^n$.

Lemma 2.22 ([15]) *Let $q > 2, m > \lceil 6n \log q + n \rceil$, there is a PPT algorithm* `SuperSamp`$(\mathbf{A}, \mathbf{C}, q)$ *which takes matrices $\mathbf{A} \in \mathbb{Z}_q^{n\times m}$ and $\mathbf{C} \in \mathbb{Z}_q^{n\times n}$ as inputs, and outputs an almost uniform matrix $\mathbf{B} \in \mathbb{Z}_q^{n\times m}$ such that $\mathbf{AB}^T = \mathbf{C}$, and a basis $\mathbf{T_B}$ of $\Lambda_q^{\perp}(\mathbf{B})$ satisfying $\|\mathbf{T}_B\| \leq m^{1.5} \cdot \omega(\sqrt{\log m})$ and $\|\widetilde{\mathbf{T}}_B\| \leq m \cdot \omega(\sqrt{\log m})$.*

Given a basis of $\Lambda_q^{\perp}(\mathbf{A})$, there is an efficient algorithm to solve the (I)SIS problem as follows.

Lemma 2.23 ([12]) *There is a PPT algorithm* SamplePre$(\mathbf{T_A}, \mathbf{A}, \mathbf{u}, s)$ *that given a basis $\mathbf{T_A}$ of $\Lambda_q^{\perp}(\mathbf{A})$ for some matrix $\mathbf{A} \in \mathbb{Z}_q^{n\times m}$, a real $s \geq \|\widetilde{\mathbf{T}}_{\mathbf{A}}\| \cdot \omega(\sqrt{\log m})$ and a vector $\mathbf{u} \in \mathbb{Z}_q^n$, outputs a vector $\mathbf{e} \in \mathbb{Z}^m$ such that the distribution of $\mathbf{e}$ is statistically close to $D_{\mathbb{Z}^m,s}$ conditioned on $\mathbf{Ae} = \mathbf{u}$.*

By utilizing a simple primitive matrix, Micciancio and Peikert [21] proposed a simple notion of lattice trapdoor, which admits many useful features of a short basis. Let $\mathbf{I}_n$ be the $n \times n$ identity matrix. We now recall the publicly known primitive matrix $\mathbf{G}_b$ in [21]. Formally, for any prime $q > 2$, integer $n, b > 1$ and $k = \lceil \log_b q \rceil$ define $\mathbf{g}_b = (1, b, \ldots, b^{k-1})^t \in \mathbb{Z}_q^k$ and $\mathbf{G}_b = \mathbf{I}_n \otimes \mathbf{g}_b^t \in \mathbb{Z}_q^{n\times nk}$, where "⊗" represents the tensor product.[2] Then, the lattice $\Lambda_q^{\perp}(\mathbf{G}_b)$ has a publicly known short basis $\mathbf{T} = \mathbf{I}_n \otimes \mathbf{T}_k \in \mathbb{Z}^{nk\times nk}$ with $\|\mathbf{T}\| \leq \max\{\sqrt{b^2+1}, \sqrt{k}\}$. Let $(q_0, q_1, \ldots, q_{k-1}) = \mathsf{BitDecomp}_q(q) \in \{0, 1\}^k$, we have

$$\mathbf{G}_b = \begin{pmatrix} \cdots \mathbf{g}_b^t \cdots & & & \\ & \cdots \mathbf{g}_b^t \cdots & & \\ & & \ddots & \\ & & & \cdots \mathbf{g}_b^t \cdots \end{pmatrix} \qquad \mathbf{T}_k = \begin{pmatrix} b & & & & q_0 \\ -1 & b & & & q_1 \\ & -1 & & & q_2 \\ & & \ddots & & \vdots \\ & & & b & q_{k-2} \\ & & & -1 & q_{k-1} \end{pmatrix}.$$

The subscript b is taken to be 2 when omitted. For any vector $\mathbf{u} \in \mathbb{Z}_q^n$, the basis $\mathbf{T} = \mathbf{I}_n \otimes \mathbf{T}_k \in \mathbb{Z}_q^{nk\times nk}$ can be used to sample short vector $\mathbf{e} \sim D_{\mathbb{Z}^{nk},s}$ satisfying $\mathbf{Ge} = \mathbf{u}$ for any $s \geq \omega(\sqrt{\log n})$ in quasilinear time. Besides, one can deterministically compute a short vector $\mathbf{v} = \mathbf{G}^{-1}(\mathbf{u}) \in \{0, 1\}^{nk}$ such that $\mathbf{Gv} = \mathbf{u}$.

[2] One can define $\mathbf{G}$ by using any base $b \geq 2$ and $\mathbf{g} = (1, b, \ldots, b^{k-1})^t$ for $k = \lceil \log_b q \rceil$.

Definition 2.9 (**G**-*trapdoor* [21]) For any integers $n, \bar{m}, q, b \in \mathbb{Z}$, $k = \lceil \log_b q \rceil$, and matrix $\mathbf{A} \in \mathbb{Z}_q^{n\times\bar{m}}$, the **G**-trapdoor for **A** is a matrix $\mathbf{R} \in \mathbb{Z}^{(\bar{m}-nk)\times nk}$ such that $\mathbf{A}\begin{bmatrix}\mathbf{R}\\ \mathbf{I}_{nk}\end{bmatrix} = \mathbf{S}\mathbf{G}_b$ for some invertible tag $\mathbf{S} \in \mathbb{Z}_q^{n\times n}$. The quality of the trapdoor is measured by its largest singular value $s_1(\mathbf{R})$.

If **R** is a **G**-trapdoor for **A**, one can obtain a **G**-trapdoor $\mathbf{R}'$ for any extension $(\mathbf{A}\|\mathbf{B})$ by padding **R** with zero rows. In particular, we have $s_1(\mathbf{R}') = s_1(\mathbf{R})$.

Lemma 2.24 ([21]) *Given any integers $n \geq 1$, $q > 2$, sufficiently large $\bar{m} = O(n \log q)$ and a tag $\mathbf{S} \in \mathbb{Z}_q^{n\times n}$, there is an efficient randomized algorithm* $\mathsf{TrapGen}(1^n, 1^{\bar{m}}, q, \mathbf{S})$ *that outputs a matrix $\mathbf{A} \in \mathbb{Z}_q^{n\times\bar{m}}$ and a **G**-trapdoor $\mathbf{R} \in \mathbb{Z}_q^{(\bar{m}-nk)\times nk}$ with quality $s_1(\mathbf{R}) \leq \sqrt{\bar{m}} \cdot \omega(\sqrt{\log n})$ such that the distribution of **A** is $\mathsf{negl}(n)$-far from uniform and $\mathbf{A}\begin{bmatrix}\mathbf{R}\\ \mathbf{I}_{nk}\end{bmatrix} = \mathbf{S}\mathbf{G}_b$, where $k = \lceil \log_b q \rceil$.*

*In addition, given a **G**-trapdoor **R** of $\mathbf{A} \in \mathbb{Z}_q^{n\times\bar{m}}$ for some invertible tag $\mathbf{S} \in \mathbb{Z}_q^{n\times n}$, any $\mathbf{U} \in \mathbb{Z}_q^{n\times n'}$ for some integer $n' \geq 1$ and real $s \geq b \cdot s_1(\mathbf{R}) \cdot \omega(\sqrt{\log n})$, there is an algorithm* $\mathsf{SampleD}(\mathbf{R}, \mathbf{A}, \mathbf{S}, \mathbf{U}, s)$ *that samples from a distribution within* $\mathsf{negl}(n)$ *statistical distance of $\mathbf{E} \sim (D_{\mathbb{Z}^{\bar{m}},s})^{n'}$ conditioned on $\mathbf{A}\mathbf{E} = \mathbf{U}$.*

As shown in [21], there exists a PPT algorithm that inverts $\mathbf{y} = \mathbf{G}_b^T\mathbf{s} + \mathbf{e}$ as long as $\|\mathbf{e}\| < \frac{q}{2\sqrt{b^2+1}}$. Moreover, if $q = b^k$, the algorithm can invert $\mathbf{y} = \mathbf{G}_b^T\mathbf{s} + \mathbf{e}$ if $\|\mathbf{e}\|_\infty < \frac{q}{2b}$. The following lemma is implicit in [21, Thm. 5.4]:

Lemma 2.25 *Let $\mathbf{I}_{nk}$ be the $nk \times nk$ identity matrix. For any matrices $\mathbf{A} \in \mathbb{Z}_q^{n\times m}$, $\mathbf{R} \in \mathbb{Z}_q^{(m-nk)\times nk}$ and invertible matrix $\mathbf{S} \in \mathbb{Z}_q^{n\times n}$ satisfying $\mathbf{A}\begin{pmatrix}\mathbf{R}\\ \mathbf{I}_{nk}\end{pmatrix} = \mathbf{S}\mathbf{G}_b$, there exists a PPT algorithm* $\mathsf{Solve}(\mathbf{A}, \mathbf{R}, \mathbf{y})$ *that given any $\mathbf{y} = \mathbf{A}^T\mathbf{s} + \begin{pmatrix}\mathbf{e}_1\\ \mathbf{e}_2\end{pmatrix} \in \mathbb{Z}_q^m$ satisfying $\|\mathbf{R}^T\mathbf{e}_1 + \mathbf{e}_2\| < \frac{q}{2\sqrt{b^2+1}}$, outputs $\mathbf{s} \in \mathbb{Z}_q^n$, where $\mathbf{e}_1 \in \mathbb{Z}^{m-nk}$ and $\mathbf{e}_2 \in \mathbb{Z}^{nk}$.*

Moreover, if $q = b^k$, the algorithm $\mathsf{Solve}(\mathbf{A}, \mathbf{R}, \mathbf{y})$ *can invert any $\mathbf{y} = \mathbf{A}^T\mathbf{s} + \begin{pmatrix}\mathbf{e}_1\\ \mathbf{e}_2\end{pmatrix} \in \mathbb{Z}_q^m$ satisfying $\|\mathbf{R}^T\mathbf{e}_1 + \mathbf{e}_2\|_\infty < \frac{q}{2b}$.*

Lemma 2.26 ([21]) *Let $\mathbf{I}_{nk}$ be the $nk \times nk$ identity matrix. For any matrices $\mathbf{A} \in \mathbb{Z}_q^{n\times m}$, $\mathbf{R} \in \mathbb{Z}_q^{(m-nk)\times nk}$ and invertible matrix $\mathbf{S} \in \mathbb{Z}_q^{n\times n}$ satisfying $\mathbf{A}\begin{pmatrix}\mathbf{R}\\ \mathbf{I}_{nk}\end{pmatrix} = \mathbf{S}\mathbf{G}_b$, there exists a PPT algorithm* DelTrap$(\mathbf{B}, \mathbf{R}, \mathbf{S}_1)$ *that given matrix $\mathbf{B} = (\mathbf{A}\|\mathbf{A}_1) \in \mathbb{Z}_q^{n\times(m+w)}$ and an invertible matrix $\mathbf{S}_1 \in \mathbb{Z}_q^{n\times n}$, outputs a **G**-trapdoor $\mathbf{R}' \in \mathbb{Z}^{m\times w}$ for **B** with tag $\mathbf{S}_1$, i.e., $\mathbf{B}\begin{pmatrix}\mathbf{R}'\\ \mathbf{I}_w\end{pmatrix} = \mathbf{S}_1\mathbf{G}_b$, such that $s_1(\mathbf{R}') \leq b(\sqrt{m} + \sqrt{w}) \cdot s_1(\mathbf{R}) \cdot \omega(\sqrt{\log n})$.*

Note that the columns of $(\mathbf{A}\|\mathbf{A}_1)$ in the above lemma can appear in any order, since this just induces a permutation of $\begin{pmatrix}\mathbf{R}'\\ \mathbf{I}_w\end{pmatrix}$'s rows [21].

References

1. Agrawal, S., Boneh, D., Boyen, X.: Efficient lattice (H)IBE in the standard model. In: Gilbert, H. (ed.) Advances in Cryptology - EUROCRYPT 2010. Lecture Notes in Computer Science, vol. 6110, pp. 553–572. Springer, Heidelberg (2010)
2. Ajtai, M.: Generating hard instances of lattice problems (extended abstract). In: Proceedings of the Twenty-Eighth Annual ACM Symposium on Theory of Computing, STOC '96, pp. 99–108. ACM (1996)
3. Ajtai, M.: Generating hard instances of the short basis problem. In: Wiedermann, J., van Emde Boas, P., Nielsen, M. (eds.) Automata, Languages and Programming, Lecture Notes in Computer Science, vol. 1644, pp. 706–706. Springer, Berlin/Heidelberg (1999)
4. Alwen, J., Peikert, C.: Generating shorter bases for hard random lattices. In: STACS, pp. 75–86 (2009)
5. Applebaum, B., Cash, D., Peikert, C., Sahai, A.: Fast cryptographic primitives and circular-secure encryption based on hard learning problems. In: Halevi, S. (ed.) Advances in Cryptology - CRYPTO 2009. Lecture Notes in Computer Science, vol. 5677, pp. 595–618. Springer, Heidelberg (2009)
6. Banaszczyk, W.: New bounds in some transference theorems in the geometry of numbers. Math. Ann. **296**, 625–635 (1993)
7. Brakerski, Z., Gentry, C., Vaikuntanathan, V.: Fully homomorphic encryption without bootstrapping. Innovations in Theoretical Computer Science, ITCS, pp. 309–325 (2012)
8. Brakerski, Z., Vaikuntanathan, V.: Fully homomorphic encryption from ring-LWE and security for key dependent messages. In: Rogaway, P. (ed.) Advances in Cryptology - CRYPTO 2011. Lecture Notes in Computer Science, vol. 6841, pp. 505–524. Springer, Heidelberg (2011)
9. Cash, D., Hofheinz, D., Kiltz, E., Peikert, C.: Bonsai trees, or how to delegate a lattice basis. In: Gilbert, H. (ed.) Advances in Cryptology - EUROCRYPT 2010. Lecture Notes in Computer Science, vol. 6110, pp. 523–552. Springer, Berlin/Heidelberg (2010)
10. Dodis, Y., Rafail, O., Reyzin, L., Smith, A.: Fuzzy extractors: how to generate strong keys from biometrics and other noisy data. SIAM J. Comput. **38**, 97–139 (2008)
11. Ducas, L., Micciancio, D.: Improved short lattice signatures in the standard model. In: Garay, J., Gennaro, R. (eds.) Advances in Cryptology - CRYPTO 2014. Lecture Notes in Computer Science, vol. 8616, pp. 335–352. Springer, Heidelberg (2014)
12. Gentry, C., Peikert, C., Vaikuntanathan, V.: Trapdoors for hard lattices and new cryptographic constructions. In: Proceedings of the 40th Annual ACM Symposium on Theory of Computing, STOC '08, pp. 197–206. ACM (2008)
13. Gordon, S., Katz, J., Vaikuntanathan, V.: A group signature scheme from lattice assumptions. In: Abe, M. (ed.) Advances in Cryptology - ASIACRYPT 2010. Lecture Notes in Computer Science, vol. 6477, pp. 395–412. Springer, Heidelberg (2010)
14. Katz, J., Vaikuntanathan, V.: Smooth projective hashing and password-based authenticated key exchange from lattices. In: Matsui, M. (ed.) Advances in Cryptology - ASIACRYPT 2009. Lecture Notes in Computer Science, vol. 5912, pp. 636–652. Springer, Heidelberg (2009)
15. Laguillaumie, F., Langlois, A., Libert, B., Stehlé, D.: Lattice-based group signatures with logarithmic signature size. In: Sako, K., Sarkar, P. (eds.) Advances in Cryptology - ASIACRYPT 2013. Lecture Notes in Computer Science, vol. 8270, pp. 41–61. Springer, Heidelberg (2013)
16. Lenstra, A., Lenstra H.W., J., Lovász, L.: Factoring polynomials with rational coefficients. Math. Ann. **261**(4), 515–534 (1982)
17. Lindner, R., Peikert, C.: Better key sizes (and attacks) for LWE-based encryption. In: Kiayias, A. (ed.) Topics in Cryptology - CT-RSA 2011. Lecture Notes in Computer Science, vol. 6558, pp. 319–339. Springer, Heidelberg (2011)
18. Lyubashevsky, V.: Lattice signatures without trapdoors. In: Pointcheval, D., Johansson, T. (eds.) Advances in Cryptology - EUROCRYPT 2012. Lecture Notes in Computer Science, vol. 7237, pp. 738–755. Springer, Heidelberg (2012)

19. Lyubashevsky, V., Peikert, C., Regev, O.: On ideal lattices and learning with errors over rings. In: Gilbert, H. (ed.) Advances in Cryptology - EUROCRYPT 2010. Lecture Notes in Computer Science, vol. 6110, pp. 1–23. Springer, Berlin/Heidelberg (2010)
20. Micciancio, D., Goldwasser, S.: Complexity of Lattice Problems: A Cryptographic Perspective, vol. 671. Springer, Netherlands (2002)
21. Micciancio, D., Peikert, C.: Trapdoors for lattices: simpler, tighter, faster, smaller. In: Pointcheval, D., Johansson, T. (eds.) Advances in Cryptology - EUROCRYPT 2012. Lecture Notes in Computer Science, vol. 7237, pp. 700–718. Springer, Heidelberg (2012)
22. Micciancio, D., Regev, O.: Worst-case to average-case reductions based on Gaussian measures. SIAM J. Comput. **37**, 267–302 (2007)
23. Peikert, C.: Public-key cryptosystems from the worst-case shortest vector problem: extended abstract. In: Proceedings of the 41st Annual ACM Symposium on Theory of Computing, STOC '09, pp. 333–342. ACM (2009)
24. Peikert, C.: An efficient and parallel gaussian sampler for lattices. In: Rabin, T. (ed.) Advances in Cryptology - CRYPTO 2010. Lecture Notes in Computer Science, vol. 6223, pp. 80–97. Springer, Heidelberg (2010)
25. Regev, O.: On lattices, learning with errors, random linear codes, and cryptography. In: Proceedings of the Thirty-Seventh Annual ACM Symposium on Theory of Computing, STOC '05, pp. 84–93. ACM (2005)
26. Vershynin, R.: Introduction to the non-asymptotic analysis of random matrices. arXiv preprint arXiv:1011.3027 (2010)
27. Zhang, J., Chen, Y., Zhang, Z.: Programmable hash functions from lattices: short signatures and IBEs with small key sizes. In: Robshaw, M., Katz, J. (eds.) Advances in Cryptology - CRYPTO 2016, pp. 303–332. Springer, Heidelberg (2016)
28. Zhang, J., Yu, Y.: Two-round PAKE from approximate SPH and instantiations from lattices. In: Takagi, T., Peyrin, T. (eds.) Advances in Cryptology – ASIACRYPT 2017, pp. 37–67. Springer (2017)
29. Zhang, J., Zhang, Z., Ding, J., Snook, M., Dagdelen, Ö.: Authenticated key exchange from ideal lattices. In: Oswald, E., Fischlin, M. (eds.) Advances in Cryptology - EUROCRYPT 2015, pp. 719–751. Springer, Heidelberg (2015)

Chapter 3
Public-key Encryption

Abstract In 1976, Diffie and Hellman first introduced the concept of public-key cryptography. Unlike the traditional symmetric-key primitive, public-key primitive has a pair of keys, namely, public key and private key, where the public key can be made freely available to anyone with the guarantee that it is computationally infeasible to compute the private key from the public key. Public-key cryptography has many nice features and solves many problems such as key distributions that are inherent in the symmetric-key cryptography. In 1978, Rivest, Shamir and Adleman proposed the first public-key encryption (PKE) which allows anyone with the public key to encrypt messages, but only allows the one with the private key to decrypt messages. In this chapter, we focus on constructing PKEs from lattices. Specifically, we will first present a chosen-plaintext secure (CPA-secure) PKE scheme from LWE due to Lindner and Peikert, which is an improved variant of the first LWE-based PKE by Regev and can be transformed into an adaptively chosen-ciphertext secure (CCA2-secure) PKE in the random oracle model following the generic Fujisaki-Okamoto transform. Then, we show how to construct an efficient CCA2-secure PKE from lattices in the standard model.

3.1 Definition

Diffie and Hellman [21] introduced the concept of Public-key cryptography in 1976. Soon afterward, Rivest, Shamir and Adleman [51] proposed the first public-key encryption (PKE) based on the hardness of integer factorization, which is known as RSA. Since then, PKE has drawn great attention from the community and has become one of the most fundamental and widely used cryptographic primitives. The basic security notion of PKE (i.e., CPA-security [29]) requires that it should be computationally infeasible for a passive adversary to obtain any useful information from a challenge ciphertext. Later, this notion was enhanced by Naor and Yung [41] to deal with "lunchtime attack". Formally, they [41] defined the security against non-adaptive chosen-ciphertext attacks (i.e., CCA1-security), where the adversary

J. Zhang and Z. Zhang, *Lattice-Based Cryptosystems*,
https://doi.org/10.1007/978-981-15-8427-5_3

can access a decryption oracle to decrypt any ciphertext of his choice before seeing the challenge ciphertext. Now, the de facto standard security notion for PKE is CCA2-security [49], where the adversary can adaptively access the decryption oracle during the whole attack period (with a restriction that the decryption oracle cannot be directly used to decrypt the challenge ciphertext). In fact, the National Institute of Standards and Technology (NIST) has considered CCA2-security as a basic security requirement for PKEs to the post-quantum cryptography standardization [44].

Definition 3.1 (*Public-key Encryption*) A public-key encryption (PKE) scheme Π_{pke} with message space $\mathcal{P}$ is a tuple $(\mathsf{KeyGen}, \mathsf{Enc}, \mathsf{Dec})$, such that

- $\mathsf{KeyGen}(1^\kappa)$ is a PPT algorithm that takes a security parameter 1^κ as input, and outputs a pair of public and secret keys $(\mathsf{pk}, \mathsf{sk})$.
- $\mathsf{Enc}(\mathsf{pk}, \mu)$ is a PPT algorithm that encrypts message $\mu \in \mathcal{P}$ under public key pk and outputs the corresponding ciphertext c.
- $\mathsf{Dec}(\mathsf{sk}, c)$ is a deterministic polynomial time algorithm that decrypts a ciphertext c using secret key sk and outputs a message μ (or $\perp$).

For correctness, we require that, for any $\mu \in \mathcal{P}$, $(\mathsf{pk}, \mathsf{sk}) \leftarrow \mathsf{KeyGen}(1^\kappa)$ and $c \leftarrow \mathsf{Enc}(\mathsf{pk}, \mu)$, the probability $\Pr[\mathsf{Dec}(\mathsf{sk}, c) \neq \mu] = \mathsf{negl}(\kappa)$, where the probability is over the random coins used in both KeyGen and Enc.

The (adaptively) chosen-ciphertext security of PKE is modeled by a game between a challenger C and an adversary $\mathcal{A}$.

KeyGen. The challenger C first computes $(\mathsf{pk}, \mathsf{sk}) \leftarrow \mathsf{KeyGen}(1^\kappa)$. Then, it gives the public key pk to the adversary $\mathcal{A}$ and keeps sk secret.

Phase 1. The adversary $\mathcal{A}$ is allowed to make any polynomial time decryption query with any (different) ciphertext c, the challenger C computes $\mu \leftarrow \mathsf{Dec}(\mathsf{sk}, c)$, and returns μ to $\mathcal{A}$.

Challenge. The adversary $\mathcal{A}$ outputs two equal-length messages (μ_0, μ_1), the challenger C chooses a bit $\delta^* \stackrel{\$}{\leftarrow} \{0, 1\}$ and computes $c^* \leftarrow \mathsf{Enc}(\mathsf{pk}, \mu_{\delta^*})$. Finally, it returns the challenge ciphertext c^* to $\mathcal{A}$.

Phase 2. The adversary is allowed to make more decryption queries with any ciphertext $c \neq c^*$. The challenger C responds as in Phase 1.

Guess. Finally, $\mathcal{A}$ outputs a guess $\delta \in \{0, 1\}$. If $\delta = \delta^*$, the challenger C outputs 1, else outputs 0.

Definition 3.2 (*CCA2-Security*) We say that a PKE scheme Π_{pke} is CCA2-secure if for any PPT adversary $\mathcal{A}$, its advantage $\text{Adv}^{\text{cca2}}_{\Pi_{\text{pke}},\mathcal{A}}(\kappa) = |\Pr[\delta = \delta^*] - \frac{1}{2}|$ in the above game is negligible in the se curity parameter κ.

We typically omit "2" in the CCA2-security, and directly use the term CCA-security for the standard notion of adaptively CCA-security. The notations of CPA-security and CCA1-security can be similarly defined by using a modified security game. Concretely, the CPA-security game does not allow the adversary to make any decryption queries, while the CCA1-security game allows the adversary to make any decryption queries before the challenge phase. Besides, one can also define the notion of one-way security for PKEs by using a modified game where the challenger simply encrypts a uniformly random message μ^* of his own choice to generate the challenge ciphertext C^*, and the adversary is asked to recover the message μ^* from C^*.

In practice, PKEs are usually used to encrypt a random session key. This functionality can also be achieved by another (possibly simpler) primitive named key encapsulation mechanism (KEM), which does not take any message in the encryption algorithm, instead it simultaneously outputs a session key ssk together with a ciphertext C that can be used to recover the session key. KEMs are equivalent to PKEs in the sense that given a secure KEM, one can obtain a secure PKE, and vice versa. For KEMs, one can define the one-wayness, CPA-security and CCA-security similarly by using a security game where the challenger first honestly generates a pair of session key ssk_0 and its corresponding ciphertext C, then it randomly chooses a random session key ssk_1 and outputs (ssk_δ, C) to the adversary by flipping a random coin δ (in the one-wayness game, the challenger only outputs C to the adversary).

We finally note that one can also define the one-wayness, CPA-security and CCA-security of a symmetric-key encryption scheme by using a security game where the challenger does not give anything to the adversary in the **KeyGen** phase.

3.2 PKE Against CPA Attacks

Ajtai and Dwork [4] proposed the first public-key cryptosystem from lattices, whose average-case security is based on the worst-case of the unique SVP (uSVP) problem on lattices. However, their scheme was of theoretical interest only [28, 43]. Latter, Goldreich, Goldwasser and Halevi [28] presented a PKE with better efficiency, but it still has very large public key size for reasonable security [42]. Hoffstein et al. [30] presented an efficient ring-based PKE whose security depends on the complexity of solving certain problems on a special lattice. In 2001, Micciancio proposed an improved variant of the GGH scheme [28] by representing the lattice basis matrix using the Hermite normal form [39]. A major breakthrough in this line was made by Regev [50], who introduced the learning with errors problem (LWE) and proposed a simple PKE scheme from LWE. Since then, Regev's PKE has become the basis of many advanced encryption schemes such as identity-based encryption [26] and fully homomorphic encryption [14]. The original PKE of Regev can only encrypt a single-bit message, which was extended by Kawachi et al. [33] to support multi-bit encryption and was improved by Lindner and Peikert [37] to obtain smaller parameter sizes. Since the Lindner-Peikert scheme [37] was the starting point of many practical lattice-based PKEs such as [7, 13] and can better illustrate the techniques of using LWE in designing public-key encryptions, we give the description of the Lindner-Peikert scheme together with its security in the following part of this section:

3.2.1 The Lindner-Peikert PKE Scheme

Let κ be the security parameter. Let $n, m, \ell, q > 0$ be integers, and $\alpha, \beta > 0$ be reals. The formal description of the Lindner-Peikert PKE scheme with parameters (n, m, ℓ, q, α) is given as follows.

- $\mathsf{KeyGen}(1^\kappa)$: randomly choose $\mathbf{A} \xleftarrow{\$} \mathbb{Z}_q^{n\times m}$, $\mathbf{S} \xleftarrow{\$} (D_{\mathbb{Z}^m,\alpha})^\ell$, $\mathbf{E} \xleftarrow{\$} (D_{\mathbb{Z}^n,\alpha})^\ell$, and compute $\mathbf{B} = \mathbf{AS} + \mathbf{E} \in \mathbb{Z}_q^{n\times\ell}$. Return the pair of public and secret keys $(\mathsf{pk}, \mathsf{sk}) = ((\mathbf{A}, \mathbf{B}), \mathbf{S})$.
- $\mathsf{Enc}(\mathsf{pk}, \mu \in \{0,1\}^\ell)$: first randomly choose $\mathbf{s}_1 \xleftarrow{\$} D_{\mathbb{Z}^n,\beta}$, $\mathbf{e}_1 \xleftarrow{\$} D_{\mathbb{Z}^m,\beta}$, and $\mathbf{e}_2 \xleftarrow{\$} D_{\mathbb{Z}^\ell,\beta}$. Then, compute

$$\mathbf{c}_1 = \mathbf{A}^T\mathbf{s}_1 + \mathbf{e}_1 \in \mathbb{Z}_q^m, \qquad \mathbf{c}_2 = \mathbf{B}^T\mathbf{s}_1 + \mathbf{e}_2 + \mu \cdot \lfloor \tfrac{q}{2} \rfloor \in \mathbb{Z}_q^\ell,$$

  where $\mu \cdot \lfloor \frac{q}{2} \rfloor \in \mathbb{Z}_q^\ell$ denotes multiplying each bit of $\mu \in \{0,1\}^\ell$ by $\lfloor \frac{q}{2} \rfloor$. Finally, return the ciphertext $C = (\mathbf{c}_1, \mathbf{c}_2) \in \mathbb{Z}_q^m \times \mathbb{Z}_q^\ell$.
- $\mathsf{Dec}(\mathsf{sk}, C = (\mathbf{c}_1, \mathbf{c}_2))$: first compute $\mathbf{u} = \mathbf{S}^T\mathbf{c}_1$ and $\mathbf{v} = \mathbf{c}_2 - \mathbf{u}$. Then, compute and return the message $\mu' = \lfloor \mathbf{v} \cdot \frac{2}{q} \rceil \bmod 2 \in \{0,1\}^\ell$.

Now, we show that the above PKE scheme is correct. Since

$$\begin{aligned} \mathbf{c}_2 = \mathbf{B}^T\mathbf{s}_1 + \mathbf{e}_2 + \mu \cdot \lfloor \tfrac{q}{2} \rfloor &= (\mathbf{AS} + \mathbf{E})^T\mathbf{s}_1 + \mathbf{e}_2 + \mu \cdot \lfloor \tfrac{q}{2} \rfloor \\ &= (\mathbf{AS})^T\mathbf{s}_1 + \mathbf{E}^T\mathbf{s}_1 + \mathbf{e}_2 + \mu \cdot \lfloor \tfrac{q}{2} \rfloor, \end{aligned}$$

and $\mathbf{u} = \mathbf{S}^T\mathbf{c}_1 = \mathbf{S}^T(\mathbf{A}^T\mathbf{s}_1 + \mathbf{e}_1) = (\mathbf{AS})^T\mathbf{s}_1 + \mathbf{S}^T\mathbf{e}_1$, we have that

$$\mathbf{v} = \mathbf{c}_2 - \mathbf{u} = \underbrace{\mathbf{E}^T\mathbf{s}_1 + \mathbf{e}_2 - \mathbf{S}^T\mathbf{e}_1}_{\hat{\mathbf{e}}} + \mu \cdot \lfloor \frac{q}{2} \rfloor.$$

In other words, if we set the parameters n, m, ℓ, q such that $\|\mathbf{v}\|_\infty < \frac{q}{4}$, we can be guaranteed that $\mu' = \lfloor \mathbf{v} \cdot \frac{2}{q} \rceil \bmod 2 = \mu$. This shows that the above scheme is correct.

3.2.2 Security

We have the following theorem for the security of the Lindner-Peikert PKE scheme [37]:

Theorem 3.1 *If the* $\mathrm{LWE}_{m,n,q,\alpha}$ *problem and the* $\mathrm{LWE}_{n,(m+\ell),q,\beta}$ *problem are hard, then the Lindner-Peikert PKE scheme is CPA-secure.*

Before giving the proof, we first recall a useful fact of the LWE problem. Formally, for any real $\alpha > 0$ and uniformly chosen $\mathbf{S} = (\mathbf{s}_1, \ldots, \mathbf{s}_\ell) \xleftarrow{\$} (D_{\mathbb{Z}_q^m,\alpha})^\ell$, let $A_{\mathbf{S},\alpha}$ be the distribution of $(\mathbf{A}, \mathbf{B} = (\mathbf{b}_1, \ldots, \mathbf{b}_\ell)) \in \mathbb{Z}_q^{n\times m} \times \mathbb{Z}_q^{n\times\ell}$, where $\mathbf{A} \xleftarrow{\$} \mathbb{Z}_q^{n\times m}$,

$\mathbf{e}_i \xleftarrow{\$} \chi_\beta$, and $\mathbf{b}_i = \mathbf{A}\mathbf{s}_i + \mathbf{e}_i$ for $i \in \{1, \ldots, \ell\}$. For any polynomially bounded $\ell = \mathsf{poly}(m)$, by a standard hybrid argument one can show that the distribution of $A_{\mathbf{S},\alpha}$ is computationally indistinguishable from the uniform distribution over $\mathbb{Z}_q^{n\times m} \times \mathbb{Z}_q^{n\times \ell}$ under the assumption that $\mathrm{LWE}_{m,n,q,\alpha}$ is hard [47].

The proof will use a sequence of games $G_0, \ldots, G_3$, with G_0 being the real CPA-security game (where the challenge ciphertext $C^* = (\mathbf{c}_1^*, \mathbf{c}_2^*, c_3^*, c_4^*)$ is honestly generated by first randomly choosing $\delta^* \xleftarrow{\$} \{0, 1\}$ and then encrypting μ_{δ^*}) and G_3 a random game (where the challenge ciphertext C^* is essentially uniformly random, and thus the adversary's advantage in game G_3 is negligible). The security is finally established by showing that games G_0 and G_3 are computationally indistinguishable in the adversary's view.

Proof Let $\mathcal{A}$ be a PPT adversary which can break the CPA-security of the Lindner-Peikert PKE scheme with advantage ϵ. Let F_i be the event that $\mathcal{A}$ correctly guesses $\delta = \delta^*$ in game G_i, where $i \in \{0, \ldots, 3\}$. By definition, the adversary's advantage $\mathrm{Adv}^{\text{ind-cpa}}_{\Pi_{\text{pke}},\mathcal{A}}(\kappa)$ in game i is exactly $|\Pr[F_i] - 1/2|$.

Game G_0. This game is the real security game. Formally, the challenger C works as follows:

KeyGen. Randomly choose $\mathbf{A} \xleftarrow{\$} \mathbb{Z}_q^{n\times m}, \mathbf{S} \xleftarrow{\$} (D_{\mathbb{Z}^m,\alpha})^\ell, \mathbf{E} \xleftarrow{\$} (D_{\mathbb{Z}^n,\alpha})^\ell$, and compute $\mathbf{B} = \mathbf{A}\mathbf{S} + \mathbf{E} \in \mathbb{Z}_q^{n\times \ell}$. Send the public key $\mathsf{pk} = (\mathbf{A}, \mathbf{B})$ to $\mathcal{A}$.

Challenge. Upon receiving two challenge messages $(\mu_0, \mu_1) \in \{0, 1\}^\ell \times \{0, 1\}^\ell$ from the adversary $\mathcal{A}$, first randomly choose $\delta^* \xleftarrow{\$} \{0, 1\}, \mathbf{s}_1^* \xleftarrow{\$} D_{\mathbb{Z}^n,\beta}, \mathbf{e}_1^* \xleftarrow{\$} D_{\mathbb{Z}^m,\beta}$, and $\mathbf{e}_2^* \xleftarrow{\$} D_{\mathbb{Z}^\ell,\beta}$. Then, compute

$$\mathbf{c}_1^* = \mathbf{A}^T\mathbf{s}_1^* + \mathbf{e}_1^* \in \mathbb{Z}_q^m, \qquad \mathbf{c}_2^* = \mathbf{B}^T\mathbf{s}_1^* + \mathbf{e}_2^* + \mu_{\delta^*} \cdot \lfloor \tfrac{q}{2} \rfloor \in \mathbb{Z}_q^\ell.$$

Finally, return the challenge ciphertext $C^* = (\mathbf{c}_1^*, \mathbf{c}_2^*)$ to $\mathcal{A}$.

By definition, we have the following lemma:

Lemma 3.1 $|\Pr[F_0] - 1/2| = \epsilon$.

Game G_1. This game is almost identical to game G_0 except that the challenger C changes the way of generating the pubic key as follows:

KeyGen. Randomly choose $\mathbf{A} \xleftarrow{\$} \mathbb{Z}_q^{n\times m}, \mathbf{B} \xleftarrow{\$} \mathbb{Z}_q^{n\times \ell}$, and send the public key $\mathsf{pk} = (\mathbf{A}, \mathbf{B})$ to the adversary $\mathcal{A}$.

Lemma 3.2 *If the* $\mathrm{LWE}_{m,n,q,\alpha}$ *problem is hard, then games* G_1 *and* G_0 *are computationally indistinguishable in* $\mathcal{A}$*'s view. Moreover,* $|\Pr[F_1] - \Pr[F_0]| \le \mathsf{negl}(\kappa)$.

Proof Recall that if the $\mathrm{LWE}_{m,n,q,\alpha}$ problem is hard, we have that the distribution of $A_{\mathbf{S},\alpha}$ is computationally indistinguishable from the uniform distribution over $\mathbb{Z}_q^{n\times m} \times \mathbb{Z}_q^{n\times \ell}$. Since the only difference between game G_0 and game G_1 is the way of generating the matrix $\mathbf{B} \xleftarrow{\$} \mathbb{Z}_q^{n\times \ell}$, an adversary that can distinguish the two games can be directly transformed into an algorithm that solves the $\mathrm{LWE}_{m,n,q,\alpha}$ problem. This completes the proof. □

Game G_2. This game is almost identical to game G_1 except that the challenger C changes the way of generating the challenge ciphertext as follows:

Challenge. Upon receiving two challenge messages $(\mu_0, \mu_1) \in \{0,1\}^\ell \times \{0,1\}^\ell$ from the adversary $\mathcal{A}$, first randomly choose $\delta^* \xleftarrow{\$} \{0,1\}$, $\mathbf{x}^* \xleftarrow{\$} \mathbb{Z}_q^m$, $\mathbf{y}^* \xleftarrow{\$} \mathbb{Z}_q^\ell$. Then, compute

$$\mathbf{c}_1^* = \mathbf{x}^* \in \mathbb{Z}_q^m, \qquad \mathbf{c}_2^* = \mathbf{y}^* + \mu_{\delta^*} \cdot \lfloor \tfrac{q}{2} \rfloor \in \mathbb{Z}_q^\ell.$$

Finally, return the challenge ciphertext $C^* = (\mathbf{c}_1^*, \mathbf{c}_2^*)$ to $\mathcal{A}$.

Lemma 3.3 *If the* $\mathrm{LWE}_{n,m+\ell,q,\beta}$ *problem is hard, then games G_2 and G_1 are computationally indistinguishable in $\mathcal{A}$'s view. Moreover,* $|\Pr[F_2] - \Pr[F_1]| \le \mathsf{negl}(\kappa)$.

Proof We only have to show that if there is a PPT adversary $\mathcal{A}$ that can distinguish games G_2 and G_1, then we can construct an algorithm $\mathcal{B}$ that solves the $\mathrm{LWE}_{n,m+\ell,q,\beta}$ problem. Formally, given an instance $(\mathbf{C}, \mathbf{d}) \in \mathbb{Z}_q^{(m+\ell)\times n} \times \mathbb{Z}_q^{m+\ell}$ of the $\mathrm{LWE}_{n,m+\ell,q,\beta}$ problem, the goal of $\mathcal{B}$ is to determine if $\mathbf{d} = \mathbf{C}\mathbf{s} + \mathbf{e}$ for some $\mathbf{s} \xleftarrow{\$} D_{\mathbb{Z}^n,\beta}$, $\mathbf{e} \xleftarrow{\$} D_{\mathbb{Z}^{m+\ell},\beta}$, or $\mathbf{d} \xleftarrow{\$} \mathbb{Z}_q^{m+\ell}$ is chosen from uniform.

Formally, $\mathcal{B}$ first parses $(\mathbf{C}, \mathbf{d})$ as

$$\mathbf{C} = \begin{pmatrix} \mathbf{A}^T \\ \mathbf{B}^T \end{pmatrix} \in \mathbb{Z}_q^{(m+\ell)\times n}, \text{ and } \mathbf{d} = \begin{pmatrix} \mathbf{x} \\ \mathbf{y} \end{pmatrix} \in \mathbb{Z}_q^{m+\ell},$$

where $\mathbf{A} \in \mathbb{Z}_q^{n\times m}$, $\mathbf{B} \in \mathbb{Z}_q^{n\times \ell}$, $\mathbf{x} \in \mathbb{Z}_q^m$ and $\mathbf{y} \in \mathbb{Z}_q^\ell$. Then, it sets the public key $\mathsf{pk} = (\mathbf{A}, \mathbf{B})$ and gives it to the adversary $\mathcal{A}$. Upon receiving a pair of challenge messages $(\mu_0, \mu_1) \in \{0,1\}^\ell \times \{0,1\}^\ell$ from the adversary $\mathcal{A}$, it randomly chooses $\delta^* \xleftarrow{\$} \{0,1\}$, sets the challenge ciphertext $\mathbf{C}^* = (\mathbf{x}, \mathbf{y} + \mu_{\delta^*} \cdot \lfloor \frac{q}{2} \rfloor)$ and sends $\mathbf{C}^*$ to $\mathcal{A}$. Finally, it returns whatever $\mathcal{A}$ outputs as its own guess.

Note that if $(\mathbf{C}, \mathbf{d}) \in \mathbb{Z}_q^{(m+\ell)\times n} \times \mathbb{Z}_q^{m+\ell}$ is a real $\mathrm{LWE}_{n,m+\ell,q,\beta}$ tuple, then the view of $\mathcal{A}$ is the same as that in game G_1, otherwise the same as that in game G_2. This means that if $\mathcal{A}$ can distinguish games G_1 and G_2, then $\mathcal{B}$ can solve the $\mathrm{LWE}_{n,(m+\ell),q,\beta}$ problem. This completes the proof. □

Game G_3. This game is almost identical to game G_2 except that the challenger C changes the way of generating the challenge ciphertext as follows:

Challenge. Upon receiving two challenge messages $(\mu_0, \mu_1) \in \{0,1\}^\ell \times \{0,1\}^\ell$ from the adversary $\mathcal{A}$, first randomly choose $\delta^* \xleftarrow{\$} \{0,1\}$, $\mathbf{x}^* \xleftarrow{\$} \mathbb{Z}_q^m$, $\mathbf{y}^* \xleftarrow{\$} \mathbb{Z}_q^\ell$. Then, compute

$$\mathbf{c}_1^* = \mathbf{x}^* \in \mathbb{Z}_q^m, \qquad \mathbf{c}_2^* = \mathbf{y}^* \in \mathbb{Z}_q^\ell.$$

Finally, return the challenge ciphertext $C^* = (\mathbf{c}_1^*, \mathbf{c}_2^*)$ to $\mathcal{A}$.

Lemma 3.4 *Games G_3 and G_2 are identical in the adversary's view. Moreover,* $\Pr[F_3] = \Pr[F_2]$ *and* $\Pr[F_3] = \frac{1}{2}$.

Proof This lemma follows from the fact that for any choice of $\mu \in \{0,1\}^\ell$, $\mathbf{c}_2^* = \mathbf{y}^* + \mu \cdot \lfloor \frac{q}{2} \rfloor$ is uniformly distributed over $\mathbb{Z}_q^\ell$ as long as $\mathbf{y}^*$ is uniformly chosen from $\mathbb{Z}_q^\ell$, and that the challenge ciphertext in game G_3 has no information about the challenge messages. □

In all, we have that $\epsilon = |\Pr[F_0] - \frac{1}{2}| \leq \mathsf{negl}(\kappa)$ by Lemmas 3.1–3.4, which completes the proof. □

3.3 Achieving CCA-Security in the Random Oracle Model

One can efficiently boost a CPA-secure PKE into a CCA-secure one [25, 48, 57]. As many existing practical PKEs following this approach, below we present the generic Fujisaki-Okamoto transform [25].

Formally, let $\Pi_{\text{pke}} = (\mathsf{KeyGen}, \mathsf{Enc}, \mathsf{Dec})$ be any CPA-secure PKE scheme with message space $\mathcal{P}$ and randomness space Ω satisfying that the sizes of $\mathcal{P}$ and Ω are larger than $2^{\omega(\log \kappa)}$, where κ is the security parameter, i.e., $|\mathcal{P}|, |\Omega| > \omega(\log \kappa)$. Let $\Pi_{\text{se}} = (\mathsf{Enc}, \mathsf{Dec})$ be a symmetric-key CCA-secure encryption scheme, where $\Pi_{\text{se}}.\mathsf{Enc}(\mathsf{ssk}, \mu)$ given a key $\mathsf{ssk} \in \{0,1\}^{\ell_1}$ and a message $\mu \in \{0,1\}^{\ell_2}$ as inputs, outputs a ciphertext C, and $\Pi_{\text{se}}.\mathsf{Dec}(\mathsf{ssk}, C)$ given a key $\mathsf{ssk} \in \{0,1\}^{\ell_1}$ and a ciphertext C as inputs, outputs a message $\mu \in \{0,1\}^{\ell_2}$. Let $\mathrm{H} : \mathcal{P} \to \Omega$ and $\mathbf{G} : \mathcal{P} \to \{0,1\}^{\ell_1}$ be two hash functions, which are modeled as random oracles. We construct a CCA-secure PKE scheme $\Pi'_{\text{pke}} = (\mathsf{KeyGen}, \mathsf{Enc}, \mathsf{Dec})$ as follows:

- $\mathsf{KeyGen}(1^\kappa)$: compute $(\mathsf{pk}, \mathsf{sk}) \leftarrow \Pi_{\text{pke}}.\mathsf{KeyGen}(1^\kappa)$, and return the pair of public and secret keys $(\mathsf{pk}, \mathsf{sk})$.
- $\mathsf{Enc}(\mathsf{pk}, \mu \in \{0,1\}^{\ell_2})$: first randomly choose $\rho \xleftarrow{\$} \mathcal{P}$, compute $r = \mathrm{H}(\rho \| \mu) \in \Omega$ and $\mathsf{ssk} = G(\rho)$. Then, use r as the random coins to run the algorithm $C_1 \leftarrow \Pi_{\text{pke}}.\mathsf{Enc}(\mathsf{pk}, \rho; r)$, and compute $C_2 = \Pi_{\text{se}}.\mathsf{Enc}(\mathsf{ssk}, \mu)$. Finally, return the ciphertext $C = (C_1, C_2)$.
- $\mathsf{Dec}(\mathsf{sk}, C = (C_1, C_2))$: first compute $\rho' = \Pi_{\text{pke}}.\mathsf{Dec}(\mathsf{sk}, C_1)$ and $\mathsf{ssk}' = \mathbf{G}(\rho')$. Then, compute $\mu' = \Pi_{\text{se}}.\mathsf{Dec}(\mathsf{ssk}', C_2)$, and recompute a ciphertext $C' = (C_1', C_2') = \Pi'_{\text{pke}}.\mathsf{Enc}(\mathsf{pk}, \mu'; \rho')$. If $C' = C$, return μ', otherwise return $\perp$.

Note that if Π_{pke} and Π_{se} is correct, then it is easy to check that the above PKE scheme Π'_{pke} is correct. Moreover, the resulting CCA-secure PKE scheme Π'_{pke} is almost as efficient as the underlying CPA-secure PKE scheme Π_{pke} because hash functions and symmetric-key encryptions are typically much faster than PKE schemes) in practice. Note that one can also remove the symmetric-key encryption from the generic transformation (and thus save an element in the ciphertext) if one only wants to obtain a key encapsulation mechanism (KEM). Furthermore, one can obtain a KEM with implicit-rejection by adding a uniformly random key of an appropriate pseudorandom function (PRF) into the secret key in the key generation algorithm, and outputting a random element produced by the PRF on input the whole ciphertext in the decryption algorithm when the check in the decryption algorithm fails. Such a KEM with implicit-rejection has a nice feature that the resulting PKE scheme is also provably CCA-secure in the quantum random oracle model [31].

As for the security, we have the following theorem:

Theorem 3.2 *If Π_{pke} is a one-way PKE, and Π_{se} is a CCA-secure symmetric-key encryption, then the above PKE scheme Π'_{pke} is secure in the random oracle model.*

We omit the proof of the above theorem, which can be found in [25].

3.4 CCA-Secure PKE in the Standard Model

In this section, we show how to construct a CCA-secure PKE in the standard model by utilizing the algebraic properties of the LWE problem.

3.4.1 Design Rational

We first note that in addition to the pseudorandomness, the LWE problem also has the following useful properties:

- **Hermite normal form (HNF):** the HNF variant [9] where the secret $\mathbf{s} \in \mathbb{Z}^n$ is chosen from $D_{\mathbb{Z}^n,\alpha q}$ is as hard as the standard LWE problem where $\mathbf{s}$ is uniformly chosen from $\mathbb{Z}_q^n$;
- **Additive Homomorphism:** given any $(\mathbf{A}, \mathbf{y} = \mathbf{A}^T\mathbf{s} + \mathbf{e}) \in \mathbb{Z}_q^n \times \mathbb{Z}_q$ and any $\mathbf{s}' \in \mathbb{Z}_q^n$, one can easily obtain an LWE tuple for the new secret $\tilde{\mathbf{s}} = \mathbf{s} + \mathbf{s}'$ by computing $(\mathbf{A}, \tilde{\mathbf{y}} = \mathbf{y} + \mathbf{A}^T\mathbf{s}' = \mathbf{A}^T\tilde{\mathbf{s}} + \mathbf{e})$;
- **Unique Witness:** if $\mathbf{e}$ is sufficiently short, then the secret $\mathbf{s} \in \mathbb{Z}_q^n$ is uniquely determined by $(\mathbf{A}, \mathbf{y} = \mathbf{A}^T\mathbf{s} + \mathbf{e}) \in \mathbb{Z}_q^{n\times m} \times \mathbb{Z}_q^m$ with overwhelming probability for certain choice of parameters (see Lemma 2.16);
- **Trapdoor Inversion:** given $(\mathbf{A}, \mathbf{y} = \mathbf{A}^T\mathbf{s} + \mathbf{e}) \in \mathbb{Z}_q^{n\times m} \times \mathbb{Z}_q^m$ for some short vector $\mathbf{e}$, if $\mathbf{A}$ is a trapdoor matrix (see Lemma 2.25), then there is a PPT inversion algorithm that outputs $\mathbf{s} \in \mathbb{Z}_q^n$.

By the Gaussian tail inequality, each coordinate of $\mathbf{s} \xleftarrow{\$} D_{\mathbb{Z}^n,\alpha q}$ is upper bounded by some constant $B > 0$ with overwhelming probability, namely, $\|\mathbf{s}\|_\infty \le B$. Thus, if $B \ll q$ holds (as is usually the case), then when we encode the secret $\mathbf{s} = (s_0, \ldots, s_{n-1})^T$ as an element of $\mathbb{Z}_q^n$, the most significant bits of each s_i in the binary representation are not "used" (i.e., the most significant bits of each s_i are always zeros with overwhelming probability). The starting point is to encode the message information into those "unused" bit-slots when using the HNF variant of LWE. In more detail, we first introduce a pair of message encode/decode algorithms ($\mathsf{encode}_d : \mathbb{Z}_d^n \to \mathbb{Z}_q^n$, $\mathsf{decode}_d : \mathbb{Z}_q^n \to \mathbb{Z}_d^n$) for some integer $d \ge 2$ (see Lemma 3.5) such that for any $\mathbf{v} \in \mathbb{Z}_d^n$ and $0 < B \ll q$, we have that $\mathsf{decode}_d(\mathbf{s} + \mathsf{encode}_d(\mathbf{v})) = \mathbf{v}$ holds as long as $\|\mathbf{s}\|_\infty \le B$. Then, given an LWE tuple $(\mathbf{A}, \mathbf{y} = \mathbf{A}^T\mathbf{s} + \mathbf{e})$ in the HNF form and any message $\mathbf{v} \in \mathbb{Z}_d^n$, we can efficiently compute $\tilde{\mathbf{y}} = \mathbf{A}^T\tilde{\mathbf{s}} + \mathbf{e} = \mathbf{y} + \mathbf{A}^T\mathsf{encode}_d(\mathbf{v})$ by the additive homomorphism, where $\tilde{\mathbf{s}} = \mathbf{s} + \mathsf{encode}_d(\mathbf{v})$. Since this can be publicly done given $(\mathbf{A}, \mathbf{y}) \in \mathbb{Z}_q^{n\times m} \times \mathbb{Z}_q^n$ and $\mathbf{v} \in \mathbb{Z}_d^n$, we have that $\tilde{\mathbf{y}} \in \mathbb{Z}_q^n$ is computationally indistinguishable from uniform distribution over $\mathbb{Z}_q^n$ by the pseudorandomness of LWE. Namely, the pair $(\mathbf{A}, \tilde{\mathbf{y}}) \in \mathbb{Z}_q^{n\times m} \times \mathbb{Z}_q^n$ computationally hides the message $\mathbf{v} \in \mathbb{Z}_d^n$.

So far, we have shown a compact (namely, from $\mathbb{Z}_q^{n\times m} \times \mathbb{Z}_q^n \times \mathbb{Z}_d^n$ to $\mathbb{Z}_q^{n\times m} \times \mathbb{Z}_q^n$) and "safe" solution to hide a message, but the solution itself is not a PKE. In particular, we have not yet shown how to recover $\mathbf{v} \in \mathbb{Z}_d^n$ from $(\mathbf{A}, \tilde{\mathbf{y}}) \in \mathbb{Z}_q^{n\times m} \times \mathbb{Z}_q^n$, which needs other properties of LWE. Specifically, by the unique witness property, for certain choices of parameters, we have that $\tilde{\mathbf{s}} \in \mathbb{Z}_q^n$ is almost fully determined by $(\mathbf{A}, \tilde{\mathbf{y}}) \in \mathbb{Z}_q^{n\times m} \times \mathbb{Z}_q^n$ when requiring that $\|\tilde{\mathbf{y}} - \mathbf{A}^T\tilde{\mathbf{s}}\|_\infty$ is small. Moreover, if $\mathbf{A} \in \mathbb{Z}_q^{n\times m}$ is a trapdoor matrix, then there exists a PPT inversion algorithm which can recover $\tilde{\mathbf{s}} \in \mathbb{Z}_q^n$ from $(\mathbf{A}, \tilde{\mathbf{y}}) \in \mathbb{Z}_q^{n\times m} \times \mathbb{Z}_q^n$ by the trapdoor inversion property. Thus, combining the above techniques with existing trapdoor generation algorithms [3, 8, 40] immediately gives us a PKE scheme:

- **Key Generation:** compute a matrix $\mathbf{A} \in \mathbb{Z}_q^{n\times m}$ and an associated trapdoor td by running the trapdoor generation algorithm, set the public key $\mathsf{pk} = \mathbf{A}$ and the secret key $\mathsf{sk} = td$;
- **Encryption:** given the public key $pk = \mathbf{A}$ and a message $\mathbf{v} \in \mathbb{Z}_d^n$, randomly choose $\mathbf{s} \xleftarrow{\$} D_{\mathbb{Z}^n,\alpha q}$, $\mathbf{e} \xleftarrow{\$} D_{\mathbb{Z}^m,\alpha q}$, compute and return the ciphertext $\mathbf{c} = \mathbf{A}^T\tilde{\mathbf{s}} + \mathbf{e} \in \mathbb{Z}_q^m$, where $\tilde{\mathbf{s}} = \mathbf{s} + \mathsf{encode}_d(\mathbf{v}) \in \mathbb{Z}_q^n$;
- **Decryption:** given the secret key sk and the ciphertext $\mathbf{c}$, compute $\tilde{\mathbf{s}} \in \mathbb{Z}_q^n$ by running the trapdoor inversion algorithm, and return $\mathbf{v} = \mathsf{decode}_d(\tilde{\mathbf{s}}) \in \mathbb{Z}_q^n$.

Usually, the matrix $\mathbf{A}$ output by the trapdoor generation algorithm [3, 8, 40] can be statistically close to uniform distribution over $\mathbb{Z}_q^{n\times m}$, which makes it easy to prove the CPA-security of the above PKE scheme under the LWE assumption. Inspired by [1, 3, 8, 47], the authors of [40] showed an efficient algorithm to generate a trapdoor matrix $\mathbf{A} \in \mathbb{Z}_q^{n\times m}$ for some hidden $tag^* \in \{0, 1\}^*$ chosen in advance. They [40] also showed a way to publicly derive a new matrix $\mathbf{A}_{tag} \in \mathbb{Z}_q^{n\times m}$ for any given $tag \in \{0, 1\}^*$ from $\mathbf{A}$ such that (1) $\mathbf{A}_{tag^*}$ is essentially a random matrix, but (2) for any $tag \neq tag^*$, the trapdoor td of $\mathbf{A}$ is also a trapdoor for $\mathbf{A}_{tag} \in \mathbb{Z}_q^{n\times m}$. By instantiating the above key generation with the trapdoor generation algorithm in [40] to generate $pk = \mathbf{A}$ for $tag^* = 0$, and using a randomly chosen $tag \in \{0, 1\}^*$ to derive a matrix $\mathbf{A}_{tag}$ from $pk = \mathbf{A}$ for each time encryption (namely, computing $\mathbf{c} = \mathbf{A}_{tag}^T\tilde{\mathbf{s}} + \mathbf{e}$ and setting the ciphertext as $(\mathbf{c}, tag)$), we get a tag-based PKE. One can easily transform it into a CCA2-secure PKE with extra overheads in both the computations and ciphertext sizes by applying the BCHK technique [11].

In the following, we give a more efficient way to achieve CCA2-security. We first observe that the matrix $\mathbf{A}_{tag}$ derived by using the technique in [40] actually consists of two parts, i.e., $\mathbf{A}_{tag} = (\mathbf{A}_1\|\mathbf{A}_2) \in \mathbb{Z}_q^{n\times(m_1+m_2)}$, where the first part $\mathbf{A}_1 \in \mathbb{Z}_q^{n\times m_1}$ is essentially a random matrix which is independent from the tag. Moreover, the vector $\mathbf{c} = \mathbf{A}_{tag}^T\tilde{\mathbf{s}} + \mathbf{e}$ can be naturally written as $\mathbf{c} = (\mathbf{c}_1, \mathbf{c}_2)$, where $\mathbf{c}_1 = \mathbf{A}_1^T\tilde{\mathbf{s}} + \mathbf{e}_1$

and $\mathbf{c}_2 = \mathbf{A}_2^T\tilde{\mathbf{s}} + \mathbf{e}_2$ for some $\mathbf{e} = (\mathbf{e}_1, \mathbf{e}_2) \in \mathbb{Z}_q^m$. This means that we can compute the first part $\mathbf{c}_1$ of $\mathbf{c} = (\mathbf{c}_1, \mathbf{c}_2)$ without knowing the information of the tag. Again, for certain choice of parameters, the partial ciphertext $\mathbf{c}_1 \in \mathbb{Z}_q^{m_1}$ fixes $\tilde{\mathbf{s}} \in \mathbb{Z}_q^n$ with overwhelming probability by the unique witness property of LWE. Since the message $\mathbf{v} \in \mathbb{Z}_d^n$ is encoded in the secret $\tilde{\mathbf{s}} = \mathbf{s} + \mathsf{encode}_d(\mathbf{v})$, we have that $\mathbf{c}_1 \in \mathbb{Z}_q^{m_1}$ also fixes $\mathbf{v} \in \mathbb{Z}_d^n$. In other words, $\mathbf{c}_1 \in \mathbb{Z}_q^{m_1}$ can be seen as a public tag which tightly binds the message $\mathbf{v} \in \mathbb{Z}_d^n$.

The basic idea is to "amplify" this binding property such that $\mathbf{c}_1$ can be seen as a "unique" tag for the whole ciphertext $\mathbf{c} = (\mathbf{c}_1, \mathbf{c}_2)$. For this, we have to modify the encryption and decryption algorithms. Formally, given a message $\mu \in \{0, 1\}^\ell$ for some integer $\ell > 0$, the encryption algorithm first chooses a random seed $\mathbf{v} \xleftarrow{\$} \mathbb{Z}_d^n$ for a pseudorandom generator $\mathsf{PRG} : \mathbb{Z}_d^n \to \{0, 1\}^{\ell+2\kappa}$ and compute $\tilde{\mathbf{s}} = \mathbf{s} + \mathsf{encode}_d(\mathbf{v})$, where κ is the security parameter. Then, it computes $\mathbf{c}_1 = \mathbf{A}_1^T\tilde{\mathbf{s}} + \mathbf{e}_1$, and simply uses $tag = \mathrm{H}(\mathbf{c}_1)$ to produce $\mathbf{A}_2 \in \mathbb{Z}_q^{n\times m_2}$ and the partial ciphertext $\mathbf{c}_2 \in \mathbb{Z}_q^{m_2}$, where $\mathrm{H} : \{0, 1\}^* \to \{0, 1\}^\kappa$ is a collision-resistant hash function. Finally, it computes $x\|y\|z = \mathsf{PRG}(\mathbf{v}) \in \{0, 1\}^{\ell+2\kappa}$, $c_3 = x \oplus \mu$, $\tau = \mathrm{H}(\mathbf{c}_2, c_3)$, $c_4 = (\tau \otimes y) \oplus z$ and sets the ciphertext $C = (\mathbf{c}_1, \mathbf{c}_2, c_3, c_4)$, where $\tau, y, z \in \{0, 1\}^\kappa$ are treated as elements in the finite field $\mathbb{F}_{2^\kappa}$, and "$\oplus$" (resp. "$\otimes$") denotes the addition (resp. multiplication) operation in $\mathbb{F}_{2^\kappa}$ (here, we also slightly abuse the notation "$\oplus$" to denote the "XOR" operation between bit strings). Accordingly, the decryption algorithm is modified to first recover $\mathbf{v} \in \mathbb{Z}_d^n$ from $(\mathbf{c}_1, \mathbf{c}_2)$, compute $x\|y\|z = \mathsf{PRG}(\mathbf{v}) \in \{0, 1\}^{\ell+2\kappa}$ and $\mu = c_3 \oplus x$. Then, it returns $\perp$ if $c_4 \neq (\tau \otimes y) \oplus z$, otherwise returns the message $\mu \in \{0, 1\}^\ell$. The above modifications finally allow us to achieve CCA-security at a very low cost. In fact, the size of C is even smaller than the CCA1-secure PKE in [40].

We now briefly explain why the above PKE scheme is CCA2-secure. Since $\mathbf{c}_1$ is independent from the message, in the security proof we can generate the first part $\mathbf{c}_1^*$ of the challenge ciphertext $C^* = (\mathbf{c}_1^*, \mathbf{c}_2^*, c_3^*, c_4^*)$ during the key generation phase, and embed $tag^* = \mathrm{H}(\mathbf{c}_1^*)$ into the public key $\mathbf{A}$ using the trapdoor generation algorithm in [40]. By doing this, we have that $\mathbf{A}_{tag^*} = (\mathbf{A}_1^*\|\mathbf{A}_2^*)$ is essentially a random matrix, and $\mathbf{c}_2^*$ can be generated from $\mathbf{c}_1^* = (\mathbf{A}_1^*)^T\tilde{\mathbf{s}}^* + \mathbf{e}_1^*$ without knowing the secret $\tilde{\mathbf{s}}^* \in \mathbb{Z}_q^n$. This allows us to show that given $(\mathbf{c}_1^*, \mathbf{c}_2^*)$, the seed $\mathbf{v}^* \in \mathbb{Z}_d^n$ is computationally indistinguishable from uniform over $\mathbb{Z}_d^n$ by the pseudorandomness of the LWE tuple $(\mathbf{A}_1^*, (\mathbf{A}_1^*)^T\mathbf{s}^* + \mathbf{e}_1^*)$, where $\tilde{\mathbf{s}}^* = \mathbf{s}^* + \mathsf{encode}_d(\mathbf{v}^*)$. Since PRG is a pseudorandom generator, we have that $x^*\|y^*\|z^* = \mathsf{PRG}(\mathbf{v}^*)$ is also computationally indistinguishable from uniform. This immediately implies that (1) $c_3^* = x^* \oplus \mu$ computationally hides the message $\mu \in \{0, 1\}^\ell$; and (2) given $C^* = (\mathbf{c}_1^*, \mathbf{c}_2^*, c_3^*, c_4^* = (\tau^* \otimes y^*) \oplus z^*)$, no PPT algorithm can output a valid c_4 for any other $\tau \neq \tau^* = \mathrm{H}(\mathbf{c}_2^*, c_3^*)$ such that $c_4 = (\tau \otimes y^*) \oplus z^*$ except with negligible probability (this is because any such pairs (c_4^*, c_4) fully determines (y^*, z^*) when $\tau \neq \tau^*$). Thus, it suffices to show how to answer the decryption query from the adversary. Note that $\mathsf{PRG}(\cdot)$ is deterministic and $\mathbf{v}^*$ is fixed by $\mathbf{c}_1^*$ for certain choices of parameters, a ciphertext $C = (\mathbf{c}_1^*, \mathbf{c}_2, c_3, c_4) \neq C^*$ output by the adversary cannot pass the ciphertext check in the decryption algorithm with overwhelming probability. Thus, any decryption query with such C can be simply responded with $\perp$ (i.e., without running the decryption algorithm). While a decryption query with $C = (\mathbf{c}_1 \neq \mathbf{c}_1^*, \mathbf{c}_2, c_3, c_4)$ can be honestly answered by decrypting C using the trapdoor information of $\mathbf{A}_{tag}$ (note that $\mathbf{A}_{tag}$ for any $tag = \mathrm{H}(\mathbf{c}_1) \neq tag^*$ is a trapdoor matrix). This finally allows us to prove the CCA2-security of the above PKE scheme.

3.4.2 *The Construction*

Before giving the construction, we first define a pair of algorithms (encode_d, decode_d), which are parameterized by positive integers (n, q, d). Formally, given any $\mathbf{v} \in \mathbb{Z}_d^n$, the algorithm $\mathsf{encode}_d : \mathbb{Z}_d^n \to \mathbb{Z}_q^n$ is defined as $\mathsf{encode}_d(\mathbf{v}) = (v_1 \cdot \lfloor \frac{q}{d} \rceil, \ldots, v_n \cdot \lfloor \frac{q}{d} \rceil)$, where $\mathbf{v} = (v_1, \ldots, v_n) \in \mathbb{Z}_d^n$. For any $\mathbf{u} \in \mathbb{Z}_q^n$, the algorithm $\mathsf{decode}_d : \mathbb{Z}_q^n \to \mathbb{Z}_d^n$ is defined as $\mathsf{decode}_d(\mathbf{u}) = (\lfloor u_1 \cdot \frac{d}{q} \rceil, \ldots, \lfloor u_n \cdot \frac{d}{q} \rceil)$, where $\mathbf{u} = (u_1, \ldots, u_n) \in \mathbb{Z}_q^n$.

Lemma 3.5 *Let n, q be positive integers, and integer $2 \leq d < \sqrt{q}$. Then, for any $\mathbf{v} \in \mathbb{Z}_d^n$, any $\mathbf{e} \in \mathbb{Z}^n$ satisfying $\|\mathbf{e}\|_\infty < \frac{q-(d-1)d}{2d}$, and $\mathbf{w} = \mathsf{encode}_d(\mathbf{v}) + \mathbf{e}$, we have that $\mathbf{v} = \mathsf{decode}_d(\mathbf{w})$ always holds.*

Proof Since both algorithms simply apply the same operations on their inputs in a coordinate-wise way, it suffices to show that for any $v \in \mathbb{Z}_d$, any $e \in \mathbb{Z}$ satisfying $|e| < \frac{q-(d-1)d}{2d}$ and $w = v \cdot \lfloor \frac{q}{d} \rceil + e$, we alway have $v = \lfloor w \cdot \frac{d}{q} \rceil$. By definition, we have $w = v \cdot (\frac{q}{d} + x) + e$ holds for some x satisfying $|x| \leq 1/2$. Thus, $w \cdot \frac{d}{q} = v + (vx + e) \cdot \frac{d}{q}$. Since $|(vx + e) \cdot \frac{d}{q}| \leq (|vx| + |e|) \cdot \frac{d}{q} < \left(\frac{d-1}{2} + \frac{q-(d-1)d}{2d}\right) \cdot \frac{d}{q} = 1/2$ holds by assumption, we have $v = \lfloor w \cdot \frac{d}{q} \rceil$. This completes the proof. □

Now, we are ready to give the construction. Formally, let κ be the security parameter. Let $n, \bar{m} > 0$ be integers, and let q be a prime or a power of prime $b \geq 2$. Let $k = \lceil \log_b q \rceil$ and $m = \bar{m} + nk$. Let (encode_d, decode_d) be the pair of encode/decode algorithms parameterized by (n, q, d) satisfying that $n \log_2 d \geq 3\kappa$. Let $\mathbb{F}_{2^\kappa}$ be a finite field of order 2^κ. Let $\mathrm{H} : \{0, 1\}^* \to \mathbb{F}_{2^\kappa} \backslash \{0\}$ be a collision-resistant hash function (Namely, we assume that the output of H does not contain the zero element in $\mathbb{F}_{2^\kappa}$ for simplicity). Let $\mathsf{FRD} : \mathbb{F}_{2^\kappa} \to \mathbb{Z}_q^{n \times n}$ be an FRD encoding. The CCA2-secure PKE with parameters $(n, \bar{m}, q, b, d, \alpha)$ is given as follows:

- $\mathsf{KeyGen}(1^\kappa)$: randomly choose $\mathbf{A} \xleftarrow{\$} \mathbb{Z}_q^{n \times \bar{m}}$, $\mathbf{R} \xleftarrow{\$} (D_{\mathbb{Z}^{\bar{m}}, \omega(\sqrt{\log n})})^{nk}$, and compute $\mathbf{B} = -\mathbf{AR}$. Return the pair of public and secret keys $(pk, sk) = ((\mathbf{A}, \mathbf{B}), \mathbf{R})$.
- $\mathsf{Enc}(pk, \mu \in \mathbb{F}_{2^\kappa})$: first randomly choose $\mathbf{s} \xleftarrow{\$} D_{\mathbb{Z}^n, \alpha q}$, $\mathbf{e}_1 \xleftarrow{\$} D_{\mathbb{Z}^{\bar{m}}, \alpha q}$, $\mathbf{e}_2 \xleftarrow{\$} D_{\mathbb{Z}^{nk}, \gamma}$, and $x, y, z \xleftarrow{\$} \mathbb{F}_{2^\kappa}$, where $\gamma = \sqrt{\|\mathbf{e}_1\|^2 + \bar{m}(\alpha q)^2} \cdot \omega(\sqrt{\log n})$. Then, interpret the bit-concatenation of $(x, y, z) \in (\mathbb{F}_{2^\kappa})^3$ as a vector $\mathbf{v} = x\|y\|z \in \mathbb{Z}_d^n$ (which can always be done since $n \log_2 d \geq 3\kappa$), and compute

$$\tilde{\mathbf{s}} = \mathbf{s} + \mathsf{encode}_d(\mathbf{v}), \qquad \mathbf{c}_1 = \mathbf{A}^T\tilde{\mathbf{s}} + \mathbf{e}_1,$$
$$\mathbf{c}_2 = (\mathbf{B} + \mathsf{FRD}(tag)\mathbf{G}_b)^T\tilde{\mathbf{s}} + \mathbf{e}_2, \qquad c_3 = x + \mu \in \mathbb{F}_{2^\kappa},$$
$$c_4 = \tau y + z \in \mathbb{F}_{2^\kappa},$$

where $tag = \mathrm{H}(\mathbf{c}_1) \in \mathbb{F}_{2^\kappa}$ and $\tau = \mathrm{H}(\mathbf{c}_2, c_3) \in \mathbb{F}_{2^\kappa}$. Finally, return the ciphertext $C = (\mathbf{c}_1, \mathbf{c}_2, c_3, c_4) \in \mathbb{Z}_q^{m_1} \times \mathbb{Z}_q^{nk} \times \mathbb{F}_{2^\kappa} \times \mathbb{F}_{2^\kappa}$.

- $\mathsf{Dec}(sk, C = (\mathbf{c}_1, \mathbf{c}_2, c_3, c_4))$: first compute $tag = \mathrm{H}(\mathbf{c}_1)$,

$$\mathbf{A}_{tag} = (\mathbf{A} \| \mathbf{B} + \mathsf{FRD}(tag)\mathbf{G}_b), \text{ and } \mathbf{u} = \begin{pmatrix} \mathbf{c}_1 \\ \mathbf{c}_2 \end{pmatrix}.$$

Then, compute $\tilde{\mathbf{s}} \leftarrow \mathsf{Solve}(\mathbf{A}_{tag}, \mathbf{R}, \mathbf{u})$, $\mathbf{v} = \mathsf{decode}_d(\tilde{\mathbf{s}}) \in \mathbb{Z}_d^n$, and parse $\mathbf{v} = \|x\|y\|z$, where $(x, y, z) \in (\mathbb{F}_{2^\kappa})^3$. Let $\mathbf{e}_1 = \mathbf{c}_1 - \mathbf{A}^T\tilde{\mathbf{s}}$ and $\mathbf{e}_2 = \mathbf{c}_2 - (\mathbf{B} + \mathsf{FRD}(tag)\mathbf{G}_b)^T\tilde{\mathbf{s}}$. Return $\perp$ if one of the following conditions holds:

  - $\|\mathbf{e}_1\| > \alpha q \sqrt{\bar{m}}$, or
  - $\|\mathbf{e}_2\| > \gamma\sqrt{nk}$ for prime q (or $\|\mathbf{e}_2\|_\infty > \gamma \cdot \omega(\sqrt{\log n})$ for $q = b^k$ a power of prime b), or
  - $c_4 \neq \mathrm{H}(\mathbf{c}_2, c_3)y + z \in \mathbb{F}_{2^\kappa}$.

Otherwise, return $\mu = c_3 - x \in \mathbb{F}_{2^\kappa}$.

3.4.3 Parameters and Correctness

Note that $\mathbf{c}_1 = \mathbf{A}^T\tilde{\mathbf{s}} + \mathbf{e}_1$ and $\mathbf{c}_2 = (\mathbf{B} + \mathsf{FRD}(tag)\mathbf{G}_b)^T\tilde{\mathbf{s}} + \mathbf{e}_2$, the algorithm $\mathsf{Solve}(\mathbf{A}_{tag}, \mathbf{R}, \mathbf{u})$ can invert $\tilde{\mathbf{s}} \in \mathbb{Z}_q^n$ if $\|\mathbf{R}^T\mathbf{e}_1 + \mathbf{e}_2\| < \frac{q}{2\sqrt{b^2+1}}$ by Lemma 2.25. Because $\tilde{\mathbf{s}} = \mathbf{s} + \mathsf{encode}_d(\mathbf{v})$, one can correctly recover $\mathbf{v} \in \mathbb{Z}_d^n$ if $\|\mathbf{s}\|_\infty < \frac{q-(d-1)d}{2d}$ by Lemma 3.5. Note that $\mathbf{s} \xleftarrow{\$} D_{\mathbb{Z}^n,\alpha q}$, $\mathbf{e}_1 \xleftarrow{\$} D_{\mathbb{Z}^{\bar{m}},\alpha q}$, and $\mathbf{e}_2 \xleftarrow{\$} D_{\mathbb{Z}^{nk},\gamma}$, we have that $\|\mathbf{s}\|_\infty \leq \alpha q \cdot \omega(\sqrt{\log n})$, $\|\mathbf{e}_1\| \leq \alpha q\sqrt{\bar{m}}$, $\|\mathbf{e}_2\|_\infty \leq \gamma \cdot \omega(\sqrt{\log n})$ and $\|\mathbf{e}_2\| \leq \gamma\sqrt{nk}$ hold with overwhelming probability by Lemma 2.3. Since $\mathbf{R} \xleftarrow{\$} (D_{\mathbb{Z}^{\bar{m}},\omega(\sqrt{\log n})})^{nk}$, the inequality $s_1(\mathbf{R}) \leq \sqrt{\bar{m}} \cdot \omega(\sqrt{\log n})$ holds with overwhelming probability by Lemma 2.6. Since $\gamma = \sqrt{\|\mathbf{e}_1\|^2 + \bar{m}(\alpha q)^2} \cdot \omega(\sqrt{\log n})$, we have $\|\mathbf{R}^T\mathbf{e}_1 + \mathbf{e}_2\| \leq \alpha q\bar{m} \cdot \omega(\sqrt{\log n})$. Besides, we need $\alpha q \geq 2\sqrt{n}$ for the hardness of the LWE problem [50]. We also need Lemmas 2.4 and 2.16 in the security proof, which require $\bar{m} \geq (n+1)\log_2 q + \omega(\log n)$ and $\|\mathbf{e}_1\|_\infty < q/8$.

In all, for the case where $b = 2$ and q is a prime, the decryption algorithm is correct if we set the parameters $\bar{m}, \alpha, q$ such that

$$\bar{m} = (n+1)\log_2 q + \omega(\log n), \quad 1/\alpha = \bar{m} \cdot \omega(\sqrt{\log n}), \quad \alpha q = 2\sqrt{n}, \tag{3.1}$$

which means that $m = \bar{m} + nk = \tilde{O}(n)$, $1/\alpha = \tilde{O}(n)$ and $q = \tilde{O}(n^{1.5})$.

To obtain better efficiency, one can set q as a power of a small prime b (e.g., $b = 3$), which allows us to use the inequality $\|\mathbf{R}^T\mathbf{e}_1 + \mathbf{e}_2\|_\infty \leq \alpha q\sqrt{\bar{m}} \cdot \omega(\sqrt{\log n})^2 < \frac{q}{2b}$

in the correctness analysis. In this case, it suffices to set the parameters $\bar{m}, \alpha, q$ such that

$$\bar{m} = (n+1)\log_2 q + \omega(\log n), \quad 1/\alpha = \sqrt{\bar{m}} \cdot \omega(\sqrt{\log n})^2, \quad \alpha q = 2\sqrt{n}, \quad (3.2)$$

which means that $m = \bar{m} + nk = \tilde{O}(n)$, $1/\alpha = \tilde{O}(\sqrt{n})$ and $q = \tilde{O}(n)$. In both cases, we can set $2 \le d \le \tilde{O}(\sqrt{n})$.

As commented in [5, 37], the requirement $\alpha q \ge 2\sqrt{n}$ used for the theoretical worst-case reduction [50] is not tight, and it is better to mainly consider concrete hardness against known attacks when choosing actual parameters. For example, one can set $n = 450$, $\bar{m} = 6690$, $m = 10740$, $q = 3^9 \approx 2^{14.27}$, $\alpha q = 1.5$ to achieve a decryption error rate less than 2^{-100}, and a security level about 131-bit by the online LWE estimator [5]. In this case, the sizes of the public key and the secret key are about $nm\lceil \log_2 q \rceil \approx 8.64$ MB, and $\bar{m}nk(\log_2(\alpha q \cdot \omega(\sqrt{\log n})) + 1) \approx 16.15$ MB, respectively. For 128-bit security, we set $\kappa = 256$, the ciphertext size for encrypting a 256-bit message is $m\lceil \log_2 q \rceil + 512$ bits ≈ 19.73 KB.

3.4.4 *Compressing the Ciphertext*

As many lattice-based PKEs in the literature (e.g., [6, 15, 45]), it is possible to discard some lower bits of the ciphertext (and thus reduce the ciphertext size) without affecting the correctness of the PKEs (because those lower bits mainly carry noise). This can be seen as a modulus switch technique. Concretely, let $\mathsf{Switch}_{q,p}(\cdot) : \mathbb{Z}_q \to \mathbb{Z}_p$ be a function defined as $\mathsf{Switch}_{q,p}(x) = \lceil p/q \cdot x \rfloor \bmod p$. It is easy to check that for any $x \in \mathbb{Z}_q$, $x - \mathsf{Switch}_{p,q}(\mathsf{Switch}_{q,p}(x)) \le \lceil \frac{q}{2p} \rfloor$. Thus, for $p < q$, one can use $\mathsf{Switch}_{q,p}(\cdot)$ to compress the ciphertext in the encryption algorithm (i.e., applying to vectors in a coordinate-wise way), and use $\mathsf{Switch}_{p,q}(\cdot)$ to approximately recover the original ciphertext in the decryption algorithm. This can be simply seen as adding a noise of size at most $\lceil \frac{q}{2p} \rfloor$ to each coordinate of the lattice vectors in the ciphertext.

We cannot simply apply $\mathsf{Switch}_{q,p}(\cdot)$ to the ciphertext in a black-box way to the above CCA2-secure PKE, since this will affect both the correctness and the security. Instead, we have to plug it into the encryption algorithm to generate the ciphertext $C = (\mathbf{c}_1, \mathbf{c}_2', c_3, c_4')$ as follows (where p is an integer, and other notations are the same as before):

$$\begin{aligned} \mathbf{c}_1 &= \mathbf{A}^T\tilde{\mathbf{s}} + \mathbf{e}_1, & tag &= \mathrm{H}(\mathbf{c}_1), \\ \mathbf{c}_2' &= \mathsf{Switch}_{q,p}((\mathbf{B} + \mathsf{FRD}(tag)\mathbf{G}_b)^T\tilde{\mathbf{s}} + \mathbf{e}_2), & c_3 &= x + \mu, \\ \tau' &= \mathrm{H}(\mathbf{c}_2', c_3), & c_4' &= \tau' y + z. \end{aligned}$$

For the choice of $(n, m, q, \alpha q) = (450, 10740, 3^9, 1.5)$, we can set $p = 8$ to compress the ciphertext from previous 19.73 KB to 13.80 KB, while still keep the decryption error rate less than 2^{-100}. Note that we do not use this technique to compress $\mathbf{c}_1$, because unlike the error in $\mathbf{c}_2$, any error in $\mathbf{c}_1$ will be sharply amplified by a factor of $s_1(\mathbf{R})$ in decryption.

3.4.5 Security

In this section, we show that the above PKE is CCA2-secure. Formally, we have the following theorem:

Theorem 3.3 *Let positive integers $n, \bar{m}, b, d, q \in \mathbb{Z}$ and real $\alpha \in (0, 1)$ satisfy Eq. (3.1) or (3.2). If $\mathrm{LWE}_{n,\bar{m},q,\alpha}$ is hard and H is a collision-resistant hash function, then the above PKE scheme is CCA2-secure in the standard model.*

The proof uses a sequence of games $G_1, \ldots, G_{11}$, with G_1 being the real CCA2-security game (where the challenge ciphertext $C^* = (\mathbf{c}_1^*, \mathbf{c}_2^*, c_3^*, c_4^*)$ is honestly generated by first randomly choosing $\delta^* \xleftarrow{\$} \{0, 1\}$ and then encrypting μ_{δ^*}) and G_{11} a random game (where the challenge ciphertext C^* is essentially uniformly random, and thus the adversary's advantage in game G_{11} is negligible). The security is established by showing that G_1 and G_{11} are computationally indistinguishable in the adversary's view. We outline the changes of game G_i with respect to its previous game G_{i-1} in Table 3.1.

Proof We now give the formal proof of Theorem 3.3. Let $\mathcal{A}$ be an adversary which can break the CCA2-security of the above PKE with advantage ϵ. Let F_i be the event that $\mathcal{A}$ correctly guesses $\delta = \delta^*$ in game $i \in \{1, \ldots, 11\}$. By definition, the adversary's advantage $\mathrm{Adv}^{\text{ind-cca2}}_{\mathcal{PKE},\mathcal{A}}(\kappa)$ in game i is exactly $|\Pr[F_i] - 1/2|$.

Game G_1. This game is the real security game as defined in Sect. 3.1. Formally, the challenger C works as follows:

KeyGen. first randomly choose $\mathbf{A} \xleftarrow{\$} \mathbb{Z}_q^{n\times\bar{m}}$, $\mathbf{R} \xleftarrow{\$} (D_{\mathbb{Z}^{\bar{m}},\omega(\sqrt{\log n})})^{nk}$, and compute $\mathbf{B} = -\mathbf{A}\mathbf{R}$. Then, return the pair of public key $\mathsf{pk} = (\mathbf{A}, \mathbf{B})$ to the adversary $\mathcal{A}$, and keeps the secret key $\mathbf{R}$ private.

Phase I. Upon receiving a decryption query $C = (\mathbf{c}_1, \mathbf{c}_2, c_3, c_4)$, first compute

$$\mathbf{A}_{tag} = (\mathbf{A}\|\mathbf{B} + \mathsf{FRD}(tag)\mathbf{G}_b), \text{ and } \mathbf{u} = \begin{pmatrix} \mathbf{c}_1 \\ \mathbf{c}_2 \end{pmatrix},$$

where $tag = \mathrm{H}(\mathbf{c}_1)$. Then, compute $\tilde{\mathbf{s}} \leftarrow \mathsf{Solve}(\mathbf{A}_{tag}, \mathbf{R}, \mathbf{u})$, $\mathbf{v} = \mathsf{decode}_d(\tilde{\mathbf{s}})$, and parse $\mathbf{v} = x\|y\|z$, where $(x, y, z) \in (\mathbb{F}_{2^\kappa})^3$. Let $\mathbf{e}_1 = \mathbf{c}_1 - \mathbf{A}^T\tilde{\mathbf{s}}$ and $\mathbf{e}_2 = \mathbf{c}_2 - (\mathbf{B} + \mathsf{FRD}(tag)\mathbf{G}_b)^T\tilde{\mathbf{s}}$. Return $\perp$ to the adversary $\mathcal{A}$ if one of the following conditions holds:

- $\|\mathbf{e}_1\| > \alpha q\sqrt{\bar{m}}$, or
- $\|\mathbf{e}_2\| > \gamma\sqrt{nk}$ for prime q (or $\|\mathbf{e}_2\|_\infty > \gamma \cdot \omega(\sqrt{\log n})$ for $q = b^k$ a power of prime b), or

Table 3.1 Outline of the game sequences for proving Theorem 3.3 (where μ_0 and μ_1 are the challenge messages. We use $\stackrel{c}{\approx}$ and $\stackrel{s}{\approx}$ to represent the computational indistinguishability and statistical indistinguishability between two games, respectively.)

Games	Changes w.r.t. Previous Game	Note
G_1	Public key: $pk = (\mathbf{A}, \mathbf{B} = -\mathbf{AR})$, Secret key: $sk = \mathbf{R}$, Challenge C^* : $C^* = (\mathbf{c}_1^*, \mathbf{c}_2^*, c_3^*, c_4^*)$, where $\mathbf{c}_1^* = \mathbf{A}^T \bar{\mathbf{s}}^* + \mathbf{e}_1^*$, $\mathbf{c}_2^* = (\mathbf{B} + \mathsf{FRD}(tag^*)\mathbf{G}_b)^T \bar{\mathbf{s}}^* + \mathbf{e}_2^*$, $c_3^* = x^* + \mu_{\delta^*} \in \mathbb{F}_{2^\kappa}$ $c_4^* = \tau^* y^* + z^* \in \mathbb{F}_{2^\kappa}$, for some $\mathbf{s}^* \stackrel{\$}{\leftarrow} D_{\mathbb{Z}^n, \alpha q}$ $\mathbf{e}_1^* \stackrel{\$}{\leftarrow} D_{\mathbb{Z}^{\bar{m}}, \alpha q}$ $\mathbf{e}_2^* \stackrel{\$}{\leftarrow} D_{\mathbb{Z}^{nk}, \sqrt{\Vert\mathbf{e}_1^*\Vert^2 + \bar{m}(\alpha q)^2} \cdot \omega(\sqrt{\log n})}$ $x^*, y^*, z^* \stackrel{\$}{\leftarrow} \mathbb{F}_{2^\kappa}$ $\mathbf{v}^* = x^* \Vert y^* \Vert z^* \in \mathbb{Z}_d^n$ $\bar{\mathbf{s}}^* = \mathbf{s}^* + \mathsf{encode}_d(\mathbf{v}^*)$, $tag^* = \mathrm{H}(\mathbf{c}_1^*)$, $\tau^* = \mathrm{H}(\mathbf{c}_2^*, c_3)$. $\delta^* \stackrel{\$}{\leftarrow} \{0, 1\}$, Dec. query C : run $\mathsf{Dec}(sk, C)$ for any $C \neq C^*$	Real game
G_2	Generate $(\mathbf{c}_1^*, \mathbf{c}_2^*, x^*, y^*, z^*)$ before giving pk to the adversary (i.e., in the KeyGen phase)	The change is conceptual: $G_2 = G_1$
G_3	Immediately return $\perp$ to the decryption query with $C = (\mathbf{c}_1, \mathbf{c}_2, c_3, c_4)$ if $\mathbf{c}_1 \neq \mathbf{c}_1^* \wedge \mathrm{H}(\mathbf{c}_1) = \mathrm{H}(\mathbf{c}_1^*)$	By the collision resistance of H: $G_3 \stackrel{c}{\approx} G_2$
G_4	Immediately return $\perp$ to the decryption query with $C = (\mathbf{c}_1, \mathbf{c}_2, c_3, c_4)$ if $(\mathbf{c}_2, c_3) \neq (\mathbf{c}_2^*, c_3^*) \wedge \mathrm{H}(\mathbf{c}_2, c_3) = \mathrm{H}(\mathbf{c}_2^*, c_3^*)$	By the collision resistance of H: $G_4 \stackrel{c}{\approx} G_3$
G_5	Immediately return $\perp$ to the decryption query with $C = (\mathbf{c}_1, \mathbf{c}_2, c_3, c_4)$ in Phase I if $\mathbf{c}_1 = \mathbf{c}_1^*$	By the high min-entropy of $\mathbf{c}_1^*$: $G_5 \stackrel{s}{\approx} G_4$
G_6	Immediately return $\perp$ to the decryption query with $C = (\mathbf{c}_1, \mathbf{c}_2, c_3, c_4)$ in Phase II if $(\mathbf{c}_1, \mathbf{c}_2, c_3) = (\mathbf{c}_1^*, \mathbf{c}_2^*, c_3^*)$ or $\mathbf{c}_1 = \mathbf{c}_1^* \wedge (\mathbf{c}_2, c_3) \neq (\mathbf{c}_2^*, c_3^*) \wedge c_4 \neq \mathrm{H}(\mathbf{c}_2, c_3)y^* + z^*$	By the unique witness of LWE (i.e., Lemma 2.16) and the definition of Dec: $G_6 \stackrel{s}{\approx} G_5$
G_7	Immediately return $\perp$ to the decryption query with $C = (\mathbf{c}_1, \mathbf{c}_2, c_3, c_4)$ in Phase II if $\mathbf{c}_1 = \mathbf{c}_1^*$	By the pseudorandomness of LWE and the definition of Dec :[a] $G_7 \stackrel{c}{\approx} G_6$
G_8	Set $pk = (\mathbf{A}, \mathbf{B} = -\mathbf{AR}' - \mathsf{FRD}(tag^*)\mathbf{G}_b)$ and use $\mathbf{R}'$ to answer the decryption query $C = (\mathbf{c}_1, \mathbf{c}_2, c_3, c_4)$ if C does not satisfy the "immediate rejection" rules in game G_7 (which means that $tag = \mathrm{H}(\mathbf{c}_1) \neq \mathrm{H}(\mathbf{c}_1^*) = tag^*$)	By the properties of trapdoor generation and inversion algorithms, and the definition of Dec: $G_8 \stackrel{s}{\approx} G_7$
G_9	Use $\mathbf{R}'$ and $\mathbf{c}_1^*$ to generate $\mathbf{c}_2^* = (-\mathbf{R}')^T \mathbf{c}_1^* + \mathbf{e}_2' =$ $(\mathbf{B} + \mathsf{FRD}(tag^*)\mathbf{G}_b)^T \bar{\mathbf{s}}^* + (-\mathbf{R}')^T \mathbf{e}_1^* + \mathbf{e}_2'$, where $\mathbf{e}_2' \stackrel{\$}{\leftarrow} D_{\mathbb{Z}^{\bar{n}k}, \alpha q \sqrt{\bar{m}} \cdot \omega(\sqrt{\log n})}$	By Lemma 2.5 and the definition of Enc: $G_9 \stackrel{s}{\approx} G_8$
G_{10}	Choose $\mathbf{c}_1^* \stackrel{\$}{\leftarrow} \mathbb{Z}_q^{\bar{m}}$ at random	By the pseudorandomness of LWE: $G_{10} \stackrel{c}{\approx} G_9$
G_{11}	Choose $\mathbf{c}_2^* \stackrel{\$}{\leftarrow} \mathbb{Z}_q^{nk}$ at random	By Lemma 2.4: $G_{11} \stackrel{s}{\approx} G_{10}$

[a] The proof of this claim is relatively involved, and we will use the proof technique of game transitions based on failure events in [55]

- $c_4 \neq \mathrm{H}(\mathbf{c}_2, c_3)y + z \in \mathbb{F}_{2^\kappa}$.

Otherwise, return $\mu = c_3 - x \in \mathbb{F}_{2^\kappa}$ to the adversary $\mathcal{A}$.

Challenge. Upon receiving two challenge messages $(\mu_0, \mu_1) \in \mathbb{F}_{2^\kappa} \times \mathbb{F}_{2^\kappa}$ from the adversary $\mathcal{A}$, first randomly choose $\delta^* \stackrel{\$}{\leftarrow} \{0,1\}$, $\mathbf{s}^* \stackrel{\$}{\leftarrow} D_{\mathbb{Z}^n,\alpha q}$, $\mathbf{e}_1^* \stackrel{\$}{\leftarrow} D_{\mathbb{Z}^{\bar{m}},\alpha q}$, $\mathbf{e}_2^* \stackrel{\$}{\leftarrow} D_{\mathbb{Z}^{nk},\gamma}$ and $x^*, y^*, z^* \stackrel{\$}{\leftarrow} \mathbb{F}_{2^\kappa}$, where $\gamma = \sqrt{\|\mathbf{e}_1^*\|^2 + \bar{m}(\alpha q)^2} \cdot \omega(\sqrt{\log n})$. Then, interpret the bit-concatenation of $(x^*, y^*, z^*) \in (\mathbb{F}_{2^\kappa})^3$ as a vector $\mathbf{v}^* = x^* \| y^* \| z^* \in \mathbb{Z}_d^n$, and compute

$$\begin{aligned} \tilde{\mathbf{s}}^* &= \mathbf{s}^* + \mathsf{encode}_d(\mathbf{v}^*), & \mathbf{c}_1^* &= \mathbf{A}^T\tilde{\mathbf{s}}^* + \mathbf{e}_1^*, \\ \mathbf{c}_2^* &= (\mathbf{B} + \mathsf{FRD}(tag^*)\mathbf{G}_b)^T\tilde{\mathbf{s}}^* + \mathbf{e}_2^*, & c_3^* &= x^* + \mu_{\delta^*} \in \mathbb{F}_{2^\kappa}, \\ c_4^* &= \tau^* y^* + z^* \in \mathbb{F}_{2^\kappa}, \end{aligned}$$

where $tag^* = \mathrm{H}(\mathbf{c}_1^*) \in \mathbb{F}_{2^\kappa}$ and $\tau^* = \mathrm{H}(\mathbf{c}_2^*, c_3^*) \in \mathbb{F}_{2^\kappa}$. Finally, return the challenge ciphertext $C^* = (\mathbf{c}_1^*, \mathbf{c}_2^*, c_3^*, c_4^*)$ to $\mathcal{A}$.

Phase II. Upon receiving a decryption query $C = (\mathbf{c}_1, \mathbf{c}_2, c_3, c_4)$, directly return $\perp$ to the adversary $\mathcal{A}$ if $C = C^*$, otherwise answer this query as in Phase I.

By definition, we have the following lemma:

Lemma 3.6 $|\Pr[F_1] - 1/2| = \epsilon$.

Game G_2. This game is similar to game G_1 except that the challenger C changes the KeyGen and Challenge phases as follows:

KeyGen. first randomly choose $\mathbf{A} \stackrel{\$}{\leftarrow} \mathbb{Z}_q^{n\times\bar{m}}$, $\mathbf{R} \stackrel{\$}{\leftarrow} (D_{\mathbb{Z}^{\bar{m}},\omega(\sqrt{\log n})})^{nk}$, $\mathbf{s}^* \stackrel{\$}{\leftarrow} D_{\mathbb{Z}^n,\alpha q}$, $\mathbf{e}_1^* \stackrel{\$}{\leftarrow} D_{\mathbb{Z}^{\bar{m}},\alpha q}$, $\mathbf{e}_2^* \stackrel{\$}{\leftarrow} D_{\mathbb{Z}^{nk},\gamma}$ and $x^*, y^*, z^* \stackrel{\$}{\leftarrow} \mathbb{F}_{2^\kappa}$, where $\gamma = \sqrt{\|\mathbf{e}_1^*\|^2 + \bar{m}(\alpha q)^2} \cdot \omega(\sqrt{\log n})$. Then, interpret the bit-concatenation of $(x^*, y^*, z^*) \in (\mathbb{F}_{2^\kappa})^3$ as a vector $\mathbf{v}^* = x^* \| y^* \| z^* \in \mathbb{Z}_d^n$, and compute

$$\begin{aligned} \tilde{\mathbf{s}}^* &= \mathbf{s}^* + \mathsf{encode}_d(\mathbf{v}^*), & \mathbf{c}_1^* &= \mathbf{A}^T\tilde{\mathbf{s}}^* + \mathbf{e}_1^*, \\ \mathbf{B} &= -\mathbf{A}\mathbf{R} & \mathbf{c}_2^* &= (\mathbf{B} + \mathsf{FRD}(tag^*)\mathbf{G}_b)^T\tilde{\mathbf{s}}^* + \mathbf{e}_2^*, \end{aligned}$$

where $tag^* = \mathrm{H}(\mathbf{c}_1^*)$. Finally, give the public key $pk = (\mathbf{A}, \mathbf{B})$ to the adversary $\mathcal{A}$, keep the secret $sk = \mathbf{R}$ and $(\mathbf{c}_1^*, \mathbf{c}_2^*, x^*, y^*, z^*)$ secret.

Challenge. Upon receiving two challenge messages $(\mu_0, \mu_1) \in \mathbb{F}_{2^\kappa} \times \mathbb{F}_{2^\kappa}$ from the adversary $\mathcal{A}$, first choose a bit $\delta^* \stackrel{\$}{\leftarrow} \{0,1\}$ and retrieve $(\mathbf{c}_1^*, \mathbf{c}_2^*, x^*, y^*, z^*)$. Then, compute $c_3^* = x^* + \mu_{\delta^*} \in \mathbb{F}_{2^\kappa}, c_4^* = \tau^* y^* + z^* \in \mathbb{F}_{2^\kappa}$, where $\tau^* = \mathrm{H}(\mathbf{c}_2^*, c_3^*)$. Finally, return the challenge ciphertext $C^* = (\mathbf{c}_1^*, \mathbf{c}_2^*, c_3^*, c_4^*)$ to $\mathcal{A}$.

Lemma 3.7 *Games G_2 and G_1 are identical in the adversary's view. Moreover,* $\Pr[F_2] = \Pr[F_1]$.

Proof This lemma follows from the fact that $(\mathbf{c}_1^*, \mathbf{c}_2^*, x^*, y^*, z^*)$ is independent from the adversary's choices of the challenge messages, and game G_2 is essentially a conceptual change of game G_1 in the adversary's view. □

Game G_3. This game is similar to game G_2 except that the challenger C immediately returns $\perp$ to the decryption query $C = (\mathbf{c}_1, \mathbf{c}_2, c_3, c_4)$ from the adversary $\mathcal{A}$ if $\mathbf{c}_1 \neq \mathbf{c}_1^* \wedge \mathrm{H}(\mathbf{c}_1) = \mathrm{H}(\mathbf{c}_1^*)$.

Lemma 3.8 *If* H *is a collision-resistant hash function, then games* G_3 *and* G_2 *are computationally indistinguishable. Moreover,* $|\Pr[F_3] - \Pr[F_2]| \leq \mathsf{negl}(\kappa)$.

Proof Let $\mathcal{E}$ be the event that the adversary makes a decryption query $C = (\mathbf{c}_1, \mathbf{c}_2, c_3, c_4)$ in Phase I such that $\mathbf{c}_1 \neq \mathbf{c}_1^* \wedge \mathrm{H}(\mathbf{c}_1) = \mathrm{H}(\mathbf{c}_1^*)$. Note that if $\mathcal{E}$ can only happen with negligible probability, then games G_3 and G_2 are computationally indistinguishable in the adversary's view. Now, we show that if there is a PPT adversary $\mathcal{A}$ that makes $\mathcal{E}$ happen with non-negligible probability, there is a PPT adversary $\mathcal{F}$ that finds a collision of H with the same probability by honestly simulating the attack environment for $\mathcal{A}$ as in game G_3. Whenever $\mathcal{A}$ outputs a ciphertext $C = (\mathbf{c}_1, \mathbf{c}_2, c_3, c_4)$ such that $\mathbf{c}_1 \neq \mathbf{c}_1^* \wedge \mathrm{H}(\mathbf{c}_1) = \mathrm{H}(\mathbf{c}_1^*)$ at some time in Phase I, $\mathcal{F}$ returns the pairs $(\mathbf{c}_1, \mathbf{c}_1^*)$ as its own output and aborts. Obviously, the probability that $\mathcal{F}$ succeeds is equal to the probability that $\mathcal{A}$ makes $\mathcal{E}$ happen. Thus, under the assumption that H is collision-resistant, the probability that $\mathcal{E}$ happens is negligible, which completes the proof. □

Game G_4. This game is similar to game G_3 except that the challenger C immediately returns $\perp$ to the decryption query $C = (\mathbf{c}_1, \mathbf{c}_2, c_3, c_4)$ from the adversary $\mathcal{A}$ if $(\mathbf{c}_2, c_3) \neq (\mathbf{c}_2^*, c_3^*) \wedge \mathrm{H}(\mathbf{c}_2, c_3) = \mathrm{H}(\mathbf{c}_2^*, c_3^*)$.

Lemma 3.9 *If* H *is a collision-resistant hash function, then games* G_4 *and* G_3 *are computationally indistinguishable. Moreover,* $|\Pr[F_4] - \Pr[F_3]| \leq \mathsf{negl}(\kappa)$.

Proof The proof is the same as that of Lemma 3.8, we omit the details. □

Game G_5. This game is similar to game G_4 except that the challenger C immediately returns $\perp$ to the decryption query $C = (\mathbf{c}_1, \mathbf{c}_2, c_3, c_4)$ from the adversary $\mathcal{A}$ in Phase I if $\mathbf{c}_1 = \mathbf{c}_1^*$.

Lemma 3.10 *Let positive integers* $n, \bar{m}, b, d, q \in \mathbb{Z}$ *and real* $\alpha \in (0, 1)$ *satisfy Eq. (3.1) or (3.2). Then, games* G_5 *and* G_4 *are statistically indistinguishable. Moreover,* $|\Pr[F_5] - \Pr[F_4]| \leq \mathsf{negl}(\kappa)$.

Proof Let $\mathcal{E}$ be the event that the adversary makes a decryption query $C = (\mathbf{c}_1, \mathbf{c}_2, c_3, c_4)$ in Phase I such that $\mathbf{c}_1 = \mathbf{c}_1^*$. Note that if $\mathcal{E}$ does not happen, then games G_5 and G_4 are identical in the adversary's view. Thus, it is enough to show that $\Pr[\mathcal{E}]$ is negligible for any (unbounded) adversary $\mathcal{A}$ making at most a polynomial number of decryption queries in Phase I. Note that in both games G_4 and G_5, the ciphertext part $\mathbf{c}_1^* = \mathbf{A}^T\tilde{\mathbf{s}}^* + \mathbf{e}_1^*$ is always generated by using $\tilde{\mathbf{s}}^* = \mathbf{s}^* + \mathsf{encode}_d(\mathbf{v}^*)$ and $\mathbf{e}_1^* \xleftarrow{\$} D_{\mathbb{Z}^{\bar{m}},\alpha q}$, where $\mathbf{s}^* \xleftarrow{\$} D_{\mathbb{Z}^n,\alpha q}$, $x^*, y^*, z^* \xleftarrow{\$} \mathbb{F}_{2^\kappa}$ and $\mathbf{v}^* = x^* \| y^* \| z^* \in \mathbb{Z}_d^n$. By the high min-entropy of the Gaussian distribution, we have that $\mathbf{c}_1^*$ has min-entropy at least κ, where κ is the security parameter. In other words, the probability that for any (unbounded) adversary to output $\mathbf{c}_1 = \mathbf{c}_1^*$ in Phase I (i.e., before seeing $\mathbf{c}_1^*$) is negligible. This means that if $\mathcal{A}$ can make $\mathcal{E}$ happen with non-negligible probability, which completes the proof. □

Game G_6. This game is similar to game G_5 except that the challenger C immediately returns $\perp$ to the decryption query $C = (\mathbf{c}_1, \mathbf{c}_2, c_3, c_4)$ from the adversary $\mathcal{A}$ in Phase II if $(\mathbf{c}_1, \mathbf{c}_2, c_3) = (\mathbf{c}_1^*, \mathbf{c}_2^*, c_3^*)$ or $\mathbf{c}_1 = \mathbf{c}_1^* \wedge (\mathbf{c}_2, c_3) \neq (\mathbf{c}_2^*, c_3^*) \wedge c_4 \neq \mathrm{H}(\mathbf{c}_2, c_3)y^* + z^*$, where $C^* = (\mathbf{c}_1^*, \mathbf{c}_2^*, c_3^*, \mathbf{c}_4^*)$ is the challenge ciphertext.

Lemma 3.11 *Let positive integers $n, \bar{m}, b, d, q \in \mathbb{Z}$ and real $\alpha \in (0, 1)$ satisfy Eq. (3.1) or (3.2), then games G_6 and G_5 are statistically indistinguishable. Moreover,* $|\Pr[F_6] - \Pr[F_5]| \leq \mathsf{negl}(\kappa)$.

Proof It suffices to show that the challenger C in game G_5 will always return $\perp$ to a decryption query $C = (\mathbf{c}_1, \mathbf{c}_2, c_3, c_4) \neq C^* = (\mathbf{c}_1^*, \mathbf{c}_2^*, c_3^*, \mathbf{c}_4^*)$ from the adversary $\mathcal{A}$ in Phase II if $(\mathbf{c}_1, \mathbf{c}_2, c_3) = (\mathbf{c}_1^*, \mathbf{c}_2^*, c_3^*)$ or $\mathbf{c}_1 = \mathbf{c}_1^* \wedge (\mathbf{c}_2, c_3) \neq (\mathbf{c}_2^*, c_3^*) \wedge c_4 \neq \mathrm{H}(\mathbf{c}_2, c_3)y^* + z^*$ in Phase II except with negligible probability. Note that given a decryption query $C = (\mathbf{c}_1, \mathbf{c}_2, c_3, c_4) \neq C^*$, the challenger C in game G_5 will first compute $tag = \mathrm{H}(\mathbf{c}_1)$, and

$$\mathbf{A}_{tag} = (\mathbf{A} \| \mathbf{B} + \mathsf{FRD}(tag)\mathbf{G}_b), \text{ and } \mathbf{u} = \begin{pmatrix} \mathbf{c}_1 \\ \mathbf{c}_2 \end{pmatrix}.$$

Then, compute $\tilde{\mathbf{s}} \leftarrow \mathsf{Solve}(\mathbf{A}_{tag}, \mathbf{R}, \mathbf{u})$, $\mathbf{v} = \mathsf{decode}_d(\tilde{\mathbf{s}})$, and parse $\mathbf{v} = x \| y \| z$, where $(x, y, z) \in (\mathbb{F}_{2^\kappa})^3$. Let $\mathbf{e}_1 = \mathbf{c}_1 - \mathbf{A}^T\tilde{\mathbf{s}}$ and $\mathbf{e}_2 = \mathbf{c}_2 - (\mathbf{B} + \mathsf{FRD}(tag)\mathbf{G}_b)^T\tilde{\mathbf{s}}$. Finally, return $\perp$ to the adversary if one of the following conditions holds:

- $\|\mathbf{e}_1\| > \alpha q\sqrt{\bar{m}}$, or
- $\|\mathbf{e}_2\| > \gamma\sqrt{nk}$ for prime q (or $\|\mathbf{e}_2\|_\infty > \gamma \cdot \omega(\sqrt{\log n})$ for $q = b^k$ a power of prime b), or
- $c_4 \neq \mathrm{H}(\mathbf{c}_2, c_3)y + z \in \mathbb{F}_{2^\kappa}$.

Otherwise, return $\mu = c_3 - x \in \mathbb{F}_{2^\kappa}$.

Clearly, the challenger C in Game G_5 will not return $\perp$ to the decryption query $C = (\mathbf{c}_1, \mathbf{c}_2, c_3, c_4) \neq C^*$ only when $\|\mathbf{e}_1\|_\infty \leq \|\mathbf{e}_1\| \leq \alpha q\sqrt{\bar{m}}$ and $c_4 = \mathrm{H}(\mathbf{c}_2, c_3)y + z$. In addition, given $\mathbf{c}_1^* = \mathbf{A}^T\tilde{\mathbf{s}}^* + \mathbf{e}_1^*$ for $\mathbf{e}_1^* \xleftarrow{\$} D_{\mathbb{Z}^{\bar{m}}, \alpha q}$, the challenger C in Game G_5 will not return $\perp$ to a decryption query $C = (\mathbf{c}_1 = \mathbf{c}_1^*, \mathbf{c}_2, c_3, c_4)$ only if $c_4 = \mathrm{H}(\mathbf{c}_2, c_3)y^* + z^*$ except with negligible probability, since in this case we always have $(\tilde{\mathbf{s}}, \mathbf{e}_1) = (\tilde{\mathbf{s}}^*, \mathbf{e}_1^*)$ with overwhelming probability by the unique witness property in Lemma 2.16, which in turn implies that $\mathbf{v} = \mathbf{v}^*$ and $(x, y, z) = (x^*, y^*, z^*)$ by the correctness of decode_d. In other words, the challenger C in Game G_5 will always return $\perp$ to a decryption query $C = (\mathbf{c}_1 = \mathbf{c}_1^*, \mathbf{c}_2, c_3, c_4) \neq C^* = (\mathbf{c}_1^*, \mathbf{c}_2^*, c_3^*, \mathbf{c}_4^*)$ from the adversary $\mathcal{A}$ in Phase II if $(\mathbf{c}_1, \mathbf{c}_2, c_3) = (\mathbf{c}_1^*, \mathbf{c}_2^*, c_3^*)$ or $\mathbf{c}_1 = \mathbf{c}_1^* \wedge (\mathbf{c}_2, c_3) \neq (\mathbf{c}_2^*, c_3^*) \wedge c_4 \neq \mathrm{H}(\mathbf{c}_2, c_3)y^* + z^*$ holds, except with negligible probability. This completes the proof. □

Game G_7. This game is similar to game G_6 except that the challenger C immediately returns $\perp$ to the decryption query $C = (\mathbf{c}_1, \mathbf{c}_2, c_3, c_4)$ from the adversary $\mathcal{A}$ in Phase II if $\mathbf{c}_1 = \mathbf{c}_1^*$.

Note that the goal is to show games G_7 and G_6 are computationally indistinguishable under the LWE assumption, but for technical reason it is difficult to do this in game G_7. Fortunately, we can still continue the game sequences by using the proof strategy (i.e., game transitions based on failure events) in [55]. Formally, for $i \in \{6, 7, 8, \ldots, 11\}$, let E_i be the failure event in game G_i that the adversary makes a decryption query with $C = (\mathbf{c}_1^*, \mathbf{c}_2, c_3, c_4)$ such that $\tau = \mathrm{H}(\mathbf{c}_2, c_3) \neq \tau^* \wedge c_4 = \tau y^* + z^*$.

Lemma 3.12 *If E_7 and E_6 do not happen, then games G_7 and G_6 are identical in the adversary's view. Moreover,* $\Pr[F_7|\neg E_7] = \Pr[F_6|\neg E_6]$ *and* $\Pr[E_7] = \Pr[E_6]$.

Proof Let $C^* = (\mathbf{c}_1^*, \mathbf{c}_2^*, c_3^*, c_4^*)$ be the corresponding challenge ciphertext, where $\tau^* = \mathrm{H}(\mathbf{c}_2^*, c_3^*)$, and $c_4^* = \tau^* y^* + z^*$ for some $y^*, z^* \in \{0, 1\}^\kappa$. Note that upon receiving a decryption query with $C = (\mathbf{c}_1, \mathbf{c}_2, c_3, c_4)$ in Phase II, the challenger C in both games will always return $\perp$ if $(\mathbf{c}_2, c_3) \neq (\mathbf{c}_2^*, c_3^*) \wedge \tau = \mathrm{H}(\mathbf{c}_2, c_3) = \tau^*$. Moreover, the challenger C in game G_6 will return $\perp$ if $(\mathbf{c}_1, \mathbf{c}_2, c_3) = (\mathbf{c}_1^*, \mathbf{c}_2^*, c_3^*)$ or $\mathbf{c}_1 = \mathbf{c}_1^* \wedge (\mathbf{c}_2, c_3) \neq (\mathbf{c}_2^*, c_3^*) \wedge c_4 \neq \mathrm{H}(\mathbf{c}_2, c_3)y^* + z^*$ holds. In other words, the only difference between games G_7 and G_6 is that the challenger C in game G_7 also returns $\perp$ to the decryption query $C = (\mathbf{c}_1^*, \mathbf{c}_2, c_3, c_4) \neq C^*$ even if $\tau = \mathrm{H}(\mathbf{c}_2, c_3) \neq \mathrm{H}(\mathbf{c}_2^*, c_3^*) = \tau^* \wedge c_4 = \tau y^* + z^*$. Clearly, if E_7 and E_6 do not happen, then both games are identical in the adversary's view. In particular, the adversary's view in game G_7 before E_7 happens is essentially identical to that in game G_6. Thus, we have $\Pr[F_7|\neg E_7] = \Pr[F_6|\neg E_6]$ and $\Pr[E_7] = \Pr[E_6]$. □

Game G_8. This game is similar to game G_7 except that the challenger C changes the KeyGen phase and handles the decryption queries as follows:

KeyGen. first randomly choose $\mathbf{A} \stackrel{\$}{\leftarrow} \mathbb{Z}_q^{n \times \bar{m}}$, $\mathbf{R}' \stackrel{\$}{\leftarrow} (D_{\mathbb{Z}^{\bar{m}}, \omega(\sqrt{\log n})})^{nk}$, $\mathbf{s}^* \stackrel{\$}{\leftarrow} D_{\mathbb{Z}^n, \alpha q}$, $\mathbf{e}_1^* \stackrel{\$}{\leftarrow} D_{\mathbb{Z}^{\bar{m}}, \alpha q}$, $\mathbf{e}_2^* \stackrel{\$}{\leftarrow} D_{\mathbb{Z}^{nk}, \gamma}$ and $x^*, y^*, z^* \stackrel{\$}{\leftarrow} \mathbb{F}_{2^\kappa}$, where $\gamma = \sqrt{\|\mathbf{e}_1^*\|^2 + \bar{m}(\alpha q)^2} \cdot \omega(\sqrt{\log n})$. Then, interpret the bit-concatenation of $(x^*, y^*, z^*) \in (\mathbb{F}_{2^\kappa})^3$ as a vector $\mathbf{v}^* = x^* \| y^* \| z^* \in \mathbb{Z}_d^n$, and compute

$$\begin{aligned} &\tilde{\mathbf{s}}^* = \mathbf{s}^* + \mathsf{encode}_d(\mathbf{v}^*), \qquad \mathbf{c}_1^* = \mathbf{A}^T \tilde{\mathbf{s}}^* + \mathbf{e}_1^*, \\ &\mathbf{B} = -\mathbf{A}\mathbf{R}' - \mathsf{FRD}(tag^*)\mathbf{G}_b, \\ &\mathbf{c}_2^* = (\mathbf{B} + \mathsf{FRD}(tag^*)\mathbf{G}_b)^T \tilde{\mathbf{s}}^* + \mathbf{e}_2^* = -(\mathbf{R}')^T \mathbf{A}^T \tilde{\mathbf{s}}^* + \mathbf{e}_2^*, \end{aligned}$$

where $tag^* = \mathrm{H}(\mathbf{c}_1^*)$. Finally, give the public key $pk = (\mathbf{A}, \mathbf{B})$ to the adversary $\mathcal{A}$, and keep $(\mathbf{R}', \mathbf{c}_1^*, \mathbf{c}_2^*, x^*, y^*, z^*)$ secret.

Decryption Query. Upon receiving a decryption query $C = (\mathbf{c}_1, \mathbf{c}_2, c_3, c_4)$ from the adversary $\mathcal{A}$, return $\perp$ to $\mathcal{A}$ if this query can be immediately responded with $\perp$ using the rules in previous games. Otherwise, first set

$$\mathbf{A}_{tag} = (\mathbf{A} \| \mathbf{B} + \mathsf{FRD}(tag)\mathbf{G}_b), \text{ and } \mathbf{u} = \begin{pmatrix} \mathbf{c}_1 \\ \mathbf{c}_2 \end{pmatrix},$$

where $tag = \mathrm{H}(\mathbf{c}_1)$. Then, compute $\tilde{\mathbf{s}} \leftarrow \mathsf{Solve}(\mathbf{A}_{tag}, \mathbf{R}', \mathbf{u})$, $\mathbf{v} = \mathsf{decode}_d(\tilde{\mathbf{s}})$, and parse $\mathbf{v} = x \| y \| z$, where $(x, y, z) \in (\mathbb{F}_{2^\kappa})^3$. Let $\mathbf{e}_1 = \mathbf{c}_1 - \mathbf{A}^T \tilde{\mathbf{s}}$ and $\mathbf{e}_2 =$

$\mathbf{c}_2 - (\mathbf{B} + \mathsf{FRD}(tag)\mathbf{G}_b)^T\tilde{\mathbf{s}}$. Return $\perp$ to the adversary $\mathcal{A}$ if one of the following conditions holds:

- $\|\mathbf{e}_1\| > \alpha q\sqrt{\bar{m}}$, or
- $\|\mathbf{e}_2\| > \gamma\sqrt{nk}$ for prime q (or $\|\mathbf{e}_2\|_\infty > \gamma \cdot \omega(\sqrt{\log n})$ for $q = b^k$ a power of prime b), or
- $c_4 \neq \mathrm{H}(\mathbf{c}_2, c_3)y + z$.

Otherwise, return $\mu = c_3 - x$ to the adversary $\mathcal{A}$.

Lemma 3.13 *Let positive integers $n, \bar{m}, b, d, q \in \mathbb{Z}$ and real $\alpha \in (0, 1)$ satisfy Eq. (3.1) or (3.2). Then, games G_8 and G_7 are statistically indistinguishable. Moreover, $|\Pr[F_8|\neg E_8] - \Pr[F_7|\neg E_7]| \leq \mathsf{negl}(\kappa)$ and $|\Pr[E_8] - \Pr[E_7]| \leq \mathsf{negl}(\kappa)$.*

Proof Note that the only differences between games G_8 and G_7 are the generation of the public key $pk = (\mathbf{A}, \mathbf{B})$ and the responses to the decryption queries. Concretely, in game G_7 the matrix $\mathbf{B} = -\mathbf{AR}$ is generated by using $\mathbf{R} \xleftarrow{\$} (D_{\mathbb{Z}^{\bar{m}},\omega(\sqrt{\log n})})^{nk}$, while in game G_7 the matrix $\mathbf{B} = -\mathbf{AR}' - \mathsf{FRD}(tag^*)\mathbf{G}_b$, where $\mathbf{R}' \xleftarrow{\$} (D_{\mathbb{Z}^{\bar{m}},\omega(\sqrt{\log n})})^{nk}$. Since $\mathbf{A} \in \mathbb{Z}_q^{n\times\bar{m}}$ is always uniformly chosen at random in both games, we have that $-\mathbf{AR}$ and $-\mathbf{AR}'$ are statistically close to uniform distribution over $\mathbb{Z}_q^{n\times nk}$ by Lemma 2.4. Namely, the public keys in games G_8 and G_7 are statistically close (and tag^* is statistically hidden in game G_8).

It suffices to show that in the adversary's view, the responses to the decryption queries are indistinguishable in games G_8 and G_7. Since for a decryption query $C = (\mathbf{c}_1, \mathbf{c}_2, c_3, c_4)$, the challenger will use the same rules to check if the query can be immediately responded with $\perp$ in both games, we only have to consider the decryption query $C = (\mathbf{c}_1, \mathbf{c}_2, c_3, c_4)$ that needs the challenger C to perform the decryption operation. By the definition of game G_7, we must have that $tag = \mathrm{H}(\mathbf{c}_1) \neq tag^*$ holds for such decryption query $C = (\mathbf{c}_1, \mathbf{c}_2, c_3, c_4)$. Note that in game G_7, the challenger has the real secret key $sk = \mathbf{R}$, and can run the decryption algorithm to handle this query. We now show that the challenger C in game G_8 can almost perfectly simulate the decryption operation. Recall that $pk = (\mathbf{A}, \mathbf{B} = -\mathbf{AR}' - \mathsf{FRD}(tag^*)\mathbf{G}_b)$, conditioned on $tag = \mathrm{H}(\mathbf{c}_1) \neq tag^*$ we have that $\mathbf{R}'$ is a valid trapdoor for $\mathbf{A}_{tag} = (\mathbf{A}\|\mathbf{B} + \mathsf{FRD}(tag)\mathbf{G}_b)$, and thus can be used to compute $\tilde{\mathbf{s}} \leftarrow \mathsf{Solve}(\mathbf{A}_{tag}, \mathbf{R}', \mathbf{u})$. Now, either there exists a tuple $(\tilde{\mathbf{s}}, \mathbf{e}_1, \mathbf{e}_2)$ such that $\|\mathbf{e}_1\| \leq \alpha q\sqrt{\bar{m}}$, $\|\mathbf{e}_2\| \leq \gamma\sqrt{nk}$ for prime q (or $\|\mathbf{e}_2\|_\infty \leq \gamma \cdot \omega(\sqrt{\log n})$ for $q = b^k$), $\mathbf{c}_1 = \mathbf{A}^T\tilde{\mathbf{s}} + \mathbf{e}_1$ and $\mathbf{c}_2 = (\mathbf{B} + \mathsf{FRD}(tag)\mathbf{G}_b)^T\tilde{\mathbf{s}} + \mathbf{e}_2$, or there does not. For the latter case, the challenger will always return $\perp$ in both games. While for the former case, the challenger C in game G_8 can recover $\tilde{\mathbf{s}}$ as long as $\|(\mathbf{R}')^T\mathbf{e}_1 + \mathbf{e}_2\| < \frac{q}{2\sqrt{b^2+1}}$ for prime q (or $\|(\mathbf{R}')^T\mathbf{e}_1 + \mathbf{e}_2\|_\infty < \frac{q}{2b}$ for $q = b^k$), which is essentially the same constraint for a correct decryption using $sk = \mathbf{R}$ in game G_7. Since both $\mathbf{R}$ and $\mathbf{R}'$ are chosen from the same Gaussian distribution, by Lemma 2.3 we have that the inequality will holds with the same overwhelming probability conditioned on $\|\mathbf{e}_1\| \leq \alpha q\sqrt{m}$ and $\|\mathbf{e}_2\| \leq \gamma\sqrt{nk}$ for prime q (or $\|\mathbf{e}_2\|_\infty \leq \gamma \cdot \omega(\sqrt{\log n})$ for $q = b^k$). By the fact that the challengers in both games will perform the same operations after obtaining $\tilde{\mathbf{s}}$, we have that the responses to such kind of decryption queries are identical in both games except with negligible probability. This finishes the proof. □

Remark 3.1 Note that the challenger in game G_8 actually does not have the "real" secret key, which implies that the adversary cannot obtain extra information about the secret key from the decryption queries (except what is obtained from the public key $pk = (\mathbf{A}, \mathbf{B})$). This fact will be used in later proofs.

Game G_9. This game is similar to game G_8 except that the challenger C changes the KeyGen phase as follows:

KeyGen. first randomly choose $\mathbf{A} \xleftarrow{\$} \mathbb{Z}_q^{n\times\bar{m}}, \mathbf{R}' \xleftarrow{\$} (D_{\mathbb{Z}^{\bar{m}},\omega(\sqrt{\log n})})^{nk}, \mathbf{s}^* \xleftarrow{\$} D_{\mathbb{Z}^n,\alpha q}$ and $\mathbf{e}_1^* \xleftarrow{\$} D_{\mathbb{Z}^{\bar{m}},\alpha q}, \mathbf{e}_2' \xleftarrow{\$} D_{\mathbb{Z}^{nk},r}$ and $x^*, y^*, z^* \xleftarrow{\$} \mathbb{F}_{2^\kappa}$, where $r = \alpha q\sqrt{\bar{m}} \cdot \omega(\sqrt{\log n})$. Then, interpret the bit-concatenation of $(x^*, y^*, z^*) \in (\mathbb{F}_{2^\kappa})^3$ as a vector $\mathbf{v}^* = x^*\|y^*\|z^* \in \mathbb{Z}_d^n$, and compute

$$\tilde{\mathbf{s}}^* = \mathbf{s}^* + \mathsf{encode}_d(\mathbf{v}^*), \qquad \mathbf{c}_1^* = \mathbf{A}^T\tilde{\mathbf{s}}^* + \mathbf{e}_1^*,$$
$$\mathbf{B} = -\mathbf{A}\mathbf{R}' - \mathsf{FRD}(tag^*)\mathbf{G}_b, \qquad \mathbf{c}_2^* = (-\mathbf{R}')^T\mathbf{c}_1^* + \mathbf{e}_2'$$

,

where $tag^* = \mathrm{H}(\mathbf{c}_1^*)$. Finally, give the public key $pk = (\mathbf{A}, \mathbf{B})$ to $\mathcal{A}$, and keep $(\mathbf{R}', \mathbf{c}_1^*, \mathbf{c}_2^*, x^*, y^*, z^*)$ secret.

Lemma 3.14 *Let positive integers $n, \bar{m}, b, d, q \in \mathbb{Z}$ and real $\alpha \in (0, 1)$ satisfy Eq. (3.1) or (3.2). Then, games G_9 and G_8 are statistically indistinguishable. Moreover,* $|\Pr[F_9|\neg E_9] - \Pr[F_8|\neg E_8]| \le \mathsf{negl}(\kappa)$ *and* $|\Pr[E_9] - \Pr[E_8]| \le \mathsf{negl}(\kappa)$.

Proof Note that the only difference between games G_9 and G_8 is the generation of $\mathbf{c}_2^*$. In game G_8, $\mathbf{c}_2^* = (\mathbf{B} + \mathsf{FRD}(tag^*)\mathbf{G}_b)^T\tilde{\mathbf{s}}^* + \mathbf{e}_2^* = -(\mathbf{R}')^T\mathbf{A}^T\tilde{\mathbf{s}}^* + \mathbf{e}_2^*$ is generated by using $\mathbf{e}_2^* \xleftarrow{\$} D_{\mathbb{Z}^{nk},\gamma}$ where $\gamma = \sqrt{\|\mathbf{e}_1^*\|^2 + \bar{m}(\alpha q)^2} \cdot \omega(\sqrt{\log n})$, while in game G_9, $\mathbf{c}_2^* = (-\mathbf{R}')^T\mathbf{c}_1^* + \mathbf{e}_2'$ is generated by using $\mathbf{e}_2' \xleftarrow{\$} D_{\mathbb{Z}^{nk},r}$. Since $\mathbf{c}_1^* = \mathbf{A}^T\tilde{\mathbf{s}}^* + \mathbf{e}_1^*$ for some $\mathbf{e}_1^* \xleftarrow{\$} D_{\mathbb{Z}^{\bar{m}},\alpha q}$, we have that $\mathbf{c}_2^* = (-\mathbf{R}')^T\mathbf{c}_1^* + \mathbf{e}_2' = (-\mathbf{R}')^T\mathbf{A}^T\tilde{\mathbf{s}}^* + \tilde{\mathbf{e}}_2$ for some $\tilde{\mathbf{e}}_2 = (-\mathbf{R}')^T\mathbf{e}_1^* + \mathbf{e}_2'$ which is distributed statistically close to $D_{\mathbb{Z}^{nk},\gamma}$ by Lemma 2.5. Thus, the distributions of $\mathbf{c}_2^*$ in games G_9 and G_8 are actually statistically close, which in turn shows that both games are statistically indistinguishable in the adversary's view. □

Game G_{10}. This game is similar to game G_9 except that the challenger C changes the KeyGen phase as follows:

KeyGen. first randomly choose $\mathbf{A} \xleftarrow{\$} \mathbb{Z}_q^{n\times\bar{m}}, \mathbf{b} \xleftarrow{\$} \mathbb{Z}_q^{m_1}, \mathbf{R}' \xleftarrow{\$} (D_{\mathbb{Z}^{\bar{m}},\omega(\sqrt{\log n})})^{nk}$, $\mathbf{e}_2' \xleftarrow{\$} D_{\mathbb{Z}^{nk},r}$ and $x^*, y^*, z^* \xleftarrow{\$} \mathbb{F}_{2^\kappa}$, where $r = \alpha q\sqrt{\bar{m}} \cdot \omega(\sqrt{\log n})$. Then, compute

$$\mathbf{c}_1^* = \mathbf{b}, \qquad \mathbf{B} = -\mathbf{A}\mathbf{R}' - \mathsf{FRD}(tag^*)\mathbf{G}_b,$$
$$\mathbf{c}_2^* = (-\mathbf{R}')^T\mathbf{c}_1^* + \mathbf{e}_2'$$

,

where $tag^* = \mathrm{H}(\mathbf{c}_1^*)$. Finally, give the public key $pk = (\mathbf{A}, \mathbf{B})$ to $\mathcal{A}$, and keep $(\mathbf{R}', \mathbf{c}_1^*, \mathbf{c}_2^*, x^*, y^*, z^*)$ secret.

Lemma 3.15 *If* $\mathrm{LWE}_{n,\bar{m},q,\alpha}$ *is hard, then games* G_{10} *and* G_9 *are computationally indistinguishable. Moreover,* $|\Pr[F_{10}|\neg E_{10}] - \Pr[F_9|\neg E_9]| \leq \mathsf{negl}(\kappa)$ *and* $|\Pr[E_{10}] - \Pr[E_9]| \leq \mathsf{negl}(\kappa)$.

Proof We prove this lemma by showing that if there is a PPT adversary $\mathcal{A}$ that distinguishes game G_{10} from G_9 with non-negligible advantage, then there is an efficient algorithm $\mathcal{B}$ that solves the $\mathrm{LWE}_{n,\bar{m},q,\alpha}$ problem with the same advantage by interacting with $\mathcal{A}$.

Formally, given an LWE challenge tuple $(\mathbf{A}, \mathbf{b}) \in \mathbb{Z}_q^{n\times\bar{m}} \times \mathbb{Z}_q^{\bar{m}}$, $\mathcal{B}$ randomly chooses $\mathbf{R}' \xleftarrow{\$} (D_{\mathbb{Z}^{\bar{m}},\omega(\sqrt{\log n})})^{nk}$, $\mathbf{e}_2' \xleftarrow{\$} D_{\mathbb{Z}^{nk},r}$ and $x^*, y^*, z^* \xleftarrow{\$} \mathbb{F}_{2^\kappa}$, where $r = \alpha q\sqrt{\bar{m}} \cdot \omega(\sqrt{\log n})$. Then, it interprets the bit-concatenation of $(x^*, y^*, z^*) \in (\mathbb{F}_{2^\kappa})^3$ as a vector $\mathbf{v}^* = x^*\|y^*\|z^* \in \mathbb{Z}_d^n$ and computes

$$\begin{aligned} \mathbf{c}_1^* &= \mathbf{b} + \mathbf{A}^T\mathsf{encode}_d(\mathbf{v}^*), \qquad \mathbf{B} = -\mathbf{A}\mathbf{R}' - \mathsf{FRD}(tag^*)\mathbf{G}_b, \\ \mathbf{c}_2^* &= (-\mathbf{R}')^T\mathbf{c}_1^* + \mathbf{e}_2' \end{aligned}$$

,

where $tag^* = \mathrm{H}(\mathbf{c}_1^*)$. Then, $\mathcal{B}$ sets the public key $pk = (\mathbf{A}, \mathbf{B})$, and keeps $(\mathbf{R}', \mathbf{c}_1^*, \mathbf{c}_2^*, x^*, y^*, z^*)$ private. Finally, $\mathcal{B}$ gives pk to the adversary $\mathcal{A}$, simulates the attack environment the same as in game G_9 and returns whatever $\mathcal{A}$ outputs as its own output.

Now, if $(\mathbf{A}, \mathbf{b}) \in \mathbb{Z}_q^{n\times\bar{m}} \times \mathbb{Z}_q^{\bar{m}}$ is a valid LWE tuple, i.e., $\mathbf{b} = \mathbf{A}^T\mathbf{s}^* + \mathbf{e}^*$ for some $\mathbf{s}^* \xleftarrow{\$} D_{\mathbb{Z}^n,\alpha q}$ and $\mathbf{e}^* \xleftarrow{\$} D_{\mathbb{Z}^{\bar{m}},\alpha q}$, then we have that $\mathbf{c}_1^* = \mathbf{b} + \mathbf{A}^T\mathsf{encode}_d(\mathbf{v}) = \mathbf{A}^T\tilde{\mathbf{s}}^* + \mathbf{e}^*$, where $\tilde{\mathbf{s}}^* = \mathbf{s}^* + \mathsf{encode}_d(\mathbf{v}^*)$. In this case, $\mathcal{B}$ perfectly simulates the attack environment in game G_9 for $\mathcal{A}$. Else if $(\mathbf{A}, \mathbf{b}) \in \mathbb{Z}_q^{n\times\bar{m}} \times \mathbb{Z}_q^{\bar{m}}$ is uniformly random, then $\mathbf{c}_1^* = \mathbf{b} + \mathbf{A}^T\mathsf{encode}_d(\mathbf{v}^*)$ is also uniformly random over $\mathbb{Z}_q^{\bar{m}}$. This means that $\mathcal{B}$ perfectly simulates the attack environment in game G_{10} for $\mathcal{A}$. Thus, if $\mathcal{A}$ can distinguish game G_{10} from G_9 with non-negligible advantage, then $\mathcal{B}$ can solve the $\mathrm{LWE}_{n,\bar{m},q,\alpha}$ problem with the same advantage. □

Game G_{11}. This game is similar to game G_{10} except that the challenger C changes the KeyGen phase as follows:

KeyGen. first randomly choose $\mathbf{A} \xleftarrow{\$} \mathbb{Z}_q^{n\times\bar{m}}$, $\mathbf{R}' \xleftarrow{\$} (D_{\mathbb{Z}^{\bar{m}},\omega(\sqrt{\log n})})^{nk}$, $\mathbf{b} \xleftarrow{\$} \mathbb{Z}_q^{\bar{m}}$, $\mathbf{d} \xleftarrow{\$} \mathbb{Z}_q^{nk}$ and $x^*, y^*, z^* \xleftarrow{\$} \mathbb{F}_{2^\kappa}$. Then, compute

$$\begin{aligned} \mathbf{c}_1^* &= \mathbf{b}, \qquad \mathbf{B} = -\mathbf{A}\mathbf{R}' - \mathsf{FRD}(tag^*)\mathbf{G}_b, \\ \mathbf{c}_2^* &= \mathbf{d}, \end{aligned}$$

where $tag^* = \mathrm{H}(\mathbf{c}_1^*)$. Finally, give the public key $pk = (\mathbf{A}, \mathbf{B})$ to $\mathcal{A}$, and keep $(\mathbf{R}', \mathbf{c}_1^*, \mathbf{c}_2^*, x^*, y^*, z^*)$ secret.

Lemma 3.16 *Let positive* $n, \bar{m}, b, d, q \in \mathbb{Z}$ *and real* $\alpha \in (0, 1)$ *satisfy Eq. (3.1) or (3.2), then games* G_{11} *and* G_{10} *are statistically indistinguishable. Moreover,* $|\Pr[F_{11}|\neg E_{11}] - \Pr[F_{10}|\neg E_{10}]| \leq \mathsf{negl}(\kappa)$ *and* $|\Pr[E_{11}] - \Pr[E_{10}]| \leq \mathsf{negl}(\kappa)$.

Proof Note that the only difference between games G_{11} and G_{10} is the generation of $\mathbf{c}_2^*$ in the challenge ciphertext. Thus, it is enough to show that $\mathbf{c}_2^*$ in game G_{11} is actually statistically close to that in game G_{10}. Note that $\mathbf{c}_1^* = \mathbf{b}$ is uniformly chosen from $\mathbb{Z}_q^{\bar{m}}$ at random in both games, and $\mathbf{c}_2^* = (-\mathbf{R}')^T \mathbf{c}_1^* + \mathbf{e}_2'$ in game G_{10}. Using the facts that $\bar{m} \geq (n+1)\log_2 q + \omega(\log n)$ and $\mathbf{R}' \xleftarrow{\$} (D_{\mathbb{Z}^{\bar{m}},\omega(\sqrt{\log n})})^{nk}$, we have that $(\mathbf{A}, \mathbf{AR}', \mathbf{b}, (\mathbf{R}')^T\mathbf{b})$ is statistically close to uniform by Lemma 2.4. In other words, $\mathbf{c}_2^*$ in game G_{10} is essentially statistically close to uniform over $\mathbb{Z}_q^{nk}$, which completes the proof. □

Lemma 3.17 $\Pr[F_{11}] = 1/2$ *and* $\Pr[E_{11}] = \mathsf{negl}(\kappa)$.

Proof Since $x^* \xleftarrow{\$} \{0,1\}^\ell$ is uniformly chosen at random, $\mu_{\delta^*} \in \{0,1\}^\ell$ is perfectly hidden in the challenge ciphertext $C^* = (\mathbf{c}_1^*, \mathbf{c}_2^*, c_3^*, c_4^*)$, where $c_3^* = x^* + \mu_{\delta^*} \in \mathbb{F}_{2^\kappa}$. Thus, $\Pr[F_{11}] = \Pr[\delta = \delta^*] = 1/2$, where $\delta \in \{0,1\}$ is output by the adversary $\mathcal{A}$ for the guess of δ^* in game G_{11}.

As for the second claim, since $y^*, z^* \in \{0,1\}^\kappa$ are uniformly chosen at random in game G_{11}, given $c_4^* = \tau^* y^* + z^* \in \mathbb{F}_{2^\kappa}$ there are still 2^κ possible choices of (y^*, z^*). Thus, for any adversary $\mathcal{A}$ with the knowledge of $c_4^* = \tau^* y^* + z^*$, the probability that it outputs $c_4 = \tau y^* + z^*$ for any $\tau \neq \tau^*$ is at most $1/2^\kappa$ (because the adversary can uniquely determine (y^*, z^*) if he can output a valid $c_4 = \tau y^* + z^*$), which means that $\Pr[E_{11}] \leq Q_{dec}/2^\kappa$, where Q_{dec} is the maximum number of decryption queries made by $\mathcal{A}$. □

In all, we have that $\Pr[F_1] \leq 1/2 + \mathsf{negl}(\kappa)$ by Lemmas 3.7–3.17. This completes the proof of Theorem 3.3. □

3.5 Background and Further Reading

One of the main problems in the area of public-key encryption is to construct CCA2-secure PKEs from primitives as weak as possible (e.g., a CPA-secure one). Using the random oracle (RO) heuristic, one can efficiently boost a CPA-secure PKE into a CCA2-secure one [25, 48, 57]. However, a scheme provably secure in the RO model may not be secure in the real world [16], and it is of great theoretical and practical interest to construct CCA2-secure PKE in the standard model. But this task becomes very challenging and highly non-trivial. In fact, Gertner et al. [27] showed that it is hard, if not impossible, to even construct a CCA1-secure PKE solely from a CPA-secure one in the standard model.

By relying on primitives with "stronger" functionality or security, there are roughly four approaches to CCA2-secure PKEs in the standard model. The first one is due to Naor and Yung [41], who showed a paradigm for transforming CPA-secure PKEs into CCA1-secure ones by using the non-interactive zero-knowledge (NIZK) proofs, which was further extended to achieve CCA2-security [22, 53]. The second one is a framework under the name of hash proof systems (HPS) or extractable HPS [20, 58], which essentially stems from high-level abstraction of some existing schemes. The third one is the BCHK transform [11] from identity-based encryption (IBE), which was later extended to the more general tag-based encryption (TBE) by Kiltz [34]. The last one follows the generic constructions from special types of injective trapdoor functions [35, 47, 52], such as lossy trapdoor functions [47] and adaptive trapdoor functions [35].

The above approaches have been shown very useful in constructing CCA2-secure PKEs from various hardness assumptions, but most of the instantiations were based on traditional number theoretic problems such as discrete logarithm and integer factorization, which are not quantum resistant [54]. Compared to the big success in the traditional setting, the progress on designing lattice-based CCA2-secure PKE in the standard model was relatively slow. For example, many practical CCA2-secure PKEs in the traditional setting were obtained by using the generic framework from HPS (e.g., [19]), but it is still hard to construct an HPS from lattices [10, 32, 46]. Moreover, it is also unclear how to obtain efficient NIZKs from lattices in the standard model [10, 36, 63].

In fact, almost all existing standard model CCA2-secure PKEs from lattices are obtained either by using the techniques from special types of injective trapdoor functions [45, 47, 56, 59] which are typically very inefficient (e.g., having large public key and ciphertext sizes due to the use of Dolev-Dwork-Naor like technique [22, 24]), or by applying the BCHK transform from IBEs/TBEs [1, 2, 18, 23, 60–62]. Based on the standard model IBE from lattices due to Agrawal et al. [1], Micciancio and Peikert [40] presented the best known standard model (tag-based) CCA1-secure PKE from lattices by using a more efficient trapdoor technique and a new message encoding. They [40] also mentioned that the CCA2-security can be achieved by using the generic BCHK transform [11], which has two modes: BCHK-SIG [17] and BCHK-MAC [12]. BCHK-SIG requires (one-time) signatures, and typically incurs noticeable overheads to both computation and storage [11]. This becomes even worse on lattices, since the resulting ciphertext should include a verification key of the (one-time) signature, which has at least one matrix [38]. In contrast, BCHK-MAC [12] makes use of message authentication codes (MAC) and commitments, and thus can reduce the extra overheads, e.g., it only adds a MAC tag and a commitment to the resulting ciphertext.

References

1. Agrawal, S., Boneh, D., Boyen, X.: Efficient lattice (H)IBE in the standard model. In: Gilbert, H. (ed.) Advances in Cryptology - EUROCRYPT 2010. Lecture Notes in Computer Science, vol. 6110, pp. 553–572. Springer, Heidelberg (2010)
2. Agrawal, S., Boneh, D., Boyen, X.: Lattice basis delegation in fixed dimension and shorter-ciphertext hierarchical IBE. In: Rabin, T. (ed.) Advances in Cryptology - CRYPTO 2010. Lecture Notes in Computer Science, vol. 6223, pp. 98–115. Springer, Heidelberg (2010)
3. Ajtai, M.: Generating hard instances of the short basis problem. In: Wiedermann, J., van Emde Boas, P., Nielsen, M. (eds.) Automata, Languages and Programming. Lecture Notes in Computer Science, vol. 1644, pp. 706–706. Springer, Berlin/Heidelberg (1999)

4. Ajtai, M., Dwork, C.: A public-key cryptosystem with worst-case/average-case equivalence. In: Proceedings of the Twenty-Ninth Annual ACM Symposium on Theory of Computing, STOC '97, pp. 284–293. ACM (1997)
5. Albrecht, M.R., Player, R., Scott, S.: On the concrete hardness of learning with errors. J. Math. Cryptol. **9**, 169–203 (2015)
6. Alkim, E., Ducas, L., Pöppelmann, T., Schwabe, P.: Newhope without reconciliation. Cryptology ePrint Archive, Report 2016/1157 (2016)
7. Alkim, E., Ducas, L., Pöppelmann, T., Schwabe, P.: Post-quantum key exchange-a new hope. In: USENIX Security Symposium, vol. 2016 (2016)
8. Alwen, J., Peikert, C.: Generating shorter bases for hard random lattices. In: STACS, pp. 75–86 (2009)
9. Applebaum, B., Cash, D., Peikert, C., Sahai, A.: Fast cryptographic primitives and circular-secure encryption based on hard learning problems. In: Halevi, S. (ed.) Advances in Cryptology - CRYPTO 2009. Lecture Notes in Computer Science, vol. 5677, pp. 595–618. Springer, Heidelberg (2009)
10. Benhamouda, F., Blazy, O., Ducas, L., Quach, W.: Hash proof systems over lattices revisited. Cryptology ePrint Archive, Report 2017/997 (2017)
11. Boneh, D., Canetti, R., Halevi, S., Katz, J.: Chosen-ciphertext security from identity-based encryption. SIAM J. Comput. **36**(5), 1301–1328 (2006)
12. Boneh, D., Katz, J.: Improved efficiency for CCA-secure cryptosystems built using identity-based encryption. In: Menezes, A. (ed.) Topics in Cryptology - CT-RSA 2005. Lecture Notes in Computer Science, vol. 3376, pp. 87–103. Springer, Berlin/Heidelberg (2005)
13. Bos, J., Ducas, L., Kiltz, E., Lepoint, T., Lyubashevsky, V., Schanck, J.M., Schwabe, P., Seiler, G., Stehle, D.: Crystals - kyber: a CCA-secure module-lattice-based KEM. In: 2018 IEEE European Symposium on Security and Privacy (Euro SP), pp. 353–367 (2018)
14. Brakerski, Z.: Fully homomorphic encryption without modulus switching from classical gapsvp. In: Safavi-Naini, R., Canetti, R. (eds.) Advances in Cryptology - CRYPTO 2012. Lecture Notes in Computer Science, vol. 7417, pp. 868–886. Springer, Heidelberg (2012)
15. Brakerski, Z., Vaikuntanathan, V.: Efficient fully homomorphic encryption from (standard) LWE. In: 2011 IEEE 52nd Annual Symposium on Foundations of Computer Science (FOCS), pp. 97–106 (2011). https://doi.org/10.1109/FOCS.2011.12
16. Canetti, R., Goldreich, O., Halevi, S.: The random oracle methodology, revisited. J. ACM **51**(4), 557–594 (2004)
17. Canetti, R., Halevi, S., Katz, J.: Chosen-ciphertext security from identity-based encryption. In: Cachin, C., Camenisch, J. (eds.) Advances in Cryptology - EUROCRYPT 2004. Lecture Notes in Computer Science, vol. 3027, pp. 207–222. Springer, Berlin/Heidelberg (2004)
18. Cash, D., Hofheinz, D., Kiltz, E., Peikert, C.: Bonsai trees, or how to delegate a lattice basis. In: Gilbert, H. (ed.) Advances in Cryptology - EUROCRYPT 2010. Lecture Notes in Computer Science, vol. 6110, pp. 523–552. Springer, Berlin/Heidelberg (2010)
19. Cramer, R., Shoup, V.: A practical public key cryptosystem provably secure against adaptive chosen ciphertext attack. In: Krawczyk, H. (ed.) Advances in Cryptology - CRYPTO '98. Lecture Notes in Computer Science, vol. 1462, pp. 13–25. Springer, Berlin/Heidelberg (1998)
20. Cramer, R., Shoup, V.: Design and analysis of practical public-key encryption schemes secure against adaptive chosen ciphertext attack. SIAM J. Comput. **33**, 167–226 (2001)
21. Diffie, W., Hellman, M.: New directions in cryptography. IEEE Trans. Inf. Theory **22**(6), 644–654 (1976)
22. Dolev, D., Dwork, C., Naor, M.: Non-malleable cryptography. SIAM J. Comput. **30**(2), 391–437 (2000)
23. Döttling, N., Garg, S., Hajiabadi, M., Masny, D.: New constructions of identity-based and key-dependent message secure encryption schemes. Cryptology ePrint Archive, Report 2017/978 (2017)
24. Dowsley, R., Hanaoka, G., Imai, H., Nascimento, A.C.: Reducing the ciphertext size of Dolev-Dwork-Naor like public key cryptosystems. Cryptology ePrint Archive **2009**, 271 (2009)

25. Fujisaki, E., Okamoto, T.: Secure integration of asymmetric and symmetric encryption schemes. J. Cryptol. **26**(1), 80–101 (2013)
26. Gentry, C., Peikert, C., Vaikuntanathan, V.: Trapdoors for hard lattices and new cryptographic constructions. In: Proceedings of the 40th Annual ACM Symposium on Theory of Computing, STOC '08, pp. 197–206. ACM (2008)
27. Gertner, Y., Malkin, T., Myers, S.: Towards a separation of semantic and CCA security for public key encryption. In: Vadhan, S.P. (ed.) Theory of Cryptography, LNCS, vol. 4392, pp. 434–455. Springer, Heidelberg (2007)
28. Goldreich, O., Goldwasser, S., Halevi, S.: Public-key cryptosystems from lattice reduction problems. In: Kaliski, B. (ed.) Advances in Cryptology - CRYPTO '97. Lecture Notes in Computer Science, vol. 1294, pp. 112–131. Springer, Berlin/Heidelberg (1997)
29. Goldwasser, S., Micali, S.: Probabilistic encryption. J. Comput. Syst. Sci. **28**(2), 270–299 (1984)
30. Hoffstein, J., Pipher, J., Silverman, J.: NTRU: a ring-based public key cryptosystem. In: Buhler, J. (ed.) Algorithmic Number Theory. Lecture Notes in Computer Science, vol. 1423, pp. 267–288. Springer, Heidelberg (1998)
31. Jiang, H., Zhang, Z., Chen, L., Wang, H., Ma, Z.: IND-CCA-secure key encapsulation mechanism in the quantum random oracle model, revisited. In: Shacham, H., Boldyreva, A. (eds.) Advances in Cryptology - CRYPTO 2018, pp. 96–125. Springer, Cham (2018)
32. Katz, J., Vaikuntanathan, V.: Smooth projective hashing and password-based authenticated key exchange from lattices. In: Matsui, M. (ed.) Advances in Cryptology - ASIACRYPT 2009. Lecture Notes in Computer Science, vol. 5912, pp. 636–652. Springer, Heidelberg (2009)
33. Kawachi, A., Tanaka, K., Xagawa, K.: Multi-bit cryptosystems based on lattice problems. In: Okamoto, T., Wang, X. (eds.) Public Key Cryptography - PKC 2007. Lecture Notes in Computer Science, vol. 4450, pp. 315–329. Springer, Heidelberg (2007)
34. Kiltz, E.: Chosen-ciphertext security from tag-based encryption. In: Halevi, S., Rabin, T. (eds.) Theory of Cryptography 2006, LNCS, vol. 3876, pp. 581–600. Springer, Heidelberg (2006)
35. Kiltz, E., Mohassel, P., O'Neill, A.: Adaptive trapdoor functions and chosen-ciphertext security. In: Gilbert, H. (ed.) EUROCRYPT 2010, LNCS, vol. 6110, pp. 673–692. Springer, Heidelberg (2010)
36. Kim, S., Wu, D.J.: Multi-theorem preprocessing NIZKs from lattices. In: Shacham, H., Boldyreva, A. (eds.) Advances in Cryptology – CRYPTO 2018, pp. 733–765. Springer International Publishing (2018)
37. Lindner, R., Peikert, C.: Better key sizes (and attacks) for LWE-based encryption. In: Kiayias, A. (ed.) Topics in Cryptology - CT-RSA 2011. Lecture Notes in Computer Science, vol. 6558, pp. 319–339. Springer, Heidelberg (2011)
38. Lyubashevsky, V., Micciancio, D.: Asymptotically efficient lattice-based digital signatures. J. Cryptol. **31**(3), 774–797 (2018)
39. Micciancio, D.: Improving lattice based cryptosystems using the Hermite normal form. In: Silverman, J. (ed.) Cryptography and Lattices. Lecture Notes in Computer Science, vol. 2146, pp. 126–145. Springer, Berlin/Heidelberg (2001)
40. Micciancio, D., Peikert, C.: Trapdoors for lattices: simpler, tighter, faster, smaller. In: Pointcheval, D., Johansson, T. (eds.) Advances in Cryptology - EUROCRYPT 2012. Lecture Notes in Computer Science, vol. 7237, pp. 700–718. Springer, Heidelberg (2012)
41. Naor, M., Yung, M.: Public-key cryptosystems provably secure against chosen ciphertext attacks. In: STOC 1990, pp. 427–437. ACM (1990). https://doi.org/10.1145/100216.100273
42. Nguyen, P.Q.: Cryptanalysis of the Goldreich-Goldwasser-Halevi cryptosystem from crypto97. In: Wiener, M. (ed.) Advances in Cryptology - CRYPTO '99. Lecture Notes in Computer Science, vol. 1666, pp. 288–304. Springer, Heidelberg (1999)
43. Nguyen, P.Q., Stern, J.: Cryptanalysis of the Ajtai-Dwork cryptosystem. In: Krawczyk, H. (ed.) Advances in Cryptology - CRYPTO '98. Lecture Notes in Computer Science, vol. 1462, pp. 223–242. Springer, Berlin/Heidelberg (1998)
44. NIST: Post-Quantum Cryptography Standardization. http://csrc.nist.gov/groups/ST/post-quantum-crypto/submission-requirements/index.html

45. Peikert, C.: Public-key cryptosystems from the worst-case shortest vector problem: extended abstract. In: Proceedings of the 41st annual ACM symposium on Theory of computing, STOC '09, pp. 333–342. ACM (2009)
46. Peikert, C., Vaikuntanathan, V., Waters, B.: A framework for efficient and composable oblivious transfer. In: Wagner, D. (ed.) CRYPTO 2008, LNCS, vol. 5157, pp. 554–571. Springer, Heidelberg (2008)
47. Peikert, C., Waters, B.: Lossy trapdoor functions and their applications. In: Proceedings of the 40th Annual ACM Symposium on Theory of Computing, STOC '08, pp. 187–196. ACM (2008)
48. Pointcheval, D.: Chosen-ciphertext security for any one-way cryptosystem. In: Imai, H., Zheng, Y. (eds.) PKC '00, LNCS, vol. 1751, pp. 129–146. Springer, Heidelberg (2000)
49. Rackoff, C., Simon, D.R.: Non-interactive zero-knowledge proof of knowledge and chosen ciphertext attack. In: Feigenbaum, J. (ed.) CRYPTO '91, LNCS, vol. 576, pp. 433–444. Springer, Heidelberg (1992)
50. Regev, O.: On lattices, learning with errors, random linear codes, and cryptography. In: Proceedings of the Thirty-seventh Annual ACM Symposium on Theory of Computing, STOC '05, pp. 84–93. ACM (2005)
51. Rivest, R.L., Shamir, A., Adleman, L.: A method for obtaining digital signatures and public-key cryptosystems. Commun. ACM **21**(2), 120–126 (1978)
52. Rosen, A., Segev, G.: Chosen-ciphertext security via correlated products. In: Reingold, O. (ed.) Theory of Cryptography. Lecture Notes in Computer Science, vol. 5444, pp. 419–436. Springer, Berlin/Heidelberg (2009)
53. Sahai, A.: Non-malleable non-interactive zero knowledge and adaptive chosen-ciphertext security. In: FOCS '99, pp. 543–553. IEEE Computer Society (1999)
54. Shor, P.: Polynomial-time algorithms for prime factorization and discrete logarithms on a quantum computer. SIAM J. Comput. **26**(5), 1484–1509 (1997)
55. Shoup, V.: Sequences of games: a taming complexity in security proofs. Cryptology ePrint Archive, Report 2004/332 (2004)
56. Steinfeld, R., Ling, S., Pieprzyk, J., Tartary, C., Wang, H.: NTRUCCA: how to strengthen NTRUEncrypt to chosen-ciphertext security in the standard model. In: Fischlin, M., Buchmann, J., Manulis, M. (eds.) PKC 2012, LNCS, vol. 7293, pp. 353–371. Springer, Heidelberg (2012)
57. Targhi, E.E., Unruh, D.: Post-quantum security of the Fujisaki-Okamoto and OAEP transforms. In: Hirt, M., Smith, A. (eds.) Theory of Cryptography, pp. 192–216. Springer, Heidelberg (2016)
58. Wee, H.: Efficient chosen-ciphertext security via extractable hash proofs. In: Rabin, T. (ed.) CRYPTO 2010, LNCS, vol. 6223, pp. 314–332. Springer, Heidelberg (2010)
59. Wee, H.: Public key encryption against related key attacks. In: Fischlin, M., Buchmann, J., Manulis, M. (eds.) PKC 2012, vol. 7293, pp. 262–279. Springer, Heidelberg (2012)
60. Yamada, S.: Adaptively secure identity-based encryption from lattices with asymptotically shorter public parameters. In: Fischlin, M., Coron, J.S. (eds.) EUROCRYPT 2016, LNCS, vol. 9666, pp. 32–62. Springer, Heidelberg (2016)
61. Yamada, S.: Asymptotically compact adaptively secure lattice IBEs and verifiable random functions via generalized partitioning techniques. In: Katz, J., Shacham, H. (eds.) CRYPTO 2017, vol. 10403, pp. 161–193. Springer, Cham (2017)
62. Zhang, J., Chen, Y., Zhang, Z.: Programmable hash functions from lattices: short signatures and IBEs with small key sizes. In: Robshaw, M., Katz, J. (eds.) Advances in Cryptology - CRYPTO 2016, pp. 303–332. Springer, Heidelberg (2016)
63. Zhang, J., Yu, Y.: Two-round PAKE from approximate SPH and instantiations from lattices. In: Takagi, T., Peyrin, T. (eds.) Advances in Cryptology – ASIACRYPT 2017, pp. 37–67. Springer (2017)
64. Zhang, J., Yu, Y., Fan, S., Zhang, Z.: Improved lattice-based CCA2-secure PKE in the standard model. SCIENCE CHINA Inf. Sci. **63**(8), 182101 (2020)

Chapter 4
Identity-Based Encryption

Abstract In real applications, public-key encryption typically requires a mechanism such as public key infrastructure (PKI) to authenticate public keys, e.g., to generate a certificate that some entity or person owns the public key. However, the use of PKI and certificates might be very heavy and cumbersome in some applications. In 1984, Shamir introduced identity-based encryption (IBE), which allows a user to use any string (e.g., emails) that uniquely identifies himself/herself as his/her own public key, to solve the problem of authenticating public keys. The first realizations of IBEs were due to Boneh and Franklin from pairings and Cocks from quadratic residues. In this chapter, we focus on constructing IBEs from lattices. Concretely, after giving some introduction in Sect. 4.1, we will present the first IBE from the learning with errors (LWE) assumption in the random oracle model due to Gentry et al. in Sect. 4.2. Then, we will show how to construct lattice-based IBEs with shorter key sizes in the standard model in Sect. 4.3.

4.1 Definition

Shamir [25] introduced identity-based encryption (IBE) in 1984, but the first realizations were due to Boneh and Franklin from pairings [6] and Cocks from quadratic residues [12]. In the lattice setting, Gentry et al. [18] proposed the first IBE scheme based on the learning with errors (LWE) assumption in the random oracle model. Later, several works [2, 10, 14, 28] were dedicated to the study of lattice-based (hierarchical) IBE schemes also in the random oracle model. There were a few works focusing on designing standard model lattice-based IBE schemes [1, 2, 10]. Concretely, the scheme in [2] was only proven to be *selective-identity* secure in the standard model. By using standard complexity leverage technique [5], one can generally transform a selective-identity secure IBE scheme into a *fully secure* one. But the resulting scheme has to suffer from a reduction loss proportional to L, where L is the number of distinct identities for the IBE system and is independent from the number Q of the adversary's private key queries in the security proof. Since L is

---

J. Zhang and Z. Zhang, *Lattice-Based Cryptosystems*,
https://doi.org/10.1007/978-981-15-8427-5_4

usually super polynomial and much larger than Q, the above generic transformation is a very unsatisfying approach [17]. In [1, 10], the authors showed how to achieve *full security* against adaptive chosen-plaintext and chosen-identity attacks, but both standard model fully secure IBE schemes in [1, 10] had large master public keys consisting of a linear number of matrices. In fact, Agrawal, Boneh and Boyen left it as an open problem to find fully secure lattice-based IBE schemes with short master public keys in the standard model [1].

Definition 4.1 (*Identity-based Encryption*) An identity-based encryption (IBE) scheme consists of four PPT algorithms $\Pi_{\text{ibe}} = (\mathsf{Setup}, \mathsf{Extract}, \mathsf{Enc}, \mathsf{Dec})$:

- $\mathsf{Setup}(\kappa)$. Taking the security parameter κ as input, the randomized key generation algorithm Setup outputs a master public key mpk and a master secret key msk, denoted as $(mpk, msk) \leftarrow \mathsf{Setup}(1^\kappa)$.
- $\mathsf{Extract}(msk, id)$. The (randomized) extract algorithm takes mpk, msk and an identity id as inputs, outputs a user private key sk_{id} for id, briefly denoted as $sk_{id} \leftarrow \mathsf{Extract}(msk, id)$.
- $\mathsf{Enc}(pk, W, M)$. The randomized encryption algorithm Enc takes mpk, id and a plaintext M as inputs, outputs a ciphertext C, denoted as $C \leftarrow \mathsf{Enc}(mpk, id, M)$.
- $\mathsf{Dec}(sk_S, C)$. The deterministic algorithm Dec takes sk_{id} and C as inputs, outputs a plaintext M, or a special symbol $\perp$, which is denoted as $M/\perp \leftarrow \mathsf{Dec}(sk_{id}, C)$.

For correctness, we require that, for all $(mpk, msk) \leftarrow \mathsf{Setup}(1^\kappa)$, $sk_{id} \leftarrow \mathsf{Extract}(msk, id)$ and any plaintext M, the equation $\mathsf{Dec}(sk_{id}, C) = M$ holds with overwhelming probability.

The standard semantic security of IBE was first introduced in [6]. We use the notion called indistinguishable from random in [1], which captures both semantic security and recipient anonymity by requiring the challenge ciphertext to be indistinguishable from a uniformly random element in the ciphertext space. Formally, consider the following game played by an adversary $\mathcal{A}$.

Setup. The challenger C first runs $\mathsf{Setup}(1^\kappa)$ with the security parameter κ. Then, it gives the adversary $\mathcal{A}$ the master public key mpk, and keeps the master secret key msk to itself.

Phase 1. The adversary is allowed to query the user private key for any identity id. The challenger C runs $\mathsf{sk}_{id} \leftarrow \mathsf{Extract}(\mathsf{msk}, id)$ and sends sk_{id} to the adversary $\mathcal{A}$. The adversary can repeat the user private key query any polynomial times for different identities.

Challenge. The adversary $\mathcal{A}$ outputs a challenge plaintext M^* and a challenge identity id^* with a restriction that id^* is not used in the user private key query in Phase 1. The challenger C chooses a uniformly random ciphertext C_0 from the ciphertext space. Then, it computes $C_1 \leftarrow \mathsf{Enc}(mpk, id^*, M^*)$. Finally, it randomly chooses a bit $b^* \xleftarrow{\$} \{0, 1\}$ and sends C_{b^*} as the challenge ciphertext to $\mathcal{A}$.

Phase 2. The adversary can adaptively make more user private key queries with any identity $id \neq id^*$. The challenger C responds as in Phase 1.

Guess. Finally, $\mathcal{A}$ outputs a guess $b \in \{0, 1\}$. If $b = b^*$, the challenger C outputs 1, else outputs 0.

The advantage of $\mathcal{A}$ in the above security game is defined as $\mathrm{Adv}_{\mathcal{IBE},\mathcal{A}}^{\text{indr-id-cpa}}(\kappa) = |\Pr[b = b^*] - \frac{1}{2}|$.

Definition 4.2 (*INDr-ID-CPA*) We say an IBE scheme Π_{ibe} is INDr-ID-CPA secure if for any PPT adversary $\mathcal{A}$, its advantage $\mathrm{Adv}_{\Pi_{\text{ibe}},\mathcal{A}}^{\text{indr-id-cpa}}(\kappa)$ is negligible in κ.

In the security game against chosen-ciphertext attacks (i.e., INDr-ID-CCA), the adversary is also allowed to make decryption queries in both Phase 1 and Phase 2, and obtain the decrypted results from any identity-ciphertext pair $(id, C) \neq (id^*, C_{b^*})$. Besides, Canetti et al. [9] also introduced a weaker security notion, known as selective-identity security, by using a modified security game, where the adversary is asked to output the challenge identity id^* before seeing the master public key in the setup phase, and is restricted to make user private key query for $id \neq id^*$ in both Phase 1 and Phase 2. The resulting security notion defined using the modified game as in Definition 4.2 is denoted as INDr-sID-CPA.

4.2 The Gentry-Peikert-Vaikuntanathan IBE Scheme

In 2008, Gentry et al. [18] showed how to use a lattice trapdoor to sample discrete Gaussian distribution over lattices, which paves the way for constructing many advanced lattice-based cryptosystems such as identity-based encryption. Actually, the first lattice-based IBE was obtained by combining the trapdoor Gaussian sampler with a dual version of Regev's public-key encryption scheme. In this following, we describe the GPV IBE scheme to demonstrate how to use a lattice trapdoor together with other techniques in constructing cryptosystems.

Formally, let κ be the security parameter. Let $n, m, q > 0$ be integers, and $\alpha, s > 0$ be reals. Let $\mathsf{TrapGen}$ and $\mathsf{SampleD}$ be the trapdoor generation and discrete Gaussian sampler algorithms given in Lemma 2.24. Let $H : \{0, 1\}^* \to \mathbb{Z}_q^n$ be a hash function, which is modeled as a random oracle. The description of the GPV IBE scheme $\Pi_{\text{ibe}} = (\mathsf{Setup}, \mathsf{Extract}, \mathsf{Enc}, \mathsf{Dec})$ with parameters (n, m, q, s, α) is given as follows:

- $\mathsf{Setup}(1^\kappa)$: given the security parameter κ as input, compute

$$(\mathbf{A}, \mathbf{R}) \leftarrow \mathsf{TrapGen}(1^n, 1^m, q, \mathbf{I}_n),$$

  where $\mathbf{I}_n$ is the identity matrix. Return the master pubic key $\mathsf{mpk} = \mathbf{A}$ and the master secret key $\mathsf{msk} = \mathbf{R}$;

- Extract(msk, id): given the master secret key $\mathsf{msk} = \mathbf{R}$ and a user identity $id \in \{0,1\}^*$, compute $\mathbf{u} = \mathrm{H}(id) \in \mathbb{Z}_q^n$ and $\mathbf{x} \leftarrow \mathsf{SampleD}(\mathbf{R}, \mathbf{A}, \mathbf{I}_n, \mathbf{u}, s)$ such that the distribution of $\mathbf{x} \in \mathbb{Z}^m$ is statistically close to $D_{\mathbb{Z}^m,s}$ conditioned on $\mathbf{A}\mathbf{x} = \mathbf{u}$, return the user secret key $\mathsf{usk}_{id} = \mathbf{x}$;
- Enc(mpk, id, μ): given the master public key $\mathsf{mpk} = \mathbf{A}$, a user identity $id \in \{0,1\}^*$ and a message bit $\mu \in \{0,1\}$, randomly choose $\mathbf{r} \stackrel{\$}{\leftarrow} D_{\mathbb{Z}^n,\alpha}$, $\mathbf{e}_1 \stackrel{\$}{\leftarrow} D_{\mathbb{Z}^m,\alpha}$ and $e_2 \stackrel{\$}{\leftarrow} D_{\mathbb{Z},\alpha}$. Then, compute $\mathbf{u} = \mathrm{H}(id)$,

$$\mathbf{c}_1 = \mathbf{A}^T\mathbf{r} + \mathbf{e}_1, \qquad c_2 = \mathbf{u}^T\mathbf{r} + e_2 + \mu \cdot \frac{q}{2}.$$

  Return the ciphertext $C = (\mathbf{c}_1, c_2) \in \mathbb{Z}_q^m \times \mathbb{Z}_q$.
- Dec(usk_{id}, C): given the user secret key $\mathsf{usk}_{id} = \mathbf{x}$ and a ciphertext $C = (\mathbf{c}_1, c_2)$, compute $v = c_2 - \mathbf{x}^T\mathbf{c}_1 \in \mathbb{Z}_q$, return the message $\mu' = \lfloor v \cdot \frac{2}{q} \rceil \bmod 2$.

Now, we show that the above IBE scheme is correct. Since $\mathbf{A}\mathbf{x} = \mathbf{u}$, we have that

$$\begin{aligned} v = c_2 - \mathbf{x}^T\mathbf{c}_1 &= \mathbf{u}^T\mathbf{r} + e_2 + \mu \cdot \tfrac{q}{2} - \mathbf{x}^T(\mathbf{A}^T\mathbf{r} + \mathbf{e}_1) \\ &= \mu \cdot \tfrac{q}{2} + e_2 - \mathbf{x}^T\mathbf{e}_1. \end{aligned}$$

Namely, if we set the parameters n, m, q such that $|e_2 - \mathbf{x}^T\mathbf{e}_1| < \frac{q}{4}$, we can be guaranteed that $\mu' = \lfloor \mathbf{v} \cdot \frac{2}{q} \rceil \bmod 2 = \mu$. This shows that the above scheme is correct.

4.2.1 Security

We have the following theorem for the security of the GPV IBE scheme [18].

Theorem 4.1 *If the* $\mathrm{LWE}_{m,n,q,\alpha}$ *problem is hard, then the GPV IBE scheme is INDr-ID-CPA-secure in the random oracle model.*

The proof will use a sequence of games $G_0, \ldots, G_6$, with G_0 being the real INDr-ID-CPA-security game (where the challenge ciphertext $C^* = (\mathbf{c}_1^*, c_2^*)$ is honestly generated by first randomly choosing $\delta^* \stackrel{\$}{\leftarrow} \{0,1\}$) and G_2 a random game (where the challenge ciphertext C^* is essentially uniformly random, and thus the adversary's advantage in game G_6 is negligible).

Proof Let $\mathcal{A}$ be a PPT adversary which can break the INDr-ID-CPA-security of the GPV IBE scheme with advantage ϵ, by making at most Q_1 queries to the random oracle H and Q_2 queries to generate user secret keys. For simplicity, we assume that the adversary makes at most a single user key generation query with the same identity id^* (otherwise, one can make the key generation query deterministic by using a pseudorandom function).

Let F_i be the event that $\mathcal{A}$ correctly guesses $\delta = \delta^*$ in game G_i, where $i \in \{0, \ldots, 6\}$. By definition, the adversary's advantage $\mathrm{Adv}^{\text{ind-cpa}}_{\Pi_{\text{pke}},\mathcal{A}}(\kappa)$ in game i is exactly $|\Pr[F_i] - 1/2|$. □

Game G_0. This game is the real security game. Formally, the challenger C initializes a list $L = \perp$ for the random oracle queries, and works as follows:

Setup. The challenger C computes $(\mathbf{A}, \mathbf{R}) \leftarrow \mathsf{TrapGen}(1^n, 1^m, q, \mathbf{I}_n)$, where $\mathbf{I}_n$ is the identity matrix. Then, it gives the master pubic key $\mathsf{mpk} = \mathbf{A}$ to the adversary and keeps the master secret key msk to itself.

Queries to H. When receiving a query $id \in \{0, 1\}^*$ to H, the challenger C first checks if there is a tuple $(id, \mathbf{u}_{id}) \in \{0, 1\}^* \times \mathbb{Z}_q^n$ in the list L, and adds a tuple $(id, \mathbf{u}_{id}) \in \{0, 1\}^* \times \mathbb{Z}_q^n$ into the list L by randomly choosing $\mathbf{u}_{id} \xleftarrow{\$} \mathbb{Z}_q^n$ if there is no such tuple in L. Then, return $\mathbf{u}_{id} \in \mathbb{Z}_q^n$.

Phase 1. The adversary is allowed to query the user private key for any identity id. The challenger C first makes a query to H with input id to obtain a vector $\mathbf{u}_{id} \in \mathbb{Z}_q^n$ on behalf of the adversary. Then, it computes $\mathbf{x} \leftarrow \mathsf{SampleD}(\mathbf{R}, \mathbf{A}, \mathbf{I}_n, \mathbf{u}_{id}, s)$ such that $\mathbf{x} \sim D_{\mathbb{Z}^m,s}$ conditioned on $\mathbf{Ax} = \mathbf{u}_{id}$, returns the user secret key $\mathsf{usk}_{id} = \mathbf{x}$ to the adversary. The adversary can repeat the user private key query any polynomial times for different identities.

Challenge. The adversary $\mathcal{A}$ outputs a challenge plaintext M^* and a challenge identity id^* with a restriction that id^* is not used in the user private key query in Phase 1. The challenger first makes a query to H with input id^* to obtain a vector $\mathbf{u}_{id^*} \in \mathbb{Z}_q^n$ on behalf of the adversary. Then, it chooses a uniformly random ciphertext $C_0 = (\mathbf{c}_{01}, c_{02}) \xleftarrow{\$} \mathbb{Z}_q^m \times \mathbb{Z}_q$. Then, it randomly chooses $\mathbf{r} \xleftarrow{\$} D_{\mathbb{Z}^n,\alpha}$, $\mathbf{e}_1 \xleftarrow{\$} D_{\mathbb{Z}^m,\alpha}$ and $e_2 \xleftarrow{\$} D_{\mathbb{Z},\alpha}$ and computes $C_1 = (\mathbf{c}_{11}, c_{12}) \xleftarrow{\$} \mathbb{Z}_q^m \times \mathbb{Z}_q$, where

$$\mathbf{c}_{11} = \mathbf{A}^T\mathbf{r} + \mathbf{e}_1, \qquad c_{12} = \mathbf{u}_{id^*}^T\mathbf{r} + e_2 + \mu \cdot \frac{q}{2}.$$

Finally, C randomly flips a bit $b^* \xleftarrow{\$} \{0, 1\}$, and returns C_{b^*} to the adversary.

Phase 2. The adversary can adaptively make more user private key queries with any identity $id \neq id^*$. The challenger C responds as in Phase 1.

Guess. Finally, $\mathcal{A}$ outputs a guess $b \in \{0, 1\}$. If $b = b^*$, the challenger C outputs 1, else outputs 0.

By definition, we have the following lemma:

Lemma 4.1 $|\Pr[F_0] - 1/2| = \epsilon$.

Game G_1. This game is almost identical to game G_0 except that the challenger C first randomly chooses an integer $k^* \xleftarrow{\$} \{1, \ldots, Q_1 + Q_2\}$ and changes the challenge phase as follows:

Challenge. The adversary $\mathcal{A}$ outputs a challenge plaintext M^* and a challenge identity id^* with a restriction that id^* is not used in the user private key query in Phase 1. The challenger first makes a query to H with input $id \in \{0, 1\}^*$ to obtain a vector $\mathbf{u}_{id} \in \mathbb{Z}_q^n$ on behalf of the adversary. If it is not the k^*-th queries to H or the adversary has used $id^* \in \{0, 1\}^*$ in a previous query to H, the challenger randomly outputs a bit and aborts, otherwise it computes the challenge ciphertext as in game G_0.

Lemma 4.2 $|\Pr[F_1] - \frac{1}{2}| \geq \frac{1}{Q_0+Q_1+1} \cdot |\Pr[F_0] - 1/2|$.

Proof This lemma follows from the fact that the adversary will use the k^*-th queries to H with probability at least $\frac{1}{Q_0+Q_1+1}$ for a uniformly and independently chosen $k^* \xleftarrow{\$} \{Q_0 + Q_1 + 1\}$. □

Game G_2. This game is almost identical to game G_1 except that the challenger C changes the way answering the queries to H as follows:

Queries to H. When receiving a query $id^* \in \{0,1\}^*$ to H, the challenger C returns $\mathbf{u}_{id}$ to the adversary if there is a tuple $(id, \mathbf{u}_{id}, *) \in \{0,1\}^* \times \mathbb{Z}_q^n$ in the list L. Otherwise, if it is the k^*-th queries to H, randomly choose $\mathbf{u}_{id} \xleftarrow{\$} \mathbb{Z}_q^n$, and add a tuple $(id, \mathbf{u}_{id}, -) \in \{0,1\}^* \times \mathbb{Z}_q^n$ into the list L. Else, randomly choose $\mathbf{x} \xleftarrow{\$} D_{\mathbb{Z}^m,s}$, set $\mathbf{u}_{id} = \mathbf{A}\mathbf{x}$, and add a tuple $(id, \mathbf{u}_{id}, \mathbf{x}) \in \{0,1\}^* \times \mathbb{Z}_q^n$ into the list L, and return $\mathbf{u}_{id} \in \mathbb{Z}_q^n$.

Lemma 4.3 *For appropriate choice of parameters n, m, q, s, games G_1 and G_0 are statistically indistinguishable in $\mathcal{A}$'s view. Moreover,* $|\Pr[F_2] - \Pr[F_1]| \leq \mathsf{negl}(\kappa)$.

Proof This lemma follows from Lemma 2.11 that the distribution of $\mathbf{u}_{id} = \mathbf{A}\mathbf{x}$ is statistically close to uniform distribution over $\mathbb{Z}_q^n$ for appropriate choice of parameters. □

Game G_3. This game is almost identical to game G_2 except that the challenger C changes the way generating the user secret key as follows:

Phase 1. The adversary is allowed to query the user private key for any identity id. The challenger C first makes a query to H with input $id \in \{0,1\}^*$ to obtain a vector $\mathbf{u}_{id} \in \mathbb{Z}_q^n$ on behalf of the adversary. Then, if it is the k^*-th query to H, the challenger outputs a random bit and aborts. Otherwise, recover the tuple $(id, \mathbf{u}_{id}, \mathbf{x})$ from the list L, and return the user secret key $\mathsf{usk}_{id} = \mathbf{x}$ to the adversary.

Lemma 4.4 *For appropriate choice of parameters n, m, q, s, games G_1 and G_0 are statistically indistinguishable in $\mathcal{A}$'s view. Moreover,* $|\Pr[F_3] - \Pr[F_2]| \leq \mathsf{negl}(\kappa)$.

Proof This lemma follows from Lemma 2.24 that the output distribution of $\mathsf{SampleD}(\mathbf{R}, \mathbf{A}, \mathbf{I}_n, \mathbf{u}_{id}, s)$ is statistically close to $\mathbf{x} \sim D_{\mathbb{Z}^m,s}$ conditioned on $\mathbf{A}\mathbf{x} = \mathbf{u}_{id}$ for appropriate choice of parameters. □

Game G_4. This game is almost identical to game G_3 except that the challenger C changes the way generating the master public key as follows:

Setup. The challenger C randomly chooses a matrix $\mathbf{A} \xleftarrow{\$} \mathbb{Z}_q^{n \times m}$ and gives the master public key $\mathsf{mpk} = \mathbf{A}$ to the adversary (note that no one knows the corresponding master secret key).

Proof Since in both games G_3 and G_4, the challenger does not need the master secret key to generate the user private key, this lemma follows from Lemma 2.24 that the distribution of the first output of $\mathsf{TrapGen}(1^n, 1^{\bar{m}}, q, \mathbf{S})$ is statistically close to uniform distribution over $\mathbb{Z}_q^{n \times m}$ for appropriate choice of parameters. □

Game G_5. This game is almost identical to game G_4 except that the challenger C changes the way generating the challenge ciphertext as follows:

Challenge. The adversary $\mathcal{A}$ outputs a challenge plaintext M^* and a challenge identity id^* with a restriction that id^* is not used in the user private key query in phase 1. The challenger first makes a query to H with input id^* to obtain a vector $\mathbf{u}_{id^*} \in \mathbb{Z}_q^n$ on behalf of the adversary. If it is not the k^*-th query to H, the challenger outputs a random bit and aborts. Otherwise, it randomly chooses $C_0 = (\mathbf{c}_{01}, c_{02}) \xleftarrow{\$} \mathbb{Z}_q^m \times \mathbb{Z}_q, \mathbf{v} \xleftarrow{\$} \mathbb{Z}_q^m, w \xleftarrow{\$} \mathbb{Z}_q$, and sets $C_1 = (\mathbf{c}_{11}, c_{12}) \xleftarrow{\$} \mathbb{Z}_q^m \times \mathbb{Z}_q$, where

$$\mathbf{c}_{11} = \mathbf{v} \qquad c_{12} = w + \mu \cdot \frac{q}{2}.$$

Finally, C randomly flips a bit $b^* \xleftarrow{\$} \{0, 1\}$ and returns C_{b^*} to the adversary.

Lemma 4.5 *If the* $\mathrm{LWE}_{n,m+1,q,\alpha}$ *problem is hard, then games* G_5 *and* G_4 *are computationally indistinguishable in* $\mathcal{A}$*'s view. Moreover,* $|\Pr[F_5] - \Pr[F_4]| \le \mathsf{negl}(\kappa)$.

Proof We only have to show that if there is a PPT adversary $\mathcal{A}$ that can distinguish games G_5 and G_4, then there is an algorithm $\mathcal{B}$ that solves the $\mathrm{LWE}_{n,m+1,q,\beta}$ problem. Formally, given an instance $(\mathbf{C}, \mathbf{d}) \in \mathbb{Z}_q^{(m+1)\times n} \times \mathbb{Z}_q^{m+1}$ of the $\mathrm{LWE}_{n,m+1,q,\alpha}$ problem, the goal of $\mathcal{B}$ is to distinguish that whether $\mathbf{d} = \mathbf{Cr} + \mathbf{e}$ for some $\mathbf{r} \xleftarrow{\$} D_{\mathbb{Z}^n,\alpha}, \mathbf{e} \xleftarrow{\$} D_{\mathbb{Z}^{m+1},\alpha}$, or $\mathbf{d} \xleftarrow{\$} \mathbb{Z}_q^{m+1}$ is chosen from uniform.

Formally, $\mathcal{B}$ first parses $(\mathbf{C}, \mathbf{d})$ as follows:

$$\mathbf{C} = \begin{pmatrix} \mathbf{A}^T \\ \mathbf{u}^T \end{pmatrix} \in \mathbb{Z}_q^{(m+1)\times n}, \text{ and } \mathbf{d} = \begin{pmatrix} \mathbf{v} \\ w \end{pmatrix} \in \mathbb{Z}_q^{m+1},$$

where $\mathbf{A} \in \mathbb{Z}_q^{n\times m}$, $\mathbf{u} \in \mathbb{Z}_q^n$, $\mathbf{x} \in \mathbb{Z}_q^m$ and $w \in \mathbb{Z}_q$. Then, it randomly chooses $k^* \xleftarrow{\$} \{1, \ldots, Q_1 + Q_2 + 1\}$ and interacts with the adversary $\mathcal{A}$ almost the same as the challenger in game G_4 except the following changes:

Setup. Set the master public key $\mathsf{mpk} = \mathbf{A}$, and give it to the adversary $\mathcal{A}$.

Queries to H. When receiving a query $id^* \in \{0, 1\}^*$ to H, if it is not the k^*-th query to H, $\mathcal{B}$ handles this query as the challenger in game G_4. Otherwise, if there is a tuple $(id, \mathbf{u}_{id}, *) \in \{0, 1\}^* \times \mathbb{Z}_q^n$ in the list L, $\mathcal{B}$ outputs a random bit and aborts, else adds a tuple $(id, \mathbf{u}, -) \in \{0, 1\}^* \times \mathbb{Z}_q^n$ into the list L, and returns $\mathbf{u}$.

Challenge. The adversary $\mathcal{A}$ outputs a challenge plaintext M^* and a challenge identity id^* with a restriction that id^* is not used in the user private key query in phase 1. The challenger first makes a query to H with input id^* to obtain a vector $\mathbf{u}_{id^*} \in \mathbb{Z}_q^n$ on behalf of the adversary. If it is not the k^*-th query to H, the challenger outputs a random bit and aborts. Otherwise, it randomly chooses $C_0 = (\mathbf{c}_{01}, c_{02}) \xleftarrow{\$} \mathbb{Z}_q^m \times \mathbb{Z}_q$, and sets $C_1 = (\mathbf{c}_{11}, c_{12}) \xleftarrow{\$} \mathbb{Z}_q^m \times \mathbb{Z}_q$, where

$$\mathbf{c}_{11} = \mathbf{v} \qquad c_{12} = w + \mu \cdot \frac{q}{2}.$$

Finally, C randomly flips a bit $b^* \xleftarrow{\$} \{0, 1\}$ and returns C_{b^*} to the adversary.

Finally, $\mathcal{B}$ returns whatever $\mathcal{A}$ outputs as its own guess.

Note that if $(\mathbf{C}, \mathbf{d}) \in \mathbb{Z}_q^{(m+1)\times n} \times \mathbb{Z}_q^{m+1}$ is a real $\mathrm{LWE}_{n,m+1,q,\beta}$ tuple, then the view of $\mathcal{A}$ is the same as that in game G_4, otherwise the same as that in game G_5. This means that if $\mathcal{A}$ can distinguish games G_4 and G_5, then $\mathcal{B}$ can solve the $\mathrm{LWE}_{n,m+1q,\beta}$ problem. This completes the proof. □

Game G_6. This game is almost identical to game G_5 except that the challenger C changes the way generating the challenge ciphertext as follows:

Challenge. The adversary $\mathcal{A}$ outputs a challenge plaintext M^* and a challenge identity id^* with a restriction that id^* is not used in the user private key query in Phase 1. The challenger first makes a query to H with input id to obtain a vector $\mathbf{u}_{id} \in \mathbb{Z}_q^n$ on behalf of the adversary. If it is not the k^*-th query to H, the challenger outputs a random bit and aborts. Otherwise, it randomly chooses $C_0 = (\mathbf{c}_{01}, c_{02}), C_1 = (\mathbf{c}_{11}, c_{12}) \xleftarrow{\$} \mathbb{Z}_q^m \times \mathbb{Z}_q$, and a bit $b \xleftarrow{\$} \{0, 1\}$. Return return C_{b^*} to the adversary.

Lemma 4.6 *Games G_6 and G_5 are identical in the adversary's view. Moreover,* $\Pr[F_6] = \Pr[F_5]$ *and* $\Pr[F_6] = \frac{1}{2}$.

Proof This lemma follows from the fact that for any choice of $\mu \in \{0, 1\}$, $c_{12} = w + \mu \cdot \lfloor \frac{q}{2} \rfloor$ is uniformly distributed over $\mathbb{Z}_q^\ell$ as long as w is uniformly chosen from $\mathbb{Z}_q$, and that C_0 and C_1 have the same distribution. □

In all, we have that $\epsilon = |\Pr[F_0] - \frac{1}{2}| \leq \mathsf{negl}(\kappa)$ by Lemmas 4.1–4.6, which completes the proof. □

4.3 Lattice-Based IBEs in the Standard Model

The first lattice-based IBEs in the standard model were given by Agrawal et al. [1] and Cash et al. [10]. In this section, we will give a generic way to construct lattice-based IBEs from lattice-based programmable hash functions in the standard model.

4.3.1 Programmable Hash Functions from Lattices

Let $\ell, \bar{m}, m, n, q, u, v \in \mathbb{Z}$ be some polynomials in the security parameter κ. By I_n we denote the set of invertible matrices in $\mathbb{Z}_q^{n\times n}$. A hash function $\mathcal{H} : \mathcal{X} \to \mathbb{Z}_q^{n\times m}$ consists of two algorithms ($\mathcal{H}$.Gen, $\mathcal{H}$.Eval). Given the security parameter κ, the probabilistic polynomial time (PPT) key generation algorithm $\mathcal{H}.\mathrm{Gen}(1^\kappa)$ outputs a key K, i.e., $K \leftarrow \mathcal{H}.\mathrm{Gen}(1^\kappa)$. For any input $X \in \mathcal{X}$, the efficiently deterministic evaluation algorithm $\mathcal{H}.\mathrm{Eval}(K, X)$ outputs a hash value $\mathbf{Z} \in \mathbb{Z}_q^{n\times m}$, i.e., $\mathbf{Z} = \mathcal{H}.\mathrm{Eval}(K, X)$. For simplicity, we write $\mathrm{H}_K(X) = \mathcal{H}.\mathrm{Eval}(K, X)$.

Definition 4.3 (*Lattice-based Programmable Hash Function*) A hash function $\mathcal{H} : \mathcal{X} \to \mathbb{Z}_q^{n\times m}$ is a $(u, v, \beta, \gamma, \delta)$-PHF if there exist a PPT trapdoor key generation algorithm $\mathcal{H}$.TrapGen and an efficiently deterministic trapdoor evaluation algorithm $\mathcal{H}$.TrapEval such that given a uniformly random $\mathbf{A} \in \mathbb{Z}_q^{n\times \bar{m}}$ and a (public) trapdoor matrix $\mathbf{B} \in \mathbb{Z}_q^{n\times m}$,[1] the following properties hold:

Syntax: The PPT algorithm $(K', td) \leftarrow \mathcal{H}.\text{TrapGen}(1^\kappa, \mathbf{A}, \mathbf{B})$ outputs a key K' together with a trapdoor td. Moreover, for any input $X \in \mathcal{X}$, the deterministic algorithm $(\mathbf{R}_X, \mathbf{S}_X) = \mathcal{H}.\text{TrapEval}(td, K', X)$ returns $\mathbf{R}_X \in \mathbb{Z}_q^{\bar{m}\times m}$ and $\mathbf{S}_X \in \mathbb{Z}_q^{n\times n}$ such that $s_1(\mathbf{R}_X) \le \beta$ and $\mathbf{S}_X \in \mathcal{I}_n \cup \{\mathbf{0}\}$ hold with overwhelming probability over the trapdoor td that is produced along with K'.

Correctness: For all possible $(K', td) \leftarrow \mathcal{H}.\text{TrapGen}(1^\kappa, \mathbf{A}, \mathbf{B})$, all $X \in \mathcal{X}$ and its corresponding $(\mathbf{R}_X, \mathbf{S}_X) = \mathcal{H}.\text{TrapEval}(td, K', X)$, we have $\text{H}_{K'}(X) = \mathcal{H}.\text{Eval}(K', X) = \mathbf{A}\mathbf{R}_X + \mathbf{S}_X\mathbf{B}$.

Statistically close trapdoor keys: For all key $(K', td) \leftarrow \mathcal{H}.\text{TrapGen}(1^\kappa, \mathbf{A}, \mathbf{B})$ and $K \leftarrow \mathcal{H}.\text{Gen}(1^\kappa)$, the statistical distance between $(\mathbf{A}, K')$ and $(\mathbf{A}, K)$ is at most γ.

Well-distributed hidden matrices: For all $(K', td) \leftarrow \mathcal{H}.\text{TrapGen}(1^\kappa, \mathbf{A}, \mathbf{B})$, any inputs $X_1, \ldots, X_u, Y_1, \ldots, Y_v \in \mathcal{X}$ such that $X_i \neq Y_j$ for any i, j, let $(\mathbf{R}_{X_i}, \mathbf{S}_{X_i}) = \mathcal{H}.\text{TrapEval}(td, K', X_i)$ and $(\mathbf{R}_{Y_i}, \mathbf{S}_{Y_i}) = \mathcal{H}.\text{TrapEval}(td, K', Y_i)$. Then, we have that

$$\Pr[\mathbf{S}_{X_1} = \cdots = \mathbf{S}_{X_u} = \mathbf{0} \wedge \mathbf{S}_{Y_1}, \ldots, \mathbf{S}_{Y_v} \in \mathcal{I}_n] \ge \delta,$$

where the probability is over the trapdoor td produced along with K'.

If γ is negligible and $\delta > 0$ is noticeable, we simply say that $\mathcal{H}$ is a (u, v, β)-PHF. Furthermore, if u (resp. v) is an arbitrary polynomial in κ, we say that $\mathcal{H}$ is a $(\mathsf{poly}, v, \beta)$-PHF (resp. $(u, \mathsf{poly}, \beta)$-PHF).

A ***weak programmable hash function*** is a relaxed version of PHF, where the $\mathcal{H}$.TrapGen algorithm additionally takes a list $X_1, \ldots, X_u \in \mathcal{X}$ as inputs such that the well-distributed hidden matrices property holds in the following sense: For all $(K', td) \leftarrow \mathcal{H}.\text{TrapGen}(1^\kappa, \mathbf{A}, \mathbf{B}, \{X_1, \ldots, X_u\})$, any inputs $Y_1, \ldots, Y_v \in \mathcal{X}$ such that $Y_j \notin \{X_1, \ldots, X_u\}$ for all j, let $(\mathbf{R}_{X_i}, \mathbf{S}_{X_i}) = \mathcal{H}.\text{TrapEval}(td, K', X_i)$ and $(\mathbf{R}_{Y_i}, \mathbf{S}_{Y_i}) = \mathcal{H}.\text{TrapEval}(td, K', Y_i)$, we have that $\Pr[\mathbf{S}_{X_1} = \cdots = \mathbf{S}_{X_u} = \mathbf{0} \wedge \mathbf{S}_{Y_1}, \ldots, \mathbf{S}_{Y_v} \in \mathcal{I}_n] \ge \delta$, where the probability is over the trapdoor td produced along with K'.

Besides, a hash function $\mathcal{H} : \mathcal{X} \to \mathbb{Z}_q^{n\times m}$ can be a (weak) (u, v, β)-PHF for different parameters u and v, since there might exist different pairs of trapdoor key generation and trapdoor evaluation algorithms for $\mathcal{H}$. If this is the case, one can easily show that the keys output by these trapdoor key generation algorithms are statistically indistinguishable by definition.

---

[1] A general trapdoor matrix $\mathbf{B}$ is used for utmost generality, but the publicly known trapdoor matrix $\mathbf{B} = \mathbf{G}$ in [23] is recommended for both efficiency and simplicity.

4.3.1.1 Constructions

In the following, we will give two concrete constructions of lattice-based PHFs.

Type-I Construction. We describe the Type-I construction of lattice-based PHFs in the following definition:

Definition 4.4 Let $\ell, n, m, q \in \mathbb{Z}$ be some polynomials in the security parameter κ. Let Enc be a deterministic encoding from $\mathcal{X}$ to $(\mathbb{Z}_q^{n\times n})^\ell$, the hash function $\mathcal{H} = (\mathcal{H}.\mathrm{Gen}, \mathcal{H}.\mathrm{Eval})$ with key space $\mathcal{K} \subseteq (\mathbb{Z}_q^{n\times m})^{\ell+1}$ is defined as follows:

- $\mathcal{H}.\mathrm{Gen}(1^\kappa)$: Randomly choose $(\mathbf{A}_0, \ldots, \mathbf{A}_\ell) \xleftarrow{\$} \mathcal{K}$, return $K = \{\mathbf{A}_i\}_{i\in\{0,\ldots,\ell\}}$.
- $\mathcal{H}.\mathrm{Eval}(K, X)$: Let $\mathsf{Enc}(X) = (\mathbf{C}_1, \ldots, \mathbf{C}_\ell)$, return $\mathbf{Z} = \mathbf{A}_0 + \sum_{i=1}^{\ell} \mathbf{C}_i\mathbf{A}_i$.

We note that the above hash function has actually been (implicitly) used to construct both signatures (e.g., [4, 8, 24]) and encryptions (e.g., [1, 23]). Let $\mathbf{I}_n$ be the $n \times n$ identity matrix. In the following theorems, we summarize several known results which were implicitly proved in [1, 8, 23].

Theorem 4.2 *Let* $\mathcal{K} = (\mathbb{Z}_q^{n\times m})^{\ell+1}$ *and* $\mathcal{X} = \{0, 1\}^\ell$. *In addition, given an input* $X = (X_1, \ldots, X_\ell) \in \mathcal{X}$, *the encoding function* $\mathsf{Enc}(X)$ *returns* $\mathbf{C}_i = (-1)^{X_i} \cdot \mathbf{I}_n$ *for* $i = \{1, \ldots, \ell\}$. *Then, for large enough integer* $\bar{m} = O(n \log q)$ *and any fixed polynomial* $v = v(\kappa) \in \mathbb{Z}$, *the instantiated hash function* $\mathcal{H}$ *of Definition 4.4 is a* $(1, v, \beta, \gamma, \delta)$-PHF *with* $\beta \le \sqrt{\ell\bar{m}} \cdot \omega(\sqrt{\log n})$, $\gamma = \mathsf{negl}(\kappa)$ *and* $\delta = \frac{1}{q^t}(1 - \frac{v}{q^t})$, *where* t *is the smallest integer satisfying* $q^t > 2v$.

Theorem 4.3 *For large enough* $\bar{m} = O(n \log q)$, *the hash function* $\mathcal{H}$ *given in Definition 4.4 is a weak* $(1, \mathsf{poly}, \beta, \gamma, \delta)$-PHF *with* $\beta \le \sqrt{\ell\bar{m}} \cdot \omega(\sqrt{\log n})$, $\gamma = \mathsf{negl}(\kappa)$, *and* $\delta = 1$ *when instantiated as follows:*

- *Let* $\mathcal{K} = (\mathbb{Z}_q^{n\times m})^2$ *(i.e.,* $\ell = 1$*) and* $\mathcal{X} = \mathbb{Z}_q^n$. *Given an input* $X \in \mathcal{X}$, *the encoding* $\mathsf{Enc}(X)$ *returns* $H(X)$ *where* $H : \mathbb{Z}_q^n \to \mathbb{Z}_q^{n\times n}$ *is an FRD encoding.*
- *Let* $\mathcal{K} = (\mathbb{Z}_q^{n\times m})^{\ell+1}$ *and* $\mathcal{X} = \{0, 1\}^\ell$. *Given an input* $X = (X_1, \ldots, X_\ell) \in \mathcal{X}$, *the encoding* $\mathsf{Enc}(X)$ *returns* $\mathbf{C}_i = X_i \cdot \mathbf{I}_n$ *for all* $i \in \{1, \ldots, \ell\}$.

Unlike the traditional PHFs [11, 20, 21] where a bigger u is usually better in constructing short signature schemes, the lattice-based PHFs seem more useful when the parameter v is bigger (e.g., a polynomial in κ). There is a simple explanation: although both notions aim at capturing some kind of partitioning proof trick, i.e., each programmed hash value contains a hidden element behaving as a trigger of some prior embedded trapdoors, for traditional PHFs the trapdoor is usually triggered when the hidden element is zero, while in the lattice setting the trapdoor is typically triggered when the hidden element is a non-zero invertible one. This also explains why previous known constructions on lattices (e.g., the instantiations in Theorems 4.2 and 4.3) are (weak) $(1, v, \beta)$-PHFs for some polynomial $v \in \mathbb{Z}$ and real $\beta \in \mathbb{R}$.

Procedure I	**Procedure II**
$\mathbf{Z} := \hat{\mathbf{A}}$	$\mathbf{R}_X := \hat{\mathbf{R}}, \mathbf{S}_X := -(-1)^c \cdot \mathbf{I}_n$
For all $z \in CF_X$	For all $z \in CF_X$
$(b_0, \ldots, b_{\mu-1}) := \mathsf{BitDecomp}_N(z)$	$(b_0, \ldots, b_{\mu-1}) := \mathsf{BitDecomp}_N(z)$
$\mathbf{B}_z := \mathbf{A}_{\mu-1} - b_{\mu-1} \cdot \mathbf{G}$	$\mathbf{B}_z := \mathbf{A}_{\mu-1} - b_{\mu-1} \cdot \mathbf{G}$
	$\mathbf{R}_z := \mathbf{R}_{\mu-1}$
	$\mathbf{S}_z := (1 - b^*_{\mu-1} - b_{\mu-1}) \cdot \mathbf{I}_n$
For $i = \mu - 2, \ldots, 0$	For $i = \mu - 2, \ldots, 0$
$\mathbf{B}_z := (\mathbf{A}_i - b_i \cdot \mathbf{G}) \cdot \mathbf{G}^{-1}(\mathbf{B}_z)$	$\mathbf{B}_z := (\mathbf{A}_i - b_i \cdot \mathbf{G}) \cdot \mathbf{G}^{-1}(\mathbf{B}_z)$
	$\mathbf{R}_z := \mathbf{R}_i \cdot \mathbf{G}^{-1}(\mathbf{B}_z) + (1 - b^*_i - b_i) \cdot \mathbf{R}_z$
	$\mathbf{S}_z := (1 - b^*_i - b_i) \cdot \mathbf{S}_z$
$\mathbf{Z} := \mathbf{Z} + \mathbf{B}_z$	$\mathbf{R}_X := \mathbf{R}_X + \mathbf{R}_z, \mathbf{S}_X := \mathbf{S}_X + \mathbf{S}_z$
Return $\mathbf{Z}$	Return $(\mathbf{R}_X, \mathbf{S}_X)$

Fig. 4.1 The Procedures Used in Definition 4.5 and Theorem 4.4

Type-II Construction. Let integers $\ell, \bar{m}, n, q, u, v, L, N$ be some polynomials in the security parameter κ, and let $k = \lceil \log_2 q \rceil$. We now exploit the nice property of the publicly known trapdoor matrix $\mathbf{B} = \mathbf{G} \in \mathbb{Z}_q^{n \times nk}$ to construct more efficient PHF from lattices for any $v = \mathsf{poly}(\kappa)$. We begin by first recalling the notion of cover-free sets. Formally, we say that set S does not cover set T if there exists at least one element $t \in T$ such that $t \notin S$. Let $CF = \{CF_X\}_{X \in [L]}$ be a family of subsets of $[N]$. The family CF is said to be v-cover-free over $[N]$ if for any subset $S \subseteq [L]$ of size at most v, then the union $\cup_{X \in S} CF_X$ does not cover CF_Y for all $Y \notin S$. Besides, we say that CF is η-uniform if every subset CF_X in the family $CF = \{CF_X\}_{X \in [L]}$ have size $\eta \in \mathbb{Z}$. Furthermore, there exists an efficient algorithm to generate cover-free sets [15, 22]. Formally,

Lemma 4.7 *There is a deterministic polynomial time algorithm that on inputs integers $L = 2^\ell$ and $v \in \mathbb{Z}$, returns an η-uniform, v-cover-free sets $CF = \{CF_X\}_{X \in [L]}$ over $[N]$, where $N \leq 16v^2\ell$ and $\eta = N/4v$.*

In the following, we use the binary representation of $[N]$ to construct lattice-based PHFs with short keys.

Definition 4.5 Let $n, q \in \mathbb{Z}$ be some polynomials in the security parameter κ. For any $\ell, v \in \mathbb{Z}$ and $L = 2^\ell$, let $N \leq 16v^2\ell$, $\eta \leq 4v\ell$ and $CF = \{CF_X\}_{X \in [L]}$ be defined as in Lemma 4.7. Let $\mu = \lceil \log_2 N \rceil$ and $k = \lceil \log_2 q \rceil$. Then, the hash function $\mathcal{H} = (\mathcal{H}.\text{Gen}, \mathcal{H}.\text{Eval})$ from $[L]$ to $\mathbb{Z}_q^{n \times nk}$ is defined as follows:

- $\mathcal{H}.\text{Gen}(1^\kappa)$: Randomly choose $\hat{\mathbf{A}}, \mathbf{A}_i \xleftarrow{\$} \mathbb{Z}_q^{n \times nk}$ for $i \in \{0, \ldots, \mu - 1\}$, return the key $K = (\hat{\mathbf{A}}, \{\mathbf{A}_i\}_{i \in \{0,\ldots,\mu-1\}})$.
- $\mathcal{H}.\text{Eval}(K, X)$: Given $K = (\hat{\mathbf{A}}, \{\mathbf{A}_i\}_{i \in \{0,\ldots,\mu-1\}})$ and integer $X \in [L]$, the algorithm performs the **Procedure I** in Fig. 4.1 to compute $\mathbf{Z} = \mathrm{H}_K(X)$.

We now show that for any prior fixed $v = \mathsf{poly}(\kappa)$, the hash function $\mathcal{H}$ given in Definition 4.5 is a $(1, v, \beta)$-PHF for some polynomially bounded $\beta \in \mathbb{R}$.

Theorem 4.4 *For any $\ell, v \in \mathbb{Z}$ and $L = 2^\ell$, let $N \leq 16v^2\ell$, $\eta \leq 4v\ell$ and $CF = \{CF_X\}_{X\in[L]}$ be defined as in Lemma 4.7. Then, for large enough $\bar{m} = O(n \log q)$, the hash function $\mathcal{H}$ in Definition 4.5 is a $(1, v, \beta, \gamma, \delta)$-PHF with $\beta \leq \mu v \ell \bar{m}^{1.5} \cdot \omega(\sqrt{\log \bar{m}})$, $\gamma = \mathsf{negl}(\kappa)$ and $\delta = 1/N$, where $\mu = \lceil \log_2 N \rceil$.*

In particular, if we set $\ell = n$ and $v = \omega(\log n)$, then $\beta = \tilde{O}(n^{2.5})$, and the key of $\mathcal{H}$ only consists of $\mu = O(\log n)$ matrices.

Proof We now construct a pair of trapdoor algorithms for $\mathcal{H}$ as follows:

- $\mathcal{H}$.TrapGen$(1^\kappa, \mathbf{A}, \mathbf{G})$: Given a uniformly random $\mathbf{A} \in \mathbb{Z}_q^{n\times \bar{m}}$ and matrix $\mathbf{G} \in \mathbb{Z}_q^{n\times nk}$ for sufficiently large $\bar{m} = O(n \log q)$, let $s \geq \omega(\sqrt{\log \bar{m}}) \in \mathbb{R}$ satisfy the requirement in Lemma 2.11. Randomly choose $\hat{\mathbf{R}}, \mathbf{R}_i \xleftarrow{\$} (D_{\mathbb{Z}^{\bar{m}},s})^{nk}$ for $i \in \{0, \ldots, \mu - 1\}$, and an integer $z^* \xleftarrow{\$} [N]$. Let $(b_0^*, \ldots, b_{\mu-1}^*) = \mathsf{BitDecomp}_N(z^*)$, and let c be the number of 1's in the vector $(b_0^*, \ldots, b_{\mu-1}^*)$. Then, compute $\hat{\mathbf{A}} = \mathbf{A}\hat{\mathbf{R}} - (-1)^c \cdot \mathbf{G}$ and $\mathbf{A}_i = \mathbf{A}\mathbf{R}_i + (1 - b_i^*) \cdot \mathbf{G}$. Finally, return the key $K' = (\hat{\mathbf{A}}, \{\mathbf{A}_i\}_{i\in\{0,\ldots,\mu-1\}})$ and the trapdoor $td = (\hat{\mathbf{R}}, \{\mathbf{R}_i\}_{i\in\{0,\ldots,\mu-1\}}, z^*)$.
- $\mathcal{H}$.TrapEval(td, K', X): Given td and an input $X \in [L]$, the algorithm first computes CF_X by Lemma 4.7. Then, let $(b_0^*, \ldots, b_{\mu-1}^*) = \mathsf{BitDecomp}_N(z^*)$, and perform the **Procedure II** in Fig. 4.1 to compute $(\mathbf{R}_X, \mathbf{S}_X)$.

Since $s \geq \omega(\sqrt{\log \bar{m}})$ and $\hat{\mathbf{R}}, \mathbf{R}_i \xleftarrow{\$} (D_{\mathbb{Z}^{\bar{m}},s})^{nk}$, each matrix in key $K' = (\hat{\mathbf{A}}, \{\mathbf{A}_i\}_{i\in\{0,\ldots,\mu-1\}})$ is statistically close to uniform over $\mathbb{Z}_q^{n\times nk}$ by Lemma 2.11. Using a standard hybrid argument, it is easy to show that the statistical distance γ between $(\mathbf{A}, K')$ and $(\mathbf{A}, K)$ is negligible, where $K \leftarrow \mathcal{H}$.Gen(1^κ). In particular, this means that z^* is statistically hidden in K'.

For correctness, we first show that $\mathbf{B}_z = \mathbf{A}\mathbf{R}_z + \mathbf{S}_z\mathbf{G}$ always holds during the computation. By definition, we have that $\mathbf{B}_z = \mathbf{A}_{\mu-1} - b_{\mu-1} \cdot \mathbf{G} = \mathbf{A}\mathbf{R}_z + \mathbf{S}_z\mathbf{G}$ holds before entering the inner loop. Assume that $\mathbf{B}_z = \mathbf{A}\mathbf{R}_z + \mathbf{S}_z\mathbf{G}$ holds before entering the j-th (i.e., $i = j$) iteration of the inner loop, we now show that the equation $\mathbf{B}_z = \mathbf{A}\mathbf{R}_z + \mathbf{S}_z\mathbf{G}$ still holds after the j-th iteration. Since $\mathbf{A}_j - b_j \cdot \mathbf{G} = \mathbf{A}\mathbf{R}_j + (1 - b_j^* - b_j) \cdot \mathbf{G}$, we have that $\mathbf{B}_z := (\mathbf{A}_j - b_j \cdot \mathbf{G}) \cdot \mathbf{G}^{-1}(\mathbf{B}_z) = \mathbf{A}\mathbf{R}_j \cdot \mathbf{G}^{-1}(\mathbf{B}_z) + (1 - b_j^* - b_j) \cdot (\mathbf{A}\mathbf{R}_z + \mathbf{S}_z\mathbf{G})$. This means that if we set $\mathbf{R}_z := \mathbf{R}_j \cdot \mathbf{G}^{-1}(\mathbf{B}_z) + (1 - b_j^* - b_j) \cdot \mathbf{R}_z$ and $\mathbf{S}_z := (1 - b_j^* - b_j) \cdot \mathbf{S}_z$, the equation $\mathbf{B}_z = \mathbf{A}\mathbf{R}_z + \mathbf{S}_z\mathbf{G}$ still holds. In particular, we have that $\mathbf{S}_z = \prod_{i=0}^{\mu-1}(1 - b_i^* - b_i) \cdot \mathbf{I}_n$ holds at the end of the inner loop. It is easy to check that $\mathbf{S}_z = \mathbf{0}$ for any $z \neq z^*$, and $\mathbf{S}_z = (-1)^c \cdot \mathbf{I}_n$ for $z = z^*$, where c is the number of 1's in the binary vector $(b_0^*, \ldots, b_{\mu-1}^*) = \mathsf{BitDecomp}_N(z^*)$. The correctness of the trapdoor evaluation algorithm follows from that fact that $\mathbf{Z} = \mathcal{H}.\mathrm{Eval}(K', X) = \hat{\mathbf{A}} + \sum_{z\in CF_X} \mathbf{B}_z = \mathbf{A}\hat{\mathbf{R}} - (-1)^c \cdot \mathbf{G} + \sum_{z\in CF_X}(\mathbf{A}\mathbf{R}_z + \mathbf{S}_z\mathbf{G}) = \mathbf{A}\mathbf{R}_X + \mathbf{S}_X\mathbf{B}$. In particular, we have that $\mathbf{S}_X = -(-1)^c \cdot \mathbf{I}_n$ if $z^* \notin CF_X$, else $\mathbf{S}_X = \mathbf{0}$.

Since $s_1(\mathbf{G}^{-1}(\mathbf{B}_z)) \leq nk$ by the fact that $\mathbf{G}^{-1}(\mathbf{B}_z) \in \{0, 1\}^{nk\times nk}$, and $s_1(\hat{\mathbf{R}})$, $s_1(\mathbf{R}_i) \leq (\sqrt{\bar{m}} + \sqrt{nk}) \cdot \omega(\sqrt{\log \bar{m}})$ by Lemma 2.6, we have that $s_1(\mathbf{R}_z) \leq \mu \bar{m}^{1.5} \cdot \omega(\sqrt{\log \bar{m}})$ holds except with negligible probability for any $z \in CF_X$. Using $|CF_X| = \eta \leq 4v\ell$, the inequality $s_1(\mathbf{R}_X) \leq \mu v \ell \bar{m}^{1.5} \cdot \omega(\sqrt{\log \bar{m}})$ holds except with negligible probability for any $X \in [L]$. Besides, for any $X_1, Y_1, \ldots, Y_v \in [L]$ such that $X_1 \neq Y_j$

for all $j \in \{1, \ldots, v\}$, there is at least one element in $CF_{X_1} \subseteq [N]$ that does not belong to the union set $\cup_{j\in\{1,\ldots,v\}} CF_{Y_j}$. This is because the family $CF = \{CF_X\}_{X\in[L]}$ is v-cover-free. Since z^* is randomly chosen from $[N]$ and is statistically hidden in the key K', the probability $\Pr[z^* \in CF_{X_1} \wedge z^* \notin \cup_{j\in\{1,\ldots,v\}} CF_{Y_j}]$ is at least $1/N$. Thus, we have that $\Pr[\mathbf{S}_{X_1} = \mathbf{0} \wedge \mathbf{S}_{Y_1} = \cdots = \mathbf{S}_{Y_v} = -(-1)^c \cdot \mathbf{I}_n \in \mathcal{I}_n] \geq \frac{1}{N}$. □

4.3.1.2 Collision-Resistance

Let $\mathcal{H} = \{\mathrm{H}_K : \mathcal{X} \to \mathcal{Y}\}_{K\in\mathcal{K}}$ be a family of hash functions with key space $\mathcal{K}$. We say that $\mathcal{H}$ is collision-resistant if for any PPT algorithm C, its advantage

$$\mathrm{Adv}^{\mathrm{cr}}_{\mathcal{H},C}(\kappa) = \Pr[K \xleftarrow{\$} \mathcal{K}; (X_1, X_2) \xleftarrow{\$} C(K, 1^\kappa) : X_1 \neq X_2 \wedge \mathrm{H}_K(X_1) = \mathrm{H}_K(X_2)]$$

is negligible in the security parameter κ.

Theorem 4.5 *Let $n, v, q \in \mathbb{Z}$ and $\bar{\beta}, \beta \in \mathbb{R}$ be polynomials in the security parameter κ. Let $\mathcal{H} = (\mathcal{H}.\mathrm{Gen}, \mathcal{H}.\mathrm{Eval})$ be a $(1, v, \beta, \gamma, \delta)$-PHF with $\gamma = \mathsf{negl}(\kappa)$ and noticeable $\delta > 0$. Then, for large enough $\bar{m}, m \in \mathbb{Z}$ and $v \geq 1$, if there exists an algorithm C breaking the collision-resistance of $\mathcal{H}$, there exists an algorithm $\mathcal{B}$ solving the $\mathrm{ISIS}_{q,\bar{m},\bar{\beta}}$ problem for $\bar{\beta} = \beta\sqrt{m} \cdot \omega(\log n)$ with probability at least $\epsilon' \geq (\epsilon - \gamma)\delta$.*

Proof If there exists an algorithm C breaking the collision-resistance of $\mathcal{H}$ with advantage ϵ, we now construct an algorithm $\mathcal{B}$ that solves the $\mathrm{ISIS}_{q,\bar{m},\bar{\beta}}$ problem. Let $\mathbf{B} \in \mathbb{Z}_q^{n\times m}$ be any trapdoor matrix that allows to efficiently sample short vector $\mathbf{v} \in \mathbb{Z}^m$ such that $\|\mathbf{v}\| \leq \sqrt{m} \cdot \omega(\log n)$ and $\mathbf{B}\mathbf{v} = \mathbf{u}'$ for any $\mathbf{u}' \in \mathbb{Z}_q^n$ (e.g., $\mathbf{B}$ is generated by using the trapdoor generation algorithm in Lemma 2.24). Formally, given an $\mathrm{ISIS}_{q,\bar{m},\bar{\beta}}$ challenge instance $(\mathbf{A}, \mathbf{u}) \in \mathbb{Z}_q^{n\times\bar{m}} \times \mathbb{Z}_q^n$. The algorithm $\mathcal{B}$ computes $(K', td) \xleftarrow{\$} \mathcal{H}.\mathrm{TrapGen}(1^\kappa, \mathbf{A}, \mathbf{B})$ and sends K' as the hash key to C. Since the statistical distance between K' and the real hash key K is at most $\gamma = \mathsf{negl}(\kappa)$, the probability that given the key K' the algorithm $C(K', 1^\kappa)$ outputs two elements $X_1 \neq X_2$ satisfying $\mathrm{H}_{K'}(X_1) = \mathrm{H}_{K'}(X_2)$ is at least $\epsilon - \gamma$. By the correctness of $\mathcal{H}$, we know that there exist two tuples $(\mathbf{R}_{X_1}, \mathbf{S}_{X_1})$ and $(\mathbf{R}_{X_2}, \mathbf{S}_{X_2})$ such that $\mathrm{H}_{K'}(X_1) = \mathbf{A}\mathbf{R}_{X_1} + \mathbf{S}_{X_1}\mathbf{B} = \mathbf{A}\mathbf{R}_{X_2} + \mathbf{S}_{X_2}\mathbf{B} = \mathrm{H}_{K'}(X_2)$. In addition, by the well-distributed hidden matrices property of $\mathcal{H}$, the probability $\Pr[\mathbf{S}_{X_1} = \mathbf{0} \wedge \mathbf{S}_{X_2} \in \mathcal{I}_n]$ is at least δ. In other words, the equation $\mathbf{A}\mathbf{R}_{X_1} = \mathbf{A}\mathbf{R}_{X_2} + \mathbf{S}_{X_2}\mathbf{B}$ holds with probability at least $(\epsilon - \gamma)\delta$. If this is the case, $\mathcal{B}$ outputs $\mathbf{x} = (\mathbf{R}_{X_1} - \mathbf{R}_{X_2})\mathbf{v}$, where $\mathbf{v} \in \mathbb{Z}_q^m$ is sampled by using the trapdoor of $\mathbf{B}$ such that $\|\mathbf{v}\| \leq \sqrt{m} \cdot \omega(\log n)$ and $\mathbf{B}\mathbf{v} = \mathbf{S}_{X_2}^{-1}\mathbf{u}$. By $\mathbf{A}\mathbf{x} = \mathbf{S}_{X_2}\mathbf{B}\mathbf{v} = \mathbf{u}$, we have that $\mathbf{x}$ is a solution of $\mathbf{A}\mathbf{x} = \mathbf{u}$. In addition, since $s_1(\mathbf{R}_{X_1}), s_1(\mathbf{R}_{X_2}) \leq \beta$ by assumption, we have $\|\mathbf{x}\| \leq \beta\sqrt{m} \cdot \omega(\log n)$. This completes the proof. □

4.3.1.3 High Min-Entropy

Let $\mathcal{H} : \mathcal{X} \to \mathbb{Z}_q^{n\times m}$ be a $(1, v, \beta, \gamma, \delta)$-PHF with $\gamma = \mathsf{negl}(\kappa)$ and noticeable $\delta > 0$. Note that the well-distributed hidden matrices property of $\mathcal{H}$ holds even for an unbounded algorithm $\mathcal{A}$ that chooses $\{X_i\}$ and $\{Z_j\}$ after seeing K'. For any noticeable $\delta > 0$, this can only happen when the decomposition $\mathrm{H}_{K'}(X) = \mathbf{A}\mathbf{R}_X + \mathbf{S}_X\mathbf{B}$ is not unique (with respect to K') and the particular pair determined by td, i.e., $(\mathbf{R}_X, \mathbf{S}_X) = \mathcal{H}.\text{TrapEval}(td, K', X)$, is information-theoretically hidden from $\mathcal{A}$. We now introduce a property called high min-entropy to formally capture this useful feature.

Definition 4.6 *(PHF with High Min-Entropy)* Let $\mathcal{H} : \mathcal{X} \to \mathbb{Z}_q^{n\times m}$ be a $(1, v, \beta, \gamma, \delta)$-PHF with $\gamma = \mathsf{negl}(\kappa)$ and noticeable $\delta > 0$. Let $\mathcal{K}$ be the key space of $\mathcal{H}$, and let $\mathcal{H}$.TrapGen and $\mathcal{H}$.TrapEval be a pair of trapdoor generation and trapdoor evaluation algorithms for $\mathcal{H}$. We say that $\mathcal{H}$ is a PHF with high min-entropy if for uniformly random $\mathbf{A} \in \mathbb{Z}_q^{n\times \bar{m}}$ and (publicly known) trapdoor matrix $\mathbf{B} \in \mathbb{Z}_q^{n\times m}$, the following conditions hold:

1. For any $(K', td) \leftarrow \mathcal{H}.\text{TrapGen}(1^\kappa, \mathbf{A}, \mathbf{B})$, $K \leftarrow \mathcal{H}.\text{Gen}(1^\kappa)$, any $X \in \mathcal{X}$ and any $\mathbf{w} \in \mathbb{Z}_q^{\bar{m}}$, the algorithm $\mathcal{H}.\text{TrapEval}(td, K, X)$ is well-defined, and the statistical distance between $(\mathbf{A}, K', (\mathbf{R}'_X)^t\mathbf{w})$ and $(\mathbf{A}, K, \mathbf{R}^t_X\mathbf{w})$ is negligible in κ, where $(\mathbf{R}'_X, \mathbf{S}'_X) = \mathcal{H}.\text{TrapEval}(td, K', X)$, and $(\mathbf{R}_X, \mathbf{S}_X) = \mathcal{H}.\text{TrapEval}$ (td, K, X).
2. For any $(K', td) \leftarrow \mathcal{H}.\text{TrapGen}(1^\kappa, \mathbf{A}, \mathbf{B})$, any $X \in \mathcal{X}$, any uniformly random $\mathbf{v} \in \mathbb{Z}_q^{\bar{m}}$, and any uniformly random $\mathbf{u} \xleftarrow{\$} \mathbb{Z}_q^m$, the statistical distance between $(\mathbf{A}, K', \mathbf{v}, (\mathbf{R}'_X)^t\mathbf{v})$ and $(\mathbf{A}, K', \mathbf{v}, \mathbf{u})$ is negligible in κ, where $(\mathbf{R}'_X, \mathbf{S}'_X) = \mathcal{H}.\text{TrapEval}(td, K', X)$.

Remark 4.1 First, since $s_1(\mathbf{R}'_X) \leq \beta$ holds with overwhelming probability, we have that $\|(\mathbf{R}'_X)^t\mathbf{w}\| \leq \beta\|\mathbf{w}\|$. Thus, the first condition implicitly implies that $\|\mathbf{R}^t_X\mathbf{w}\| \leq \beta\|\mathbf{w}\|$ holds with overwhelming probability for any $K \leftarrow \mathcal{H}.\text{Gen}(1^\kappa)$, $X \in \mathcal{X}$, and $(\mathbf{R}_X, \mathbf{S}_X) = \mathcal{H}.\text{TrapEval}$ (td, K, X). Second, we note that the well-distributed hidden matrices property of PHF only holds when the information (except that is already leaked via the key K') of the trapdoor td is hidden. This means that it provides no guarantee when some information of $\mathbf{R}_X$ for any $X \in \mathcal{X}$ (which is usually related to the trapdoor td) is given public. However, for a PHF with high min-entropy, this property still holds when the information of $\mathbf{R}^t_X\mathbf{v}$ for a uniformly random vector $\mathbf{v}$ is leaked.

For appropriate choices of parameters, the work [1] implicitly showed that the Type-I PHF construction satisfied the high min-entropy property. Now, we show that the Type-II PHF construction also has the high min-entropy property.

Theorem 4.6 *Let integers $n, \bar{m}, q$ be some polynomials in the security parameter κ, and let $k = \lceil \log_2 q \rceil$. For any $\ell, v \in \mathbb{Z}$ and $L = 2^\ell$, let $N \leq 16v^2\ell$, $\eta \leq 4v\ell$ and $CF = \{CF_X\}_{X\in[L]}$ be defined as in Lemma 4.7. Then, for large enough $\bar{m} =$*

$O(n \log q)$, the hash function $\mathcal{H} : [L] \to \mathbb{Z}_q^{n \times nk}$ given in Definition 4.5 (and proved in Theorem 4.4) is a PHF with high min-entropy.

Proof For any $\mathbf{w} \in \mathbb{Z}_q^{\bar{m}}$, let $f_{\mathbf{w}} : \mathbb{Z}_q^{\bar{m} \times nk} \to \mathbb{Z}_q^{nk}$ be the function defined by $f_{\mathbf{w}}(\mathbf{X}) = \mathbf{X}^t \mathbf{w} \in \mathbb{Z}_q^{nk}$. By the definition of $\mathcal{H}$.TrapGen in Theorem 4.4, for any $(K', td) \leftarrow \mathcal{H}.\mathrm{TrapGen}(1^\kappa, \mathbf{A}, \mathbf{G})$, we have $td = (\hat{\mathbf{R}}, \{\mathbf{R}_i\}_{i \in \{0,\ldots,\mu-1\}}, z^*)$. Denote $I = \{f_{\mathbf{w}}(\hat{\mathbf{R}}), \{f_{\mathbf{w}}(\mathbf{R}_i)\}_{i \in \{0,\ldots,\mu-1\}})\}$. First, it is easy to check that the algorithm $\mathcal{H}.\mathrm{TrapEval}(td, K, X)$ is well-defined for any $K \in \mathcal{K} = \mathbb{Z}_q^{n \times nk}$ and $X \in \mathcal{X}$. In addition, given $I = \{f_{\mathbf{w}}(\hat{\mathbf{R}}), \{f_{\mathbf{w}}(\mathbf{R}_i)\}_{i \in \{0,\ldots,\mu-1\}})\}$ and (K, X, z^*) as inputs, there exists a public algorithm that computes $\mathbf{R}_X^t \mathbf{w}$ by simulating the **Procedure II** in Theorem 4.4, where $(\mathbf{R}_X, \mathbf{S}_X) = \mathcal{H}.\mathrm{TrapEval}(td, K, X)$. To prove that $\mathcal{H}$ satisfies the first condition of high min-entropy, it suffices to show that K' is statistically close to uniform over $(\mathbb{Z}_q^{n \times nk})^{\mu+1}$ conditioned on I and z^* (recall that the real key K of $\mathcal{H}$ is uniformly distributed over $(\mathbb{Z}_q^{n \times nk})^{\mu+1}$ by Definition 4.5). Since each matrix in the key K' always has a form of $\mathbf{A}\tilde{\mathbf{R}} + b\mathbf{G}$ for some randomly chosen $\tilde{\mathbf{R}} \xleftarrow{\$} (D_{\mathbb{Z}^{\bar{m}},s})^{nk}$, and a bit $b \in \{0, 1\}$ depending on a random $z^* \xleftarrow{\$} [N]$. Using a standard hybrid argument, it is enough to show that conditioned on $\mathbf{A}$ and $f_{\mathbf{w}}(\tilde{\mathbf{R}})$, $\mathbf{A}\tilde{\mathbf{R}}$ is statistically close to uniform over $\mathbb{Z}_q^{n \times nk}$.

Let $f'_{\mathbf{w}} : \mathbb{Z}_q^{\bar{m}} \to \mathbb{Z}_q$ be defined by $f'_{\mathbf{w}}(\mathbf{x}) = \mathbf{x}^t \mathbf{w}$, and let $\tilde{\mathbf{R}} = (\mathbf{r}_1, \ldots, \mathbf{r}_{nk})$. Then, $f_{\mathbf{w}}(\tilde{\mathbf{R}}) = (f'_{\mathbf{w}}(\mathbf{r}_1), \ldots, f'_{\mathbf{w}}(\mathbf{r}_{nk}))^t \in \mathbb{Z}_q^{nk}$. By Lemma 2.3, the guessing probability $\gamma(\mathbf{r}_i)$ is at most $2^{1-\bar{m}}$ for all $i \in \{1, \ldots, nk\}$. By the generalized leftover hash lemma in [13], conditioned on $\mathbf{A}$ and $f'_{\mathbf{w}}(\mathbf{r}_i) \in \mathbb{Z}_q$, the statistical distance between $\mathbf{A}\mathbf{r}_i \in \mathbb{Z}_q^n$ and uniform over $\mathbb{Z}_q^n$ is at most $\frac{1}{2} \cdot \sqrt{2^{1-\bar{m}} \cdot q^n \cdot q}$, which is negligible if we set $\bar{m} = O(n \log q) > (n + 1) \log q + \omega(\log n)$. Using a standard hybrid argument, we have that conditioned on $\mathbf{A}$ and $f_{\mathbf{w}}(\tilde{\mathbf{R}})$, the matrix $\mathbf{A}\tilde{\mathbf{R}} = (\mathbf{A}\mathbf{r}_1 \| \ldots \| \mathbf{A}\mathbf{r}_{nk})$ is statistically close to uniform over $\mathbb{Z}_q^{n \times nk}$.

Since for any input X and $(\mathbf{R}_X, \mathbf{S}_X) = \mathcal{H}.\mathrm{TrapEval}(td, K', X)$, we always have that $\mathbf{R}_X = \hat{\mathbf{R}} + \tilde{\mathbf{R}}$ for some $\tilde{\mathbf{R}}$ that is independent from $\hat{\mathbf{R}}$. Let $\mathbf{R}_X^t \mathbf{v} = \hat{\mathbf{R}}^t \mathbf{v} + \tilde{\mathbf{R}}^t \mathbf{v} = \hat{\mathbf{u}} + \tilde{\mathbf{u}}$. Then, in order to prove that $\mathcal{H}$ satisfies the second condition of high min-entropy property, it suffices to show that given K' and $\mathbf{v}$, the element $\hat{\mathbf{u}} = \hat{\mathbf{R}}^t \mathbf{v}$ is uniformly random. Since $\hat{\mathbf{R}} \xleftarrow{\$} (D_{\mathbb{Z}^{\bar{m}},s})^{nk}$ for $s \geq \omega(\sqrt{\log \bar{m}})$ is only used to generate the matrix $\hat{\mathbf{A}} = \mathbf{A}\hat{\mathbf{R}} - (-1)^c \cdot \mathbf{G}$ in the key K', we have that for large enough $\bar{m} = O(n \log q)$, the pair $(\mathbf{A}\hat{\mathbf{R}}, \hat{\mathbf{u}}^t = \mathbf{v}^t \hat{\mathbf{R}})$ is statistically close to uniform over $\mathbb{Z}_q^{n \times nk} \times \mathbb{Z}_q^{nk}$ by the fact in Lemma 2.11.[2] Thus, $\mathbf{R}_X^t \mathbf{v} = \hat{\mathbf{R}}^t \mathbf{v} + \tilde{\mathbf{R}}^t \mathbf{v}$ is uniformly distributed over $\mathbb{Z}_q^{nk}$. This completes the proof of Theorem 4.6. □

---

[2]This is because one can first construct a new uniformly random matrix $\mathbf{A}'$ by appending the row vector $\mathbf{v}^t$ to the rows of $\mathbf{A}$, and then apply the fact in Lemma 2.11.

4.3.2 A Generic IBE from Lattice-Based PHFs

Let integers n, m', v, β, q be polynomials in the security parameter κ, and let $k = \lceil \log_2 q \rceil$. Let $\mathcal{H} = (\mathcal{H}.\text{Gen}, \mathcal{H}.\text{Eval})$ be any $(1, v, \beta)$-PHF with high min-entropy from $\{0,1\}^n$ to $\mathbb{Z}_q^{n \times m'}$. Let $\mathcal{H}$.TrapGen and $\mathcal{H}$.TrapEval be a pair of trapdoor generation and trapdoor evaluation algorithm of $\mathcal{H}$ that satisfies the conditions in Definition 4.6. For convenience, we set both the user identity space and the message space as $\{0,1\}^n$. Let integers $\bar{m} = O(n \log q)$, $m = \bar{m} + m'$, $\alpha \in \mathbb{R}$, and large enough $s > \max(\beta, \sqrt{m}) \cdot \omega(\sqrt{\log n})$ be the system parameters. The generic IBE scheme $\Pi_{\text{ibe}} = (\mathsf{Setup}, \mathsf{Extract}, \mathsf{Enc}, \mathsf{Dec})$ is defined as follows:

$\mathsf{Setup}(1^\kappa)$: Given a security parameter κ, compute $(\mathbf{A}, \mathbf{R}) \leftarrow \mathsf{TrapGen}(1^n, 1^{\bar{m}}, q, \mathbf{I}_n)$ such that $\mathbf{A} \in \mathbb{Z}_q^{n \times \bar{m}}$, $\mathbf{R} = \mathbb{Z}_q^{(\bar{m}-nk) \times nk}$. Randomly choose $\mathbf{U} \xleftarrow{\$} \mathbb{Z}_q^{n \times n}$, and compute $K \leftarrow \mathcal{H}.\text{Gen}(1^\kappa)$. Finally, return $(mpk, msk) = ((\mathbf{A}, K, \mathbf{U}), \mathbf{R})$.

$\mathsf{Extract}(msk, id \in \{0,1\}^n)$: Given msk and a user identity id, compute $\mathbf{A}_{id} = (\mathbf{A} \| \mathrm{H}_K(id)) \in \mathbb{Z}_q^{n \times m}$, where $\mathrm{H}_K(id) = \mathcal{H}.\text{Eval}(K, id) \in \mathbb{Z}_q^{n \times m'}$. Then, compute $\mathbf{E}_{id} \leftarrow \mathsf{SampleD}(\mathbf{R}, \mathbf{A}_{id}, \mathbf{I}_n, \mathbf{U}, s)$, and return $sk_{id} = \mathbf{E}_{id} \in \mathbb{Z}^{m \times n}$.

$\mathsf{Enc}(mpk, id \in \{0,1\}^n, M \in \{0,1\}^n)$: Given mpk, id and plaintext M, compute $\mathbf{A}_{id} = (\mathbf{A} \| \mathrm{H}_K(id)) \in \mathbb{Z}_q^{n \times m}$, where $\mathrm{H}_K(id) = \mathcal{H}.\text{Eval}(K, id) \in \mathbb{Z}_q^{n \times m'}$. Then, randomly choose $\mathbf{s} \xleftarrow{\$} \mathbb{Z}_q^n$, $\mathbf{x}_0 \xleftarrow{\$} D_{\mathbb{Z}^n, \alpha q}$, $\mathbf{x}_1 \xleftarrow{\$} D_{\mathbb{Z}^{\bar{m}}, \alpha q}$, and compute $(K', td) \leftarrow \mathcal{H}.\mathsf{TrapGen}(1^\kappa, \mathbf{A}, \mathbf{B})$ for some trapdoor matrix $\mathbf{B} \in \mathbb{Z}_q^{n \times m'}$, $(\mathbf{R}_{id}, \mathbf{S}_{id}) = \mathcal{H}.\mathsf{TrapEval}(td, K, id)$. Finally, compute and return the ciphertext $\mathbf{C} = (\mathbf{c}_0, \mathbf{c}_1)$, where

$$\mathbf{c}_0 = \mathbf{U}^t \mathbf{s} + \mathbf{x}_0 + \frac{q}{2} M, \qquad \mathbf{c}_1 = \mathbf{A}_{id}^t \mathbf{s} + \begin{pmatrix} \mathbf{x}_1 \\ \mathbf{R}_{id}^t \mathbf{x}_1 \end{pmatrix} = \begin{pmatrix} \mathbf{A}^t \mathbf{s} + \mathbf{x}_1 \\ \mathrm{H}_K(id)^t \mathbf{s} + \mathbf{R}_{id}^t \mathbf{x}_1 \end{pmatrix}.$$

$\mathsf{Dec}(sk_{id}, \mathbf{C})$: Given $sk_{id} = \mathbf{E}_{id}$ and a ciphertext $\mathbf{C} = (\mathbf{c}_0, \mathbf{c}_1)$ under identity id, compute $\mathbf{b} = \mathbf{c}_0 - \mathbf{E}_{id}^t \mathbf{c}_1 \in \mathbb{Z}_q^n$. Then, treat each coordinate of $\mathbf{b} = (b_1, \ldots, b_n)^t$ as an integer in $\mathbb{Z}$, and set $M_i = 1$ if $|b_i - \lfloor \frac{q}{2} \rfloor| \le \lfloor \frac{q}{4} \rfloor$, else $M_i = 0$, where $i \in \{1, \ldots, n\}$. Finally, return the plaintext $M = (M_0, \ldots, M_n)^t$.

By Lemma 2.24, we have that $s_1(\mathbf{R}) \le O(\sqrt{\bar{m}}) \cdot \omega(\sqrt{\log n})$. For large enough $s \ge \sqrt{m} \cdot \omega(\sqrt{\log n})$, by the correctness of $\mathsf{SampleD}$ we know that $\mathbf{A}_{id}\mathbf{E}_{id} = \mathbf{U}$ and $\|\mathbf{E}_{id}\| \le s\sqrt{m}$ hold with overwhelming probability. In this case, $\mathbf{c}_0 - \mathbf{E}_{id}^t \mathbf{c}_1 = \mathbf{c}_0 - \mathbf{E}_{id}^t \left(\mathbf{A}_{id}^t \mathbf{s} + \hat{\mathbf{x}}\right) = \mathbf{c}_0 - \mathbf{U}^t \mathbf{s} - \mathbf{E}_{id}^t \hat{\mathbf{x}} = \frac{q}{2} M + \mathbf{x}_0 - \mathbf{E}_{id}^t \hat{\mathbf{x}}$, where $\hat{\mathbf{x}} = \begin{pmatrix} \mathbf{x}_1 \\ \mathbf{R}_{id}^t \mathbf{x}_1 \end{pmatrix}$. Now, we estimate the size of $\|\mathbf{x}_0 - \mathbf{E}_{id}^t \hat{\mathbf{x}}\|_\infty$. Since $\mathbf{x}_0 \xleftarrow{\$} D_{\mathbb{Z}^n, \alpha q}$, $\mathbf{x}_1 \xleftarrow{\$} D_{\mathbb{Z}^{\bar{m}}, \alpha q}$, we have that $\|\mathbf{x}_0\|, \|\mathbf{x}_1\| \le \alpha q \sqrt{m}$ holds with overwhelming probability by Lemma 2.3. In addition, using the fact that $s_1(\mathbf{R}_{id}^t \mathbf{x}_1) \le \beta \cdot \|\mathbf{x}_1\|$, we have that $\|\hat{\mathbf{x}}\| \le \alpha q \sqrt{m(\beta^2+1)}$. Thus, we have that $\|\mathbf{E}_{id}^t \hat{\mathbf{x}}\|_\infty \le \alpha q m s \sqrt{\beta^2+1}$, and $\|\mathbf{x}_0 - \mathbf{E}_{id}^t \hat{\mathbf{x}}\|_\infty \le 2\alpha q m s \sqrt{\beta^2+1}$. This means that the decryption algorithm is correct if we set parameters such that $2\alpha q m s \sqrt{\beta^2+1} < \frac{q}{4}$ holds. For instance, we can set the parameters as follows: $m = 4n^{1+\psi}$, $s = \beta \cdot \omega(\sqrt{\log n})$, $q = \beta^2 m^2 \cdot \omega(\sqrt{\log n})$, $\alpha = (\beta^2 m^{1.5} \cdot \omega(\sqrt{\log n}))^{-1}$, where real $\psi \in \mathbb{R}$ satisfies $\log q < n^\psi$.

4.3.3 Security

Theorem 4.7 *Let $n, q, m' \in \mathbb{Z}$ and $\alpha, \beta \in \mathbb{R}$ be polynomials in the security parameter κ. For large enough $v = \mathsf{poly}(n)$, let $\mathcal{H} = (\mathcal{H}.\mathrm{Gen}, \mathcal{H}.\mathrm{Eval})$ be any $(1, v, \beta, \gamma, \delta)$-PHF with high min-entropy from $\{0, 1\}^n$ to $\mathbb{Z}_q^{n\times m'}$, where $\gamma = \mathsf{negl}(\kappa)$ and $\delta > 0$ is noticeable. Then, if there exists a PPT adversary $\mathcal{A}$ breaking the* INDr-ID-CPA security of Π_{ibe} with non-negligible advantage ϵ and making at most $Q < v$ user private key queries, there exists an algorithm $\mathcal{B}$ solving the $\mathrm{LWE}_{q,\alpha}$ problem with advantage at least $\epsilon' \geq \epsilon\delta/3 - \mathsf{negl}(\kappa)$.

Proof In the following, we use a sequence of games from Game G_0 to Game G_5, where Game G_0 is exactly the real security game as in Definition 4.2 where the challenger honestly encrypts the challenge plaintext, while Game G_5 is a random game where the challenge ciphertext is independent from the challenge plaintext. The security is established by showing that if $\mathcal{A}$ can succeed in Game G_0 with non-negligible advantage ϵ, then it can also succeed in Game G_5 with non-negligible advantage, which is contradictory to the fact that Game G_5 is a random game. Let $\mathcal{H}$.TrapGen and $\mathcal{H}$.TrapEval be a pair of trapdoor generation and trapdoor evaluation algorithm of $\mathcal{H}$ that satisfies the conditions in Definition 4.6. For simplicity, we fix the trapdoor matrix $\mathbf{B} = \mathbf{G} \in \mathbb{Z}_q^{n\times nk}$ throughout the proof. One can extend the proof to any other general trapdoor matrix $\mathbf{B}$ that allows to efficiently sample short vector $\mathbf{v}$ satisfying $\mathbf{Bv} = \mathbf{u}$ for any $\mathbf{u} \in \mathbb{Z}_q^n$, by using the trapdoor delegation techniques in [1].

Game G_0. The challenger C honestly simulates the INDr-ID-CPA security game for $\mathcal{A}$ as follows:

Setup. First compute $(\mathbf{A}, \mathbf{R}) \leftarrow \mathsf{TrapGen}(1^n, 1^{\bar{m}}, q, \mathbf{I}_n)$ such that $\mathbf{A} \in \mathbb{Z}_q^{n\times \bar{m}}$, $\mathbf{R} = \mathbb{Z}_q^{(\bar{m}-nk)\times nk}$. Then, randomly choose $\mathbf{U} \xleftarrow{\$} \mathbb{Z}_q^{n\times n}$, and compute $K \leftarrow \mathcal{H}.\mathrm{Gen}(1^\kappa)$. Finally, send the master public key $mpk = (\mathbf{A}, K, \mathbf{U})$ to the adversary $\mathcal{A}$, and keep the master secret key $\mathbf{R}$ private.

Phase 1. Upon receiving the user private key query with identity $id \in \{0, 1\}^n$, compute the hash value $\mathbf{A}_{id} = (\mathbf{A}\|\mathrm{H}_K(id)) \in \mathbb{Z}_q^{n\times m}$, where $\mathrm{H}_K(id) = \mathcal{H}.\mathrm{Eval}(K, id) \in \mathbb{Z}_q^{n\times nk}$. Then, compute $\mathbf{E}_{id} \leftarrow \mathsf{SampleD}(\mathbf{R}, \mathbf{A}_{id}, \mathbf{I}_n, \mathbf{U}, s)$, and send the user private key $sk_{id} = \mathbf{E}_{id} \in \mathbb{Z}^{m\times n}$ to the adversary $\mathcal{A}$.

Challenge. At some time, the adversary $\mathcal{A}$ outputs a challenge identity id^* and a plaintext $M^* \in \{0, 1\}^n$ with the restriction that it never obtains the user private key of id^* in Phase 1. The challenger C first randomly chooses $\mathbf{C}_0 \xleftarrow{\$} \mathbb{Z}_q^n \times \mathbb{Z}_q^m$, $\mathbf{s} \xleftarrow{\$} \mathbb{Z}_q^n$, $\mathbf{x}_0 \xleftarrow{\$} D_{\mathbb{Z}^n,\alpha q}$, and $\mathbf{x}_1 \xleftarrow{\$} D_{\mathbb{Z}^{\bar{m}},\alpha q}$. Then, it computes $(K', td) \leftarrow \mathcal{H}.\mathsf{TrapGen}(1^\kappa, \mathbf{A}, \mathbf{G})$, $(\mathbf{R}_{id^*}, \mathbf{S}_{id^*}) = \mathcal{H}.\mathsf{TrapEval}(td, K, id^*)$ and sets $\mathbf{C}_1 = (\mathbf{c}_0^*, \mathbf{c}_1^*)$, where

$$\mathbf{c}_0^* = \mathbf{U}^t\mathbf{s} + \mathbf{x}_0 + \frac{q}{2}M_{b^*}, \quad \mathbf{c}_1^* = \mathbf{A}_{id^*}^t\mathbf{s} + \begin{pmatrix} \mathbf{x}_1 \\ \mathbf{R}_{id^*}^t\mathbf{x}_1 \end{pmatrix} = \begin{pmatrix} \mathbf{A}^t\mathbf{s} + \mathbf{x}_1 \\ \mathrm{H}_K(id^*)^t\mathbf{s} + \mathbf{R}_{id^*}^t\mathbf{x}_1 \end{pmatrix},$$

where $\mathbf{A}_{id^*} = (\mathbf{A}\|\mathrm{H}_K(id^*)) \in \mathbb{Z}_q^{n\times m}$ and $\mathrm{H}_K(id^*) = \mathcal{H}.\mathrm{Eval}(K, id^*) \in \mathbb{Z}_q^{n\times nk}$. Finally, it randomly chooses a bit $b^* \xleftarrow{\$} \{0, 1\}$, and sends the challenge ciphertext $\mathbf{C}_{b^*}$ to the adversary $\mathcal{A}$.

Phase 2. $\mathcal{A}$ can adaptively make more user private key queries with any identity $id \neq id^*$. The challenger C responds as in Phase 1.

Guess. Finally, $\mathcal{A}$ outputs a guess $b \in \{0, 1\}$. If $b = b^*$, the challenger C outputs 1, else outputs 0.

Denote F_i be the event that C outputs 1 in Game G_i for $i \in \{0, 1, \ldots, 5\}$.

Lemma 4.8 $|\Pr[F_0] - \frac{1}{2}| = \epsilon$.

Proof This lemma immediately follows from the fact that C honestly simulates the attack environment for $\mathcal{A}$, and outputs 1 if and only if $b = b^*$. □

Game G_1. This game is identical to Game G_0 except that the challenger C changes the setup and the challenge phases as follows:

Setup. First compute $(\mathbf{A}, \mathbf{R}) \leftarrow \mathsf{TrapGen}(1^n, 1^{\bar{m}}, q, \mathbf{I}_n)$ such that $\mathbf{A} \in \mathbb{Z}_q^{n\times \bar{m}}$, $\mathbf{R} = \mathbb{Z}_q^{(\bar{m}-nk)\times nk}$. Then, randomly choose $\mathbf{U} \xleftarrow{\$} \mathbb{Z}_q^{n\times n}$, and compute $(K', td) \leftarrow \mathcal{H}.\mathrm{TrapGen}(1^\kappa, \mathbf{A}, \mathbf{G})$. Finally, send $mpk = (\mathbf{A}, K', \mathbf{U})$ to the adversary $\mathcal{A}$, and keep the master secret key $\mathbf{R}$ and the trapdoor td private.

Challenge. This phase is the same as in Game G_2 except that the challenger C directly uses the pair (K', td) produced in the setup phase to generate the ciphertext $\mathbf{C}_1 = (\mathbf{c}_0^*, \mathbf{c}_1^*)$.

Lemma 4.9 *If* $\mathcal{H}$ *is a PHF with high min-entropy, then* $|\Pr[F_1] - \Pr[F_0]| \leq \mathsf{negl}(\kappa)$.

Proof By the first condition of high min-entropy in Definition 4.6, for any $K \xleftarrow{\$} \mathcal{H}.\mathrm{Gen}(1^\kappa)$, $(K', td) \leftarrow \mathcal{H}.\mathrm{TrapGen}(1^\kappa, \mathbf{A}, \mathbf{G})$, any $id^* \in \{0, 1\}^n$ and any $\mathbf{x}_1 \in \mathbb{Z}_q^{\bar{m}}$, we have that the statistical distance between $(\mathbf{A}, K', (\mathbf{R}'_{id^*})^t\mathbf{x}_1)$ and $(\mathbf{A}, K, \mathbf{R}^t_{id^*}\mathbf{x}_1)$ is negligible, where $(\mathbf{R}'_{id^*}, \mathbf{S}'_{id^*}) = \mathcal{H}.\mathsf{TrapEval}(td, K', id^*)$ and $(\mathbf{R}_{id^*}, \mathbf{S}_{id^*}) = \mathcal{H}.\mathsf{TrapEval}(td, K, id^*)$. This means that the master public key mpk and the ciphertext $\mathbf{C}_1 = (\mathbf{c}_0^*, \mathbf{c}_1^*)$ in Game G_1 are statistically close to that in Game G_0. Thus, we have $|\Pr[F_1] - \Pr[F_0]| \leq \mathsf{negl}(\kappa)$. □

Game G_2. This game is identical to Game G_1 except that C changes the guess phase as follows:

Guess. Finally, the adversary $\mathcal{A}$ outputs a guess $b \in \{0, 1\}$. Let $id_1, \ldots, id_Q$ be all the identities in the user private queries, and let id^* be the challenge identity. Denote $I^* = \{id_1, \ldots, id_Q, id^*\}$, the challenger C first defines the following function:

$$\tau(\hat{td}, \hat{K}, I^*) = \begin{cases} 0, & \text{if } \hat{\mathbf{S}}_{id^*} = \mathbf{0}, \text{ and } \hat{\mathbf{S}}_{id_i} \text{ is invertible for all } i \in \{1, \ldots, Q\} \\ 1, & \text{otherwise,} \end{cases}$$

where $(\hat{\mathbf{R}}_{id^*}, \hat{\mathbf{S}}_{id^*}) = \mathcal{H}.\text{TrapEval}(\hat{td}, \hat{K}, id^*)$ and $(\hat{\mathbf{R}}_{id_i}, \hat{\mathbf{S}}_{id_i}) = \mathcal{H}.\text{TrapEval}(\hat{td}, \hat{K}, id_i)$. Then, C proceeds the following steps:

1. **Abort check**: Let (td, K') be produced in the setup phase when generating the master public key $mpk = (\mathbf{A}, K', \mathbf{U})$, the challenger C computes the value of $\tau(td, K', I^*)$. If $\tau(td, K', I^*) = 1$, the challenger C aborts the game, and outputs a uniformly random bit.
2. **Artificial abort**: Fixing $I^* = \{id_1, \ldots, id_Q, id^*\}$, let p be the probability $p = \Pr[\tau(\hat{td}, \hat{K}, I^*) = 0]$ over the random choice of $(\hat{td}, \hat{K})$. Then, the challenger C samples $O(\epsilon^2 \log(\epsilon^{-1})\delta^{-1} \log(\delta^{-1}))$ times the probability p by independently running $(\hat{td}, \hat{K}) \leftarrow \mathcal{H}.\text{TrapGen}(1^\kappa, \mathbf{A}, \mathbf{G})$ and evaluating $\tau(\hat{td}, \hat{K}, I^*)$ to compute an estimate p'.[3] Let δ be the parameter for the well-distributed hidden matrices property of $\mathcal{H}$, if $p' > \delta$, the challenger C aborts with probability $\frac{p'-\delta}{p'}$, and outputs a uniformly random bit.

Finally, if $b = b^*$, the challenger C outputs 1, else outputs 0.

Remark 4.2 As in [1, 3, 10, 16, 26], this seemingly meaningless artificial abort stage is necessary for later refinements. Looking ahead, in the following games the challenger C can continue the simulation only when the identities $id_1, \ldots, id_Q, id^*$ will not cause an abort (in the abort check stage). Since the success probability of the adversary $\mathcal{A}$ might be correlated with the probability that C aborts, it becomes complicated when we try to rely the success probability of C (in solving the underlying LWE problem) on the success probability of the adversary $\mathcal{A}$ (in attacking the IBE scheme). In [26], Waters introduced the artificial abort to force the probability that C aborts to be independent of $\mathcal{A}$'s particular queries. In certain cases, Bellare and Ristenpart [3] showed that the artificial abort can be avoided. Because the construction uses general lattice-based PHFs as a "black-box", we opt for the Waters approach and introduce an artificial abort. Besides, we clarify that there is no artificial abort involved in the generic signature scheme because any valid forgery can be publicly checked by the challenger C. Similar argument can be found in [26].

For $i \in \{2, 3, 4, 5\}$, let $\tilde{p}_i$ be the probability that C does not abort in the abort check stage in Game i, and let p_i be the probability in the artificial abort stage of Game i defined by $p_i = \Pr[\tau(\hat{td}, \hat{K}, I^*) = 0]$. Since the adversary might obtain some information of td from the challenge ciphertext $\mathbf{C}_{b^*}$, the probability $\tilde{p}_i$ might not be equal to the probability p. However, we will show later that the two probabilities can be very close under the LWE assumption. Formally, let Γ_i be the absolute difference between $\tilde{p}_i$ and p_i (i.e., $\Gamma_i = |\tilde{p}_i - p_i|$), we have the following lemma:

Lemma 4.10 *If $\mathcal{H}$ is a $(1, v, \beta, \gamma, \delta)$-PHF and $Q \leq v$, then* $|\Pr[F_2] - \frac{1}{2}| \geq \frac{1}{2}\epsilon(\delta - \Gamma_2)$.

[3]In general, the sampling procedure generally makes the running time of C dependent on the success advantage ϵ of $\mathcal{A}$, but for concrete PHFs (e.g., the construction in Theorem 4.4), it is possible to directly calculate the probability p.

So as not to interrupt the game sequences, we defer the proof of Lemma 4.10 to the end of Game G_5.

Game G_3. This game is identical to Game G_2 except that the challenger C changes the way of generating the user private keys and the challenge ciphertext as follows:

Phase 1. Upon receiving the user private key query with identity $id \in \{0, 1\}^n$, first compute $\mathbf{A}_{id} = (\mathbf{A} \| \mathrm{H}_{K'}(id)) \in \mathbb{Z}_q^{n \times m}$, where $\mathrm{H}_{K'}(id) = \mathcal{H}.\mathrm{Eval}(K', id) \in \mathbb{Z}_q^{n \times nk}$. Then, compute $(\mathbf{R}_{id}, \mathbf{S}_{id}) = \mathcal{H}.\mathrm{TrapEval}(td, K', id)$. If $\mathbf{S}_{id}$ is not invertible, the challenger C outputs a uniformly random bit and aborts the game. Otherwise, compute $\mathbf{E}_{id} \leftarrow \mathsf{SampleD}(\mathbf{R}_{id}, \mathbf{A}_{id}, \mathbf{S}_{id}, \mathbf{U}, s)$ and send $sk_{id} = \mathbf{E}_{id} \in \mathbb{Z}^{m \times n}$ to $\mathcal{A}$.

Challenge. This phase is the same as in Game G_2 except that the challenger directly aborts and outputs a uniformly random bit if $\mathbf{S}_{id^*} \neq \mathbf{0}$, where $(\mathbf{R}_{id^*}, \mathbf{S}_{id^*}) = \mathcal{H}.\mathrm{TrapEval}(td, K', id^*)$. Note that if $\mathbf{S}_{id^*} = \mathbf{0}$, we have $\mathrm{H}_{K'}(id^*) = \mathbf{A}\mathbf{R}_{id^*}$ and $\mathbf{c}_1^* = \begin{pmatrix} \mathbf{b}_1 \\ \mathbf{R}_{id^*}^t \mathbf{b}_1 \end{pmatrix}$ for some vector $\mathbf{b}_1 = \mathbf{A}^t \mathbf{s} + \mathbf{x}_1 \in \mathbb{Z}_q^{\bar{m}}$.

Phase 2. $\mathcal{A}$ can adaptively make more user private key queries with any identity $id \neq id^*$. The challenger C responds as in Phase 1.

Lemma 4.11 *If $\mathcal{H}$ is a $(1, v, \beta, \gamma, \delta)$-PHF and $Q \leq v$, then* $\Pr[F_3] = \Pr[F_2]$ *and* $\Gamma_3 = \Gamma_2$.

Proof Note that both stages of the abort check and the artificial abort in Game G_3 and Game G_2 are identical. By the fact that the same abort conditions as in the abort check stage are examined when generating the user private keys and the ciphertext $\mathbf{C}_1 = (\mathbf{c}_0^*, \mathbf{c}_1^*)$, the challenger C in Game G_3 will abort with the same probability as that in Game G_2. Besides, if C does not abort in Game G_3, we have that $\mathbf{S}_{id^*} = 0$ and $\mathbf{S}_{id}$ is invertible for any id in the user private key queries. In this case, C can use the $\mathsf{SampleD}$ algorithm to successfully generate the user private keys by the fact that $s_1(\mathbf{R}_{id}) \leq \beta$ and $s > \max(\beta, \sqrt{m}) \cdot \omega(\sqrt{\log n})$. Thus, if C does not abort during the game, then Game G_3 is identical to Game G_2 in the adversary $\mathcal{A}$'s view. In all, we have that $\Pr[F_4] = \Pr[F_3]$ and $\Gamma_3 = \Gamma_2$ hold. □

Game G_4. This game is identical to Game G_3 except that the challenger C changes the setup and the challenge phases as follows:

Setup. First randomly choose $\mathbf{A} \xleftarrow{\$} \mathbb{Z}_q^{n \times \bar{m}}$, $\mathbf{U} \xleftarrow{\$} \mathbb{Z}_q^{n \times n}$, and compute $(K', td) \leftarrow \mathcal{H}.\mathrm{TrapGen}(1^\kappa, \mathbf{A}, \mathbf{G})$. Then, send $mpk = (\mathbf{A}, K', \mathbf{U})$ to the adversary $\mathcal{A}$, and keep the trapdoor td private.

Challenge. This phase is the same as in Game G_3 except that the challenger generates the ciphertext $\mathbf{C}_1 = (\mathbf{c}_0^*, \mathbf{c}_1^*)$ as follows: randomly choose vector $\mathbf{b}_0 \xleftarrow{\$} \mathbb{Z}_q^n$, $\mathbf{b}_1 \xleftarrow{\$} \mathbb{Z}_q^{\bar{m}}$, and compute

$$\mathbf{c}_0^* = \mathbf{b}_0 + \frac{q}{2} M_{b^*}, \qquad \mathbf{c}_1^* = \begin{pmatrix} \mathbf{b}_1 \\ \mathbf{R}_{id^*}^t \mathbf{b}_1 \end{pmatrix},$$

where $(\mathbf{R}_{id^*}, \mathbf{S}_{id^*}) = \mathcal{H}.\text{TrapEval}(td, K', id^*)$.

Lemma 4.12 *If the advantage of any PPT algorithm $\mathcal{B}$ in solving the* $\text{LWE}_{q,\alpha}$ *problem is at most ϵ', then we have that* $|\Pr[F_4] - \Pr[F_3]| \leq \epsilon'$ *and* $|\Gamma_4 - \Gamma_3| \leq \epsilon'$ *hold.*

Proof We construct an algorithm $\mathcal{B}$ for the $\text{LWE}_{q,\alpha}$ as follows. Given the $\text{LWE}_{q,\alpha}$ challenge instance $(\hat{\mathbf{U}}, \hat{\mathbf{b}}_0) \in \mathbb{Z}_q^{n\times n} \times \mathbb{Z}_q^n$ and $(\hat{\mathbf{A}}, \hat{\mathbf{b}}_1) \in \mathbb{Z}_q^{n\times \bar{m}} \times \mathbb{Z}_q^{\bar{m}}$. $\mathcal{B}$ simulates the security game for the adversary $\mathcal{A}$ the same as in Game G_3 except that it replaces $(\mathbf{A}, \mathbf{U})$ in the setup phase and $(\mathbf{b}_0, \mathbf{b}_1)$ in the challenge phase with $(\hat{\mathbf{A}}, \hat{\mathbf{U}})$ and $(\hat{\mathbf{b}}_0, \hat{\mathbf{b}}_1)$, respectively.

It is easy to check that if $(\hat{\mathbf{U}}, \hat{\mathbf{b}}_0) \in \mathbb{Z}_q^{n\times n} \times \mathbb{Z}_q^n$ and $(\hat{\mathbf{A}}, \hat{\mathbf{b}}_1) \in \mathbb{Z}_q^{n\times \bar{m}} \times \mathbb{Z}_q^{\bar{m}}$ are valid LWE tuples, then $\mathcal{A}$ is in Game G_3, otherwise $\mathcal{A}$ is in Game G_4. This means that $\mathcal{B}$ is a valid LWE distinguisher, which implies that both $|\Pr[F_4] - \Pr[F_3]| \leq \epsilon'$ and $|\Gamma_4 - \Gamma_3| \leq \epsilon'$ hold by the assumption. □

Game G_5. This game is identical to Game G_4 except that the challenger C makes the following changes:

Setup. First compute $(\mathbf{A}, \mathbf{R}) \leftarrow \mathsf{TrapGen}(1^n, 1^{\bar{m}}, q, \mathbf{I}_n)$ such that $\mathbf{A} \in \mathbb{Z}_q^{n\times \bar{m}}$, $\mathbf{R} = \mathbb{Z}_q^{(\bar{m}-nk)\times nk}$. Then, the challenger C randomly chooses matrix $\mathbf{U} \xleftarrow{\$} \mathbb{Z}_q^{n\times n}$, computes $K \xleftarrow{\$} \mathcal{H}.\text{Gen}(1^\kappa)$ and $(K', td) \xleftarrow{\$} \mathcal{H}.\text{TrapGen}(1^\kappa, \mathbf{A}, \mathbf{G})$. Finally, send the master public key $mpk = (\mathbf{A}, K, \mathbf{U})$ to the adversary $\mathcal{A}$, and keep $(\mathbf{R}, K', td)$ private.

Phase 1. Upon receiving the user private key query with identity $id \in \{0, 1\}^n$, the challenger computes $\mathbf{A}_{id} = (\mathbf{A} \| \mathrm{H}_K(id)) \in \mathbb{Z}_q^{n\times m}$, where $\mathrm{H}_K(id) = \mathcal{H}.\text{Eval}(K, id) \in \mathbb{Z}_q^{n\times nk}$. Then, compute $\mathbf{E}_{id} \leftarrow \mathsf{SampleD}(\mathbf{R}, \mathbf{A}_{id}, \mathbf{I}_n, \mathbf{U}, s)$, and send $sk_{id} = \mathbf{E}_{id} \in \mathbb{Z}^{m\times n}$ to $\mathcal{A}$.

Challenge. This phase is the same as in Game G_4 except that the challenger generates the ciphertext $\mathbf{C}_1 = (\mathbf{c}_0^*, \mathbf{c}_1^*)$ by randomly choosing $\mathbf{c}_0^* \xleftarrow{\$} \mathbb{Z}_q^n$ and $\mathbf{c}_1^* \xleftarrow{\$} \mathbb{Z}_q^m$.

Phase 2. $\mathcal{A}$ can adaptively make more user private key queries with any identity $id \neq id^*$. The challenger C responds as in Phase 1.

Lemma 4.13 *If $\mathcal{H}$ is a $(1, v, \beta, \mathsf{negl}(\kappa), \delta)$-PHF with min-entropy, then we have* $|\Pr[F_5] - \Pr[F_4]| \leq \mathsf{negl}(\kappa)$ *and* $|\Gamma_5 - \Gamma_4| \leq \mathsf{negl}(\kappa)$.

Proof Since $\mathcal{H}$ is a $(1, v, \beta, \mathsf{negl}(\kappa), \delta)$-PHF, the statistical distance between the master public key mpk in Game G_4 and that in Game G_5 is negligible by the property of $\mathsf{TrapGen}$. By the property of $\mathsf{SampleD}$, the distribution of user private keys in Game G_5 is almost identical to that in Game G_4. Since both $\mathbf{b}_0, \mathbf{b}_1$ in Game G_4 are uniformly random over $\mathbb{Z}_q^n$ and $\mathbb{Z}_q^{\bar{m}}$, the challenge ciphertext in Game G_4 is statistically close to that in Game G_5 by the second condition of high min-entropy in Definition 4.6. Finally, using the fact that both Game G_4 and Game 5 implement the same abort strategy in the guess phase, we have that Game G_4 is negligibly close to Game G_5 in the adversary $\mathcal{A}$'s view. Thus, we have that both $|\Pr[F_5] - \Pr[F_4]| \leq \mathsf{negl}(\kappa)$ and $|\Gamma_5 - \Gamma_4| \leq \mathsf{negl}(\kappa)$ hold. □

Lemma 4.14 $\Pr[F_5] = \frac{1}{2}$ *and* $\Gamma_5 = 0$.

Proof The first claim follows from the fact that $\mathbf{C}_1$ is uniformly random. Since both the master pubic key mpk and the challenge ciphertext are independent from the random choice of td in Game G_5, the challenger C can actually compute $(K', td) \leftarrow \mathcal{H}.\text{TrapGen}(1^\kappa, \mathbf{A}, \mathbf{G})$ in the guess phase, and use (K', td) in the abort check stage. By the definition of Γ_5, we have $\Gamma_5 = 0$. This completes the proof. □

By Lemmas 4.10 and 4.11, we have $|\Pr[F_3] - \frac{1}{2}| \geq \frac{1}{2}\epsilon(\delta - \Gamma_3)$. By Lemmas 4.13 and 4.14, we have $\Pr[F_4] \leq \frac{1}{2} + \mathsf{negl}(\kappa)$ and $\Gamma_4 \leq \mathsf{negl}(\kappa)$. By the fact that $|\Pr[F_4] - \Pr[F_3]| \leq \epsilon'$ and $|\Gamma_4 - \Gamma_3| \leq \epsilon'$ in Lemma 4.12, we have $\frac{1}{2}\epsilon(\delta - \epsilon') - \mathsf{negl}(\kappa) \leq |\Pr[F_3] - \frac{1}{2}| \leq \epsilon' - \mathsf{negl}(\kappa)$. This shows that $\epsilon' \geq \frac{\epsilon\delta}{3} - \mathsf{negl}(\kappa)$ holds, which completes the proof of Theorem 4.7. □

Proof of Lemma 4.10 Let $\mathcal{QID} = (\{0, 1\}^n)^{Q+1}$ be the set of all $Q + 1$ tuples of identities. Let $Q(I)$ be the event that the adversary $\mathcal{A}$ uses the first Q identities in $I = \{id_1, \ldots, id_Q, id^*\} \in \mathcal{QID}$ for user private key queries, and the last one for the challenge identity. Let $F_i(I) \subseteq Q(I)$ be the event that the challenger C outputs 1 in Game G_i when $Q(I)$ happens, where $i \in \{1, 2\}$. Let $\mathcal{E}$ be the event that C aborts in Game G_2. Then, by the definition we have the following facts:

$$\begin{aligned}
&\textstyle\sum_{I \in \mathcal{QID}} \Pr[Q(I)] = 1 \\
&\Pr[F_i] = \textstyle\sum_{I \in \mathcal{QID}} \Pr[F_i(I)] \\
&\Pr[F_i] = \Pr[F_i \wedge \mathcal{E}] + \Pr[F_i \wedge \neg\mathcal{E}] \\
&\Pr[Q(I)] = \Pr[Q(I) \wedge \mathcal{E}] + \Pr[Q(I) \wedge \neg\mathcal{E}]
\end{aligned}$$

.

Besides, by the description of Game G_2, we have that $\Pr[F_2(I) \wedge \mathcal{E}] = \frac{1}{2}\Pr[Q(I) \wedge \mathcal{E}]$ and $\Pr[F_2(I) \wedge \neg\mathcal{E}] = \Pr[F_1(I) \wedge \neg\mathcal{E}] = \Pr[F_1(I)]\Pr[\neg\mathcal{E}|Q(I)]$ hold. By a simple calculation, we have

$$\begin{aligned}
|\Pr[F_2] - \tfrac{1}{2}| &= |\textstyle\sum_{I \in \mathcal{QID}}(\Pr[F_2(I) \wedge \mathcal{E}] + \Pr[F_2(I) \wedge \neg\mathcal{E}]) - \tfrac{1}{2}| \\
&= |\textstyle\sum_{I \in \mathcal{QID}}(\Pr[F_2(I) \wedge \neg\mathcal{E}] - \tfrac{1}{2}\Pr[Q(I) \wedge \neg\mathcal{E}])| \\
&= |\textstyle\sum_{I \in \mathcal{QID}}(\Pr[F_1(I)] - \tfrac{1}{2}\Pr[Q(I)])\Pr[\neg\mathcal{E}|Q(I)]|.
\end{aligned}$$

Since $\Pr[F_i(I)] \leq \Pr[Q(I)]$, we have that $|\Pr[F_1(I)] - \frac{1}{2}\Pr[Q(I)]| \leq \frac{1}{2}\Pr[Q(I)]$ holds. Note that the term $\Pr[F_1(I)] - \frac{1}{2}\Pr[Q(I)]$ can be either positive or negative. Let $\mathcal{QID}^+$ (resp. $\mathcal{QID}^-$) be the set of identities such that $\Pr[F_1(I) - \frac{1}{2}\Pr[Q(I)]$ is positive (resp. negative), and let $\eta(I) = \Pr[\neg\mathcal{E}|Q(I)]$. In addition, let $\eta_{max} = \max_{I \in \mathcal{QID}} \eta(I)$ and $\eta_{min} = \min_{I \in \mathcal{QID}} \eta(I)$. Then, we have

$$
\begin{aligned}
|\Pr[F_2]-\tfrac{1}{2}| &= |\textstyle\sum_{I\in QI\mathcal{D}^+}(\Pr[F_1(I)]-\tfrac{1}{2}\Pr[Q(I)])\eta(I) \\
&\quad +\textstyle\sum_{I\in QI\mathcal{D}^-}(\Pr[F_1(I)]-\tfrac{1}{2}\Pr[Q(I)])\eta(I)| \\
&\geq \eta_{min}|\textstyle\sum_{I\in QI\mathcal{D}}(\Pr[F_1(I)]-\tfrac{1}{2}\Pr[Q(I)])|-\tfrac{1}{2}(\eta_{max}-\eta_{min}) \\
&= \eta_{min}|\Pr[F_1]-\tfrac{1}{2}|-\tfrac{1}{2}(\eta_{max}-\eta_{min}).
\end{aligned}
$$

By the definition of $\tilde{p}_2$ and p_2 in Game G_2, we have $\eta(I)=\Pr[\neg\mathcal{E}|Q(I)]=\tilde{p}_2\frac{\delta}{p'}$, where p' is an estimate of p_2. Since the challenger C always samples $O(\epsilon^2\log(\epsilon^{-1})\delta^{-1}\log(\delta^{-1}))$ times the probability p_2 to compute p', we have that $\Pr[p'>p_2(1+\frac{\epsilon}{8})]<\delta\frac{\epsilon}{8}$ and $\Pr[p'<p_2(1-\frac{\epsilon}{8})]<\delta\frac{\epsilon}{8}$ hold by the Chernoff bounds. Then, we have

$$
\begin{aligned}
\eta_{max} &\leq (1-\delta\tfrac{\epsilon}{8})\tilde{p}_2\frac{\delta}{p_2(1-\frac{\epsilon}{8})} \\
\eta_{min} &\geq (1-\delta\tfrac{\epsilon}{8})\tilde{p}_2\frac{\delta}{p_2(1+\frac{\epsilon}{8})} \geq \frac{7\delta\tilde{p}_2}{9p_2} \\
\eta_{max}-\eta_{min} &\leq (1-\delta\tfrac{\epsilon}{8})\frac{\epsilon\delta\tilde{p}_2}{4(1-\frac{\epsilon^2}{64})p_2} \leq \frac{16\epsilon\delta\tilde{p}_2}{63p_2}.
\end{aligned}
$$

By Lemmas 4.8 and 4.9, $|\Pr[F_1]-\frac{1}{2}|\geq\epsilon-\mathsf{negl}(\kappa)$ holds. Then, we have

$$
\begin{aligned}
|\Pr[F_2]-\tfrac{1}{2}| &\geq \eta_{min}|\Pr[F_1]-\tfrac{1}{2}|-\tfrac{1}{2}(\eta_{max}-\eta_{min}) \\
&\geq \frac{7\delta\tilde{p}_2}{9p_2}(\epsilon-\mathsf{negl}(\kappa))-\frac{8\epsilon\delta\tilde{p}_2}{63p_2} \\
&\geq \frac{\epsilon\delta(p_2-\Gamma_2)}{2p_2} \geq \tfrac{1}{2}\epsilon(\delta-\Gamma_2),
\end{aligned}
$$

where the last two inequalities are due to the fact that $\Gamma_2=|\tilde{p}_2-p_2|$ and $p_2\geq\delta$. This completes the proof of Lemma 4.10. □

4.3.4 Instantiations

By instantiating $\mathcal{H}$ in the generic scheme Π_{ibe} with the Type-I PHF construction, we recover the fully secure IBE scheme due to Agrawal et al. [1]. Besides, if $\mathcal{H}$ is replaced by a weak $(1,v,\beta)$-PHF with high min-entropy, we can further show that the resulting scheme is INDr-sID-CPA secure and subsumes the selectively secure IBE scheme in [1]. Formally,

Corollary 4.1 *Let $n,m',q\in\mathbb{Z}$ and $\alpha,\beta\in\mathbb{R}$ be polynomials in the security parameter κ. For large enough $v=\mathsf{poly}(n)$, let $\mathcal{H}=(\mathcal{H}.\mathrm{Gen},\mathcal{H}.\mathrm{Eval})$ be any weak $(1,v,\beta,\gamma,\delta)$-PHF with high min-entropy from $\{0,1\}^n$ to $\mathbb{Z}_q^{n\times m'}$, where $\gamma=\mathsf{negl}(\kappa)$ and $\delta>0$ is noticeable. Then, under the $\mathrm{LWE}_{q,\alpha}$ assumption, the generic IBE scheme Π_{ibe} is INDr-sID-CPA secure.*

By instantiating the generic IBE scheme Π_{ibe} with the Type-II PHF in Definition 4.5, we can obtain a fully secure IBE scheme with master public key containing $O(\log n)$ number of matrices. Let Π^*_{ibe} be the instantiated scheme.

Corollary 4.2 *If there exists a PPT adversary $\mathcal{A}$ breaking the* INDr-ID-CPA *security of Π^*_{ibe} with non-negligible advantage ϵ and making at most $Q = \mathsf{poly}(\kappa)$ user private key queries, then there exists an algorithm $\mathcal{B}$ solving the* $\text{LWE}_{q,\alpha}$ *problem with advantage at least $\epsilon' \geq \frac{\epsilon}{48nQ^2} - \mathsf{negl}(\kappa)$.*

4.4 Background and Further Reading

Gentry et al. [18] proposed the first IBE scheme based on the learning with errors (LWE) assumption in the random oracle model. Later, several works [2, 10, 14, 28] were dedicated to the study of lattice-based (hierarchical) IBE schemes also in the random oracle model. There were a few works focusing on designing standard model lattice-based IBE schemes [1, 2, 10]. Concretely, the scheme in [2] was only proven to be *selective-identity* secure in the standard model. By using standard complexity leverage technique [5], one can generally transform a selective-identity secure IBE scheme into a *fully secure* one. But the resulting scheme has to suffer from a reduction loss proportional to L, where L is the number of distinct identities for the IBE system and is independent from the number Q of the adversary's private key queries in the security proof. Since L is usually super polynomial and much larger than Q, the above generic transformation is a very unsatisfying approach [17]. In [1, 10], the authors showed how to achieve *full security* against adaptive chosen-plaintext and chosen-identity attacks, but both standard model fully secure IBE schemes in [1, 10] had large master public keys consisting of a linear number of matrices.

Hofheinz and Kiltz [21] introduced the notion of PHF based on group hash functions, and gave a concrete (2, 1)-PHF instantiation. Then, the work [20] constructed a $(u, 1)$-PHF for any $u \geq 1$ by using cover-free sets. Later, Yamada et al. [27] reduced the key size from $O(u^2\ell)$ in [20] to $O(u\sqrt{\ell})$ by combining the two-dimensional representation of cover-free sets with the bilinear groups, where ℓ was the bit size of the inputs. At CRYPTO 2012, Hanaoka et al. [19] showed that it was impossible to construct *algebraic* $(u, 1)$-PHF over prime order groups in a black-box way such that its key has less than u group elements.[4] Later, Freire et al. [16] got around the impossibility result of [19] and constructed a $(\mathsf{poly}, 1)$-PHF by adapting PHFs to the multilinear maps setting. Despite its great theoretical interests, the current state of multilinear maps might be a big obstacle in any attempt to securely and efficiently instantiate the PHFs in [16]. More recently, Catalano et al. [11] introduced a variant of traditional PHF called asymmetric PHF over bilinear maps, and used it to construct (homomorphic) signature schemes with short verification keys.

[4]Informally, an algorithm is algebraic if there is way to compute the representation of a group element output by the algorithm in terms of its input group elements [7].

All the above PHF constructions [11, 16, 20, 21, 27] seem specific to groups with nice properties, which might constitute a main barrier to instantiate them from lattices. The notion of lattice-based PHFs introduced in this chapter is an analogue of the traditional PHFs, which not only captures the partitioning proof tricks that appeared in lattice-based constructions [1, 8, 23], but also provides a modular way to design other primitives from lattices.

References

1. Agrawal, S., Boneh, D., Boyen, X.: Efficient lattice (H)IBE in the standard model. In: Gilbert, H. (ed.) Advances in Cryptology - EUROCRYPT 2010. Lecture Notes in Computer Science, vol. 6110, pp. 553–572. Springer, Heidelberg (2010)
2. Agrawal, S., Boneh, D., Boyen, X.: Lattice basis delegation in fixed dimension and shorter-ciphertext hierarchical IBE. In: Rabin, T. (ed.) Advances in Cryptology - CRYPTO 2010. Lecture Notes in Computer Science, vol. 6223, pp. 98–115. Springer, Heidelberg (2010)
3. Bellare, M., Ristenpart, T.: Simulation without the artificial abort: simplified proof and improved concrete security for Waters' IBE scheme. In: Joux, A. (ed.) EUROCRYPT 2009, LNCS, vol. 5479, pp. 407–424. Springer (2009). https://doi.org/10.1007/978-3-642-01001-9_24
4. Böhl, F., Hofheinz, D., Jager, T., Koch, J., Seo, J., Striecks, C.: Practical signatures from standard assumptions. In: Johansson, T., Nguyen, P. (eds.) Advances in Cryptology - EUROCRYPT 2013. Lecture Notes in Computer Science, vol. 7881, pp. 461–485. Springer, Heidelberg (2013)
5. Boneh, D., Boyen, X.: Efficient selective-ID secure identity-based encryption without random oracles. In: Cachin, C., Camenisch, J. (eds.) Advances in Cryptology - EUROCRYPT 2004. Lecture Notes in Computer Science, vol. 3027, pp. 223–238. Springer, Berlin/Heidelberg (2004)
6. Boneh, D., Franklin, M.: Identity-based encryption from the Weil pairing. In: Kilian, J. (ed.) Advances in Cryptology - CRYPTO 2001. Lecture Notes in Computer Science, vol. 2139, pp. 213–229. Springer, Berlin/Heidelberg (2001)
7. Boneh, D., Venkatesan, R.: Breaking RSA may not be equivalent to factoring. In: Nyberg, K. (ed.) Advances in Cryptology - EUROCRYPT '98. Lecture Notes in Computer Science, vol. 1403, pp. 59–71. Springer, Heidelberg (1998)
8. Boyen, X.: Lattice mixing and vanishing trapdoors: a framework for fully secure short signatures and more. In: Nguyen, P., Pointcheval, D. (eds.) Public Key Cryptography - PKC 2010. Lecture Notes in Computer Science, vol. 6056, pp. 499–517. Springer, Heidelberg (2010)
9. Canetti, R., Halevi, S., Katz, J.: Chosen-ciphertext security from identity-based encryption. In: Cachin, C., Camenisch, J. (eds.) Advances in Cryptology - EUROCRYPT 2004. Lecture Notes in Computer Science, vol. 3027, pp. 207–222. Springer, Berlin/Heidelberg (2004)
10. Cash, D., Hofheinz, D., Kiltz, E., Peikert, C.: Bonsai trees, or how to delegate a lattice basis. In: Gilbert, H. (ed.) Advances in Cryptology - EUROCRYPT 2010. Lecture Notes in Computer Science, vol. 6110, pp. 523–552. Springer, Berlin/Heidelberg (2010)
11. Catalano, D., Fiore, D., Nizzardo, L.: Programmable hash functions go private: constructions and applications to (homomorphic) signatures with shorter public keys. In: Gennaro, R., Robshaw, M. (eds.) CRYPTO 2015, LNCS, vol. 9216, pp. 254–274. Springer (2015). https://doi.org/10.1007/978-3-662-48000-7_13
12. Cocks, C.: An identity based encryption scheme based on quadratic residues. In: Honary, B. (ed.) Cryptography and Coding. Lecture Notes in Computer Science, vol. 2260, pp. 360–363. Springer, Heidelberg (2001)
13. Dodis, Y., Rafail, O., Reyzin, L., Smith, A.: Fuzzy extractors: how to generate strong keys from biometrics and other noisy data. SIAM J. Comput. **38**, 97–139 (2008)

14. Ducas, L., Lyubashevsky, V., Prest, T.: Efficient identity-based encryption over NTRU lattices. In: Sarkar, P., Iwata, T. (eds.) ASIACRYPT 2014, LNCS, vol. 8874, pp. 22–41. Springer (2014)
15. Erdös, P., Frankl, P., Füredi, Z.: Families of finite sets in which no set is covered by the union of r others. Isr. J. Math. **51**(1–2), 79–89 (1985)
16. Freire, E., Hofheinz, D., Paterson, K., Striecks, C.: Programmable hash functions in the multilinear setting. In: Canetti, R., Garay, J. (eds.) CRYPTO 2013, LNCS, vol. 8042, pp. 513–530. Springer (2013). https://doi.org/10.1007/978-3-642-40041-4_28
17. Gentry, C.: Practical identity-based encryption without random oracles. In: Vaudenay, S. (ed.) Advances in Cryptology - EUROCRYPT 2006. Lecture Notes in Computer Science, vol. 4004, pp. 445–464. Springer, Heidelberg (2006)
18. Gentry, C., Peikert, C., Vaikuntanathan, V.: Trapdoors for hard lattices and new cryptographic constructions. In: Proceedings of the 40th Annual ACM Symposium on Theory of Computing, STOC '08, pp. 197–206. ACM (2008)
19. Hanaoka, G., Matsuda, T., Schuldt, J.: On the impossibility of constructing efficient key encapsulation and programmable hash functions in prime order groups. In: Safavi-Naini, R., Canetti, R. (eds.) CRYPTO 2012, LNCS, vol. 7417, pp. 812–831. Springer (2012). https://doi.org/10.1007/978-3-642-32009-5_47
20. Hofheinz, D., Jager, T., Kiltz, E.: Short signatures from weaker assumptions. In: Lee, D., Wang, X. (eds.) ASIACRYPT 2011, LNCS, vol. 7073, pp. 647–666. Springer (2011). https://doi.org/10.1007/978-3-642-25385-0_1
21. Hofheinz, D., Kiltz, E.: Programmable hash functions and their applications. In: Wagner, D. (ed.) Advances in Cryptology - CRYPTO 2008. Lecture Notes in Computer Science, vol. 5157, pp. 21–38. Springer, Heidelberg (2008)
22. Kumar, R., Rajagopalan, S., Sahai, A.: Coding constructions for blacklisting problems without computational assumptions. In: CRYPTO '99, pp. 609–623. Springer (1999)
23. Micciancio, D., Peikert, C.: Trapdoors for lattices: simpler, tighter, faster, smaller. In: Pointcheval, D., Johansson, T. (eds.) Advances in Cryptology - EUROCRYPT 2012. Lecture Notes in Computer Science, vol. 7237, pp. 700–718. Springer, Heidelberg (2012)
24. Nguyen, P., Zhang, J., Zhang, Z.: Simpler efficient group signatures from lattices. In: Katz, J. (ed.) Public-Key Cryptography - PKC 2015. Lecture Notes in Computer Science, vol. 9020, pp. 401–426. Springer, Heidelberg (2015)
25. Shamir, A.: Identity-based cryptosystems and signature schemes. In: Blakley, G., Chaum, D. (eds.) Advances in Cryptology. Lecture Notes in Computer Science, vol. 196, pp. 47–53. Springer, Berlin/Heidelberg (1984)
26. Waters, B.: Efficient identity-based encryption without random oracles. In: Cramer, R. (ed.) Advances in Cryptology - EUROCRYPT 2005. Lecture Notes in Computer Science, vol. 3494, pp. 114–127. Springer, Heidelberg (2005)
27. Yamada, S., Hanaoka, G., Kunihiro, N.: Two-dimensional representation of cover free families and its applications: short signatures and more. In: Dunkelman, O. (ed.) CT-RSA 2012, LNCS, vol. 7178, pp. 260–277. Springer (2012). https://doi.org/10.1007/978-3-642-27954-6_17
28. Zhandry, M.: Secure identity-based encryption in the quantum random oracle model. In: Safavi-Naini, R., Canetti, R. (eds.) CRYPTO 2012, LNCS, vol. 7417, pp. 643–662. Springer (2012)
29. Zhang, J., Chen, Y., Zhang, Z.: Programmable hash functions from lattices: short signatures and IBEs with small key sizes. In: Robshaw, M., Katz, J. (eds.) Advances in Cryptology - CRYPTO 2016, pp. 303–332. Springer, Heidelberg (2016)

Chapter 5
Attribute-Based Encryption

Abstract Sahai and Waters first introduced attribute-based encryption (ABE) as an extension of identity-based encryption (IBE), which allows the sender (who generates the ciphertext) to set a policy describing who can decrypt his ciphertext. Note that in an IBE system, a user is associated with a unique identity and can only decrypt the ciphertexts corresponding to his own identity. This means that the sender must have full knowledge of the receiver's identity and that a ciphertext can only be decrypted by a single user. In comparison, ABE can provide much more flexible access control, in particular, the sender does not necessarily know which particular user can decrypt his ciphertext, and a ciphertext can be decrypted by multiple users. In this chapter, we focus on constructing ABEs from lattices. Specifically, we will first present a ciphertext policy attribute-based encryption (CP-ABE) scheme from lattices, which only supports the simple and-gate policy. Then, we propose a lattice-based CP-ABE scheme supporting flexible threshold access policies on literal (or boolean) attributes. We also show how to extend the second scheme to support multi-valued attributes without increasing the master public key and ciphertext size.

5.1 Definition

Sahai and Waters [23] introduced the notion of attribute-based encryption as an extension of identity-based encryption (IBE), where users' secret keys are produced by a trust authority according to a set of attributes. In an ABE system, a user's secret keys and ciphertexts are labeled with sets of descriptive attributes and a particular key can decrypt a particular ciphertext only if there is a match between the attributes of the ciphertext and the user's key. Goyal et al. [13] further extended the idea of ABE and introduced two variants: key policy attribute-based encryption (KP-ABE) and ciphertext policy attribute-based encryption (CP-ABE). In a KP-ABE system, the ciphertext is associated with a set of descriptive attributes, while the private key of a party is associated with an access policy which is defined over a set of attributes

J. Zhang and Z. Zhang, *Lattice-Based Cryptosystems*,
https://doi.org/10.1007/978-981-15-8427-5_5

and specifies which type of ciphertexts the key can decrypt. A CP-ABE system can be seen as a complementary form to KP-ABE system, where the private keys are associated with a set of attributes, while a policy defined over a set of attributes is attached to the ciphertext. A ciphertext can be decrypted by a party if the attributes associated with its private keys satisfy the ciphertext's policy. In general, a ciphertext policy scheme is more flexible than a key policy scheme (if both of them support the same type access policies), since users can set access policies when encrypting messages instead of the authority setting policies when extracting users' secret keys. Actually, in many applications, we just care about what attributes a user has when extracting his secret key, rather than how a user will use his attributes.

In this chapter, we only focus on CP-ABE. Formally, let $\mathcal{R}$ be an attribute set, let $S \subseteq \mathcal{R}$ be a subset of $\mathcal{R}$ and let W be an access policy defined over $\mathcal{R}$. If the attribute set S satisfies the policy W, we briefly denote $S \vdash W$. For example, let $\mathcal{R} =$ {Lecturer, Professor, PhD, Male}, $S_1 =$ {Professor, Female}, $S_2 =$ {Lecturer, Male, PhD}, $W =$ "Lecturer $\vee$ Professor $\wedge$ Male".

Definition 5.1 (*Attribute-based Encryption*) A ciphertext policy attribute-based encryption (CP-ABE) scheme $\Pi_{\text{abe}} = \{\mathsf{Setup}, \mathsf{Extract}, \mathsf{Enc}, \mathsf{Dec}\}$ consists of four algorithms:

- $\mathsf{Setup}(\kappa, \mathcal{R})$. Given a security parameter κ and an attribute set $\mathcal{R}$, the algorithm returns a master public key mpk and a master secret key msk. The master public key is used for encryption. The master secret key, held by the central authority, is used to extract users' secret keys.
- $\mathsf{Extract}(\mathsf{msk}, S)$. The algorithm takes as input the master secret key msk and an attribute set $S \subseteq \mathcal{R}$, returns a user secret key sk_S.
- $\mathsf{Enc}(\mathsf{mpk}, W, M)$. Given the master public key mpk, an access structure W, and a message M, returns the ciphertext C.
- $\mathsf{Dec}(sk_S, C)$. The algorithm takes a secret key sk_S and a ciphertext C as input, it first checks whether the attribute set of sk_S satisfies the access structure W in C. If not, the algorithm returns $\perp$. Otherwise, it decrypts C and returns the result.

For correctness, we require that, for any message $M \in \{0, 1\}^*$, any access structure W, and any attribute $S \subseteq \mathcal{R}$ that $S \vdash W$, the equation $\mathsf{Dec}(\mathsf{Enc}(\mathsf{mpk}, W, M), sk_S) = M$ holds with overwhelming probability.

We now review the security model for CP-ABE in [7, 14], in which the attacker specifies the challenge access structure before the setup phase. The formal description of this model is given below:

Init. The adversary chooses the challenge access structure W^* and gives it to the challenger.

Setup. The challenger runs the Setup algorithm, gives mpk to the adversary and keeps the master secret key msk secret.

Key Extraction Query. The adversary can adaptively make a number of key generation queries on attribute sets S except that he is not allowed to query an attribute set S that satisfies W^*.

Challenge. At some time, the adversary outputs two messages M_0, M_1, and $|M_0| = |M_1|$. The challenger randomly chooses one bit $b \in \{0, 1\}$, computes $C^* = \texttt{Enc}(pk, W^*, M_b)$, and returns C^* to the adversary.

Guess. The adversary makes more key extraction queries on any attribute set S with a restriction that S doesn't satisfy W^*. Finally, the adversary will output a bit b'.

The advantage of an adversary $\mathcal{A}$ in the above IND-sCPA game is defined as

$$\texttt{Adv}_{\Pi_{\text{abe}},\mathcal{A}}^{\text{ind-scpa}}(\kappa) = |\Pr[b = b'] - \frac{1}{2}|$$

Definition 5.2 A CP-ABE scheme Π_{abe} is said to be secure against selective chosen-plaintext attack (sCPA) if the advantage $\texttt{Adv}_{\Pi_{\text{abe}},\mathcal{A}}^{\text{ind-scpa}}(\kappa)$ is a negligible function in κ for all polynomial time adversary $\mathcal{A}$.

In literatures, there are two methods to deal with attributes. The first is called literal attribute. Namely, this kind of attributes has no values and can be directly used in boolean formulas. We can say a user has or has not some attributes. For example, a user can either be a "student" or not a "student".

The second is multi-valued attribute, where an attribute can take multiple values. Many users may have the same attribute, but two different users can have different values. Actually, an attribute in such a system has meanings only if a specified value is assigned to it. In respect to attribute "age", we won't say a user A has an attribute "age" (since it makes no sense), instead we say user A's age is 20.

5.2 A CP-ABE Supporting And-Gate Policy

In this section, we investigate ciphertext policy attribute-based encryption (CP-ABE), which supports and-gates on positive and negative attributes.

5.2.1 Design Rational

In our considered setting, each attribute is associated with two types of attributes, namely, positive attribute and negative attribute. And if a user has attribute i, we say he has positive attribute i. Otherwise, we say he has negative attribute i. Actually, positive attribute i and negative attribute i are two different attributes, and denoted by i^+ and i^-, respectively. Each user in this system has one and only one of the two attributes, since a user either has attribute i or doesn't. For instance, for a real attribute system which has four attributes $\{att_1, att_2, att_3, att_4\}$, we extend these four attributes into $\{att_1^+, att_1^-, att_2^+, att_2^-, att_3^+, att_3^-, att_4^+, att_4^-\}$ in the system. If a user has attributes $\{att_1, att_3\}$ in the real world, we implicitly define his attributes set as $\{att_1^+, att_2^-, att_3^+, att_4^-\}$. Moreover, all access structures are organized by AND gates in this setting. For example, a user can decrypt a ciphertext if he has all the positive attributes and doesn't have any negative attributes, which are specified in the ciphertext's policy. For instance, a ciphertext encrypted under access structure $W = (att_1^+ \text{ AND } att_2^- \text{ AND } att_3^+)$ can only be decrypted by those who have attributes att_1, att_3 and doesn't have attributes att_2, and we don't care about whether he has att_4.

The basic idea of the construction is that, in the positive and negative setting, each user in this system has an "identity" (i.e., the set of his positive and negative attributes), which is unique in the sense of attribute sets, thus we can use this "identity" to do some things as we do in IBE systems. Specifically, we associate each (positive or negative) attribute with a matrix, actually a matrix uniquely defines a lattice by a well-known definition in lattice cryptography [4]. Thus a user's "identity" uniquely defines a set of lattices. When we generate secret keys for a user with attribute set S, we use his lattices set determined by S to share a public vector, which is used for encryption, by utilizing a trapdoor of these lattices, and the secret key for each attribute in S is a short vector in a lattice (strictly, a coset defined by the lattice) determined by the attribute. As two users with different attribute sets have different "identities", they share the same public vector in two different methods (i.e., in two different lattice sets). The security of this method is guaranteed by inhomogeneous small integer solution (ISIS) problem [20], which was shown to be as hard as some lattice hard problems.

5.2.2 The Construction

In this section, we present the first CP-ABE scheme in which the access structures are and-gates on positive and negative attributes. Basically, each negative attribute is considered as a new attribute [7]. Namely, if a user has attribute set $S \subseteq \mathcal{R}$ in the real system, we consider all of his attributes in S as positive attributes, and the other attributes in $\mathcal{R}\backslash S$ are implicitly considered as his negative ones. Hence, each user in the system actually has $|\mathcal{R}|$ attributes. Without loss of generality, we denote $\mathcal{R} = \{1, \ldots, |\mathcal{R}|\}$.

The construction is defined below, which is parameterized by modulus q, dimension m, Gaussian parameter s, and α that determines the error distribution $\chi = D_{\mathbb{Z},\alpha}$. Usually, all these parameters are functions of security parameter n, and all of these will be instantiated later. All the additions here are performed in $\mathbb{Z}_q$.

$\mathsf{Setup}(n, m, q, \mathcal{R})$: Given positive integers n, m, q, and an attribute set $\mathcal{R}$, first compute $(\mathbf{B}_0, \mathbf{T}_{\mathbf{B}_0}) \leftarrow \mathsf{TrapGen}(1^n, 1^m, q, \mathbf{I}_n)$. Then for each $i \in \mathcal{R}$, randomly choose $\mathbf{B}_{i^+} \xleftarrow{\$} \mathbb{Z}_q^{n\times m}$, $\mathbf{B}_{i^-} \xleftarrow{\$} \mathbb{Z}_q^{n\times m}$. Next, randomly choose a vector $\mathbf{u} \xleftarrow{\$} \mathbb{Z}_q^n$, and set public key $\mathsf{mpk} = (\mathbf{B}_0, \{\mathbf{B}_{i^+}, \mathbf{B}_{i^-}\}_{i\in\mathcal{R}}, \mathbf{u})$, and master secret key $\mathsf{msk} = (\mathsf{mpk}, \mathbf{T}_{\mathbf{B}_0})$. Finally, return $(\mathsf{mpk}, \mathsf{msk})$.

Extract(msk, S): Given the master secret key msk and a user's attribute set $S \subseteq \mathcal{R}$, for each $i \in \mathcal{R}$, if $i \in S$, define $\tilde{\mathbf{B}}_i = \mathbf{B}_{i^+}$, else define $\tilde{\mathbf{B}}_i = \mathbf{B}_{i^-}$. Then for each $i \in \mathcal{R}$, randomly choose $\mathbf{e}_i \leftarrow D_{\mathbb{Z}^m,s}$, and compute $\mathbf{y} = \mathbf{u} - \sum_{i \in R} \tilde{\mathbf{B}}_i \mathbf{e}_i$. Finally, compute $\mathbf{e}_0 \leftarrow$ SampleD($\mathbf{T}_{\mathbf{B}_0}, \mathbf{B}_0, \mathbf{I}_n, \mathbf{y}, s$), and return secret key $\mathbf{sk}_S = [\mathbf{e}_0; \ldots; \mathbf{e}_{|\mathcal{R}|}]$.

Observe that, if let $\mathbf{D} = [\mathbf{B}_0 \| \tilde{\mathbf{B}}_1 \| \ldots \| \tilde{\mathbf{B}}_{|\mathcal{R}|}]$, we have $\mathbf{D} \cdot \mathbf{sk}_S = \mathbf{u}$.

Enc(mpk, W, M): Given the master public key mpk $= (\{\mathbf{B}_{i^+}, \mathbf{B}_{i^-}\}_{i \in \mathcal{R}}, \mathbf{u})$, an access structure W, and a message bit $M \in \{0, 1\}$, denote $S^+(S^-)$ as the set of positive (negative) attributes in W, and $S' = S^+ \cup S^-$. Then for each $i \in S'$, if $i \in S^+$, define $\tilde{\mathbf{B}}_i = \mathbf{B}_{i^+}$, else, define $\tilde{\mathbf{B}}_i = \mathbf{B}_{i^-}$. Next, randomly choose $\mathbf{s} \xleftarrow{\$} \mathbb{Z}_q^n$ and compute:

- $z = \mathbf{u}^T \mathbf{s} + x_z + M \lfloor \frac{q}{2} \rfloor$, where $x_z \xleftarrow{\$} \chi$,
- $\mathbf{c}_0 = \mathbf{B}_0^T \mathbf{s} + \mathbf{x}_0$, where $\mathbf{x}_0 \xleftarrow{\$} \chi^m$,
- $\mathbf{c}_i = \tilde{\mathbf{B}}_i^T \mathbf{s} + \mathbf{x}_i$ for each $i \in S'$, where $\mathbf{x}_i \xleftarrow{\$} \chi^m$,
- $\mathbf{c}_{i^+} = \mathbf{B}_{i^+}^T \mathbf{s} + \mathbf{x}_{i^+}$ and $\mathbf{c}_{i^-} = \mathbf{B}_{i^-}^T \mathbf{s} + \mathbf{x}_{i^-}$ for each $i \in \mathcal{R} \backslash S'$, where $\mathbf{x}_{i^+}, \mathbf{x}_{i^-} \xleftarrow{\$} \chi^m$.

Finally, return ciphertext $C = (W, z, \mathbf{c}_0, \{\mathbf{c}_i\}_{i \in S'}, \{\mathbf{c}_{i^+}, \mathbf{c}_{i^-}\}_{i \in \mathcal{R} \backslash S'})$.

Dec(C, $\mathbf{sk}$): Given the ciphertext C and the secret key $\mathbf{sk} = [\mathbf{e}_0; \ldots; \mathbf{e}_{|\mathcal{R}|}]$, let S be the attribute set associated to $\mathbf{sk}$, if S doesn't satisfy W, then return $\perp$. Otherwise $S \vdash W$. Define $S^+(S^-)$ as the set of positive (negative) attributes in W, and $S' = S^+ \cup S^-$. Obliviously, $S^+ \subset S$ and $S^- \cap S = \emptyset$. Parse C into $(W, z, \mathbf{c}_0, \{\mathbf{c}_i\}_{i \in S'}, \{\mathbf{c}_{i^+}, \mathbf{c}_{i^-}\}_{i \in \mathcal{R} \backslash S'})$. Then let $\mathbf{y}_i = \mathbf{c}_i$ for each $i \in S' \cup \{0\}$, and for each $i \in \mathcal{R} \backslash S'$, if $i \in S$, let $\mathbf{y}_i = \mathbf{c}_{i^+}$, else let $\mathbf{y}_i = \mathbf{c}_{i^-}$. Define $\mathbf{y} = [\mathbf{y}_0; \mathbf{y}_1; \ldots; \mathbf{y}_{|\mathcal{R}|}]$, and compute $a = \mathbf{sk}^T \mathbf{y} = \mathbf{u}^T \mathbf{s} + x', b = z - a = x_z - x' + M \lfloor \frac{q}{2} \rfloor$. Finally, If $|b - \lfloor \frac{q}{2} \rfloor| \leq \lfloor \frac{q}{4} \rfloor$ in $\mathbb{Z}$, return 1, otherwise return 0.

Let $\mathbf{D}$ be the matrix determined by the attribute set in $\mathbf{sk}$, thus $\mathbf{D} \cdot \mathbf{sk} = \mathbf{u}$. By the method we choose vector $\mathbf{y}$, we have $\mathbf{y} = \mathbf{D}^T \mathbf{s} + \mathbf{x}_y$, where $\mathbf{s} \in \mathbb{Z}_q^n, \mathbf{x}_y \in \chi^{m(|\mathcal{R}|+1)}$ are chosen in the encryption. Thus, $a = \mathbf{sk}^T \mathbf{y} = \mathbf{sk}^T (\mathbf{D}^T \mathbf{s} + \mathbf{x}_y) = \mathbf{u}^T \mathbf{s} + \mathbf{sk}^T \mathbf{x}_y = \mathbf{u}^T \mathbf{s} + x'$. And if $|x_z - x'| \leq q/5$ holds (with overwhelming probability), it is easy to check that the decryption algorithm always outputs plaintext M correctly.

5.2.3 Security

Theorem 5.1 *Let m, s, q, α be defined as above, and let $\chi = \bar{\Psi}_\alpha$. Then if $LWE_{q,\chi}$ is hard, the CP-ABE scheme is secure against selective chosen-ciphertext attack (sCPA).*

In particularly, if there exists an adversary $\mathcal{A}$ that breaks the sCPA security of the scheme with advantage ϵ, then there exists an algorithm $\mathcal{B}$ solves $LWE_{q,\chi}$ with probability ϵ.

Proof Suppose there exists a polynomial time adversary $\mathcal{A}$ that breaks the sCPA security of the CP-ABE scheme with advantage ϵ and makes at most q key extraction queries. We construct an algorithm $\mathcal{B}$ that solves the LWE problem with probability negligible to ϵ. Note that algorithm $\mathcal{B}$ has an oracle $\mathcal{O}(\cdot)$, and he wants to decide whether the samples output by $\mathcal{O}(\cdot)$ are from $\mathrm{A}_{\mathbf{s},\chi}$ or uniform. $\mathcal{B}$ runs adversary $\mathcal{A}$ and simulates $\mathcal{A}$'s view in the sCPA security experiment as follows:

Init. Adversary $\mathcal{A}$ chooses a challenge access structure W^* and gives it to $\mathcal{B}$. Let $S^+(S^-)$ be the set of positive (negative) attributes in W^*, and let $S' = S^+ \cup S^-$.

Setup. After receiving W^*, $\mathcal{B}$ computes:

- $\mathcal{B}$ obtains $(\mathbf{B}_0, \mathbf{v}_0) \in \mathbb{Z}_q^{n\times m} \times \mathbb{Z}_q^m$ and $(\mathbf{u}, v_u) \in \mathbb{Z}_q^n \times \mathbb{Z}_q$ from $\mathcal{O}(\cdot)$.
- For each $i \in \mathcal{R}\backslash S'$, $\mathcal{B}$ obtains $(\mathbf{B}_{i^+}, \mathbf{v}_{i^+}), (\mathbf{B}_{i^-}, \mathbf{v}_{i^-}) \in \mathbb{Z}_q^{n\times m} \times \mathbb{Z}_q^m$ from $\mathcal{O}(\cdot)$.
- For each $i \in S^+$, $\mathcal{B}$ obtains $(\mathbf{B}_{i^+}, \mathbf{v}_{i^+}) \in \mathbb{Z}_q^{n\times m} \times \mathbb{Z}_q^m$ from $\mathcal{O}(\cdot)$, then computes $(\mathbf{B}_{i^-}, \mathbf{T}_{\mathbf{B}_{i^-}}) \leftarrow \mathsf{TrapGen}(1^n, 1^m, q, \mathbf{I}_n)$.
- For each $i \in S^-$, $\mathcal{B}$ obtains $(\mathbf{B}_{i^-}, \mathbf{v}_{i^-}) \in \mathbb{Z}_q^{n\times m} \times \mathbb{Z}_q^m$ from $\mathcal{O}(\cdot)$, then computes $(\mathbf{B}_{i^+}, \mathbf{T}_{\mathbf{B}_{i^+}}) \leftarrow \mathsf{TrapGen}(1^n, 1^m, q, \mathbf{I}_n)$.

Finally, $\mathcal{B}$ sets $pk = (\mathbf{B}_0, \{\mathbf{B}_{i^+}, \mathbf{B}_{i^-}\}_{i\in R}, \mathbf{u})$, and keeps $(\{\mathbf{T}_{\mathbf{B}_{i^-}}, \mathbf{v}_{i^+}\}_{i\in S^+}, \{\mathbf{T}_{\mathbf{B}_{i^+}}, \mathbf{v}_{i^-}\}_{i\in S^-}, \{\mathbf{v}_{i^+}, \mathbf{v}_{i^-}\}_{i\in\mathcal{R}\backslash S'})$ secret.

Key Extraction Query. After receiving a query with attribute set $S \subseteq \mathcal{R}$. If $S \vdash W^*$, $\mathcal{B}$ simply outputs $\perp$. Otherwise, for each $i \in \mathcal{R}$, if $i \in S$, $\mathcal{B}$ lets $\tilde{\mathbf{B}}_i = \mathbf{B}_{i^+}$, else lets $\tilde{\mathbf{B}}_i = \mathbf{B}_{i^-}$. Since S doesn't satisfy W^*, namely, $S^+ \cap S \neq S^+$ or $S^- \cap S \neq \emptyset$, there must exists a $j \in \mathcal{R}$, such that $\tilde{\mathbf{B}}_j$ is generated by `TrapGen`. Hence, $\mathcal{B}$ knows its trapdoor $\mathbf{T}_{\tilde{\mathbf{B}}_j}$. Let $\mathbf{D} = [\mathbf{B}_0\|\tilde{\mathbf{B}}_1\| \ldots \|\tilde{\mathbf{B}}_n\|]$, $\mathcal{B}$ computes $\mathbf{e}_S \leftarrow \mathsf{SampleD}(\mathbf{T}_{\tilde{\mathbf{B}}_j}, \mathbf{D}, \mathbf{I}_n, \mathbf{u}, s)$, and returns $sk_S = \mathbf{e}_S$ to $\mathcal{A}$.

Challenge. When $\mathcal{A}$ submits $M_0, M_1 \in \{0, 1\}$, $\mathcal{B}$ randomly chooses $b \in \{0, 1\}$ and computes $z = v_u + M_b\lfloor\frac{q}{2}\rfloor$ and $\mathbf{c}_0 = \mathbf{v}_0$. For each $i \in S^+$, let $\mathbf{c}_i = \mathbf{v}_{i^+}$. For each $i \in S^-$, let $\mathbf{c}_i = \mathbf{v}_{i^-}$. For each $i \in \mathcal{R}\backslash S'$, let $\mathbf{c}_{i^+} = \mathbf{v}_{i^+}$ and $\mathbf{c}_{i^-} = \mathbf{v}_{i^-}$. Finally, $\mathcal{B}$ returns $C^* = (W, z, \mathbf{c}_0, \{\mathbf{c}_i\}_{i\in S'}, \{\mathbf{c}_{i^+}, \mathbf{c}_{i^-}\}_{i\in\mathcal{R}\backslash S'})$.

$\mathcal{A}$ can make more key extraction queries on attribute set S that doesn't satisfy W^*. Eventually, $\mathcal{A}$ outputs a bit b' as a guess for b. if $b' = b$, $\mathcal{B}$ outputs 1, else outputs 0.

Note that $\mathcal{B}$ answers the key extraction queries almost the same as the challenger does in the real game by Lemma 2.24. On one hand, if $\mathcal{O}(\cdot)$ is a LWE oracle for some $\mathbf{s}^*$, C^* is a valid ciphertext, thus the distribution of $\mathcal{A}$'s view is statistically close to that in the real game. On the other hand, if $\mathcal{O}(\cdot)$ is chosen from uniform, then the ciphertext z is uniform on $\mathbb{Z}_q$, thus the probability that $\mathcal{A}$ guesses the right b is exactly 1/2. So if $\mathcal{A}$ can break the system, $\mathcal{B}$ can break the LWE assumption, which yields the claim. □

5.3 A CP-ABE Supporting Flexible Threshold Policy

In this section, we first construct a ciphertext policy attribute-based encryption from lattices, which supports flexible threshold access policies on literal attributes [27].

Then, we extend it to support multi-valued attributes without increasing the master public key and ciphertext size.

5.3.1 Design Rational

At a high level, we will use the Shamir secret sharing scheme to generate users' secret keys in a manner similar to [2], together with two more techniques to achieve "flexible threshold policy" and to reduce the size of public key and master secret key. First, we introduce default attributes into the system. We assume all the users in the system have some common default attributes, and the constructions benefit substantially from those extra default attributes in two aspects. On the one hand, with the help of default attributes, the system can support flexible threshold access policies. That's to say, both integers (i.e., t and k) of a t-out-of-k threshold chosen for the encryption algorithm are variable, and a user who encrypts a message can set these two integers arbitrarily and independently to determine who can decrypt his ciphertext. For instance, if a system with total ℓ attributes supports a maximum threshold of d ($\leq \ell$), we add d default attributes to the system and assume all the users in the system have all the default attributes. Thus a system user has at least $d + 1$ attributes (including one or more normal attributes). We extract a user's secret key by using the Shamir secret sharing scheme with a degree d polynomial, and each attribute is associated with a unique share. If a sender wants anyone who has t ($\leq \min(d, k)$) of k normal attributes (that the sender chooses) can decrypt his ciphertext, he first chooses $d - t + 1$ default attributes, then encrypts the message by using all the $k + d - t + 1$ attributes. Since all users have d default attributes, anyone satisfies the sender's access policy must have at least $d + 1$ of the $k + d - t + 1$ attributes (i.e., $d - t + 1$ default attributes together with at least t normal attributes), and then he can use any $d + 1$ shares to decrypt the ciphertext correctly. On the other hand, the security proof makes good use of those default attributes. Though the default attributes increase the system's parameter size, the parameters are still shorter than those in [2, 3, 26].

5.3.2 The Construction

In this section, we present a CP-ABE scheme, which supports flexible threshold access policies. For convenience, we assume the system has ℓ attributes, denoted by $\mathcal{R} = \{1, \ldots, \ell\}$. Each ciphertext C in the system is associated with an access policy (W, t), where $W \subset \mathcal{R}$ is an attribute set, and integer t is a threshold which is up bounded by a system parameter d[1] (i.e., $1 \leq t \leq d$). The access policy (W, t)

[1] d is determined by concrete applications, and one would like to set $d = \ell$ for all possible applications.

specifies that anyone who at least has t attributes in W can decrypt the ciphertext C correctly. We also denote $D = ((\ell + d)!)^2$.

The scheme is parameterized by modulus q, dimension m, Gaussian parameter s and α that determines the error distribution $\chi = D_{\mathbb{Z},\alpha}$. Let $\mathbf{G}$ be the public primitive matrix with base $b = 2$ (i.e., $\mathbf{G} = \mathbf{G}_2 \in \mathbb{Z}_q^{n\times nk}$) as defined in Sect. 2.5. Usually, all these parameters are functions of security parameter n, and all of these will be instantiated later. All the additions here are performed in $\mathbb{Z}_q$.

$\mathsf{Setup}(n, m, q, d, \mathcal{R})$: Given positive integers n, m, q, d, and an attribute set $\mathcal{R}$, the algorithm first chooses d default attributes $\mathcal{D} = \{\ell + 1, \ldots, \ell + d\}$ and denotes $\mathcal{R}' = \mathcal{R} \cup \mathcal{D}$. Then it computes $(\mathbf{A}_0, \mathbf{T}_{\mathbf{A}_0}) \leftarrow \mathsf{TrapGen}(1^n, 1^m, q, \mathbf{I}_n)$, and randomly chooses $\mathbf{u} = (u_1, \ldots, u_n)^T \xleftarrow{\$} \mathbb{Z}_q^n$. Let $k = \lfloor \log_2 q \rceil$. For each $i \in \mathcal{R}'$, randomly choose $\mathbf{A}_i \xleftarrow{\$} \mathbb{Z}_q^{n\times nk}$. Finally, it returns public key $\mathsf{mpk} = (\mathbf{A}_0, \{\mathbf{A}_i\}_{i\in\mathcal{R}'}, \mathbf{u})$, and master secret key $\mathsf{msk} = \mathbf{T}_{\mathbf{A}_0}$.

$\mathsf{Extract}(\mathsf{msk}, S)$: Given the master secret key msk and a user's attribute set $S \subseteq \mathcal{R}$, let $S' = S \cup \mathcal{D}$. For $j = 1, \ldots, n$, randomly choose degree d polynomial $p_j(x) \in \mathbb{Z}_q[x]$, such that $p_j(0) = u_j$. For each $i \in S'$, let $\hat{\mathbf{u}}_i = (p_1(i), \ldots, p_n(i))^T$, and $\mathbf{E}_i = (\mathbf{A}_0 \| \mathbf{A}_i + \mathbf{G}) \in \mathbb{Z}_q^{n\times(m+nk)}$, compute $\mathbf{e}_i \leftarrow \mathsf{SampleD}(\mathbf{T}_{\mathbf{A}_0}, \mathbf{E}_i, \mathbf{I}_n, \hat{\mathbf{u}}_i, s)$ (i.e., $\mathbf{E}_i\mathbf{e}_i = \hat{\mathbf{u}}_i$). Return secret key $sk = (\{\mathbf{e}_i\}_{i\in S'})$.
Note that, for any subset $J \subseteq S'$ with $|J| = d + 1$, we have $\mathbf{u} = \sum_{j\in J} L_j \cdot \hat{\mathbf{u}}_j$, where the Lagrangian coefficient $L_j = \frac{\prod_{i\in J, i\neq j} -i}{\prod_{i\in J, i\neq j}(j-i)}$.

$\mathsf{Enc}(\mathsf{mpk}, (W, t), M)$: Given the master public key $\mathsf{mpk} = (\mathbf{A}_0, \mathbf{B}, \{\mathbf{A}_i\}_{i\in\mathcal{R}'}, \mathbf{u})$, an attribute set W, an integer $1 \le t \le \min\{|W|, d\}$, and a message bit $M \in \{0, 1\}$, let $W' = W \cup \{\ell + 1, \ldots, \ell + d + 1 - t\}$. Choose $\mathbf{s} \xleftarrow{\$} \mathbb{Z}_q^n$ randomly, compute $c_0 = \mathbf{u}^T\mathbf{s} + Dx_0 + M\lfloor \frac{q}{2} \rfloor$, and $\mathbf{c}' = \mathbf{A}_0^T\mathbf{s} + D\mathbf{x}$, where $x_0 \xleftarrow{\$} \chi$ and $\mathbf{x} \xleftarrow{\$} \chi^m$. For each $i \in W'$, randomly choose $\mathbf{R}_i \in \{-1, 1\}^{m\times m}$, compute $\mathbf{c}_i = (\mathbf{A}_i + \mathbf{B})^T\mathbf{s} + D\mathbf{R}_i^T\mathbf{x}$. Return ciphertext $C = (c_0, \mathbf{c}', \{\mathbf{c}_i\}_{i\in W'})$.

$\mathsf{Dec}(C, sk)$: Given the ciphertext C and the secret key sk, let S be the attribute set associated to sk, (W, t) be the access structure associated to C. If $|S \cap W| < t$, then return $\perp$. Otherwise, Let $S' = S \cup \mathcal{D}$ and $W' = W \cup \{\ell + 1, \ldots, \ell + d + 1 - t\}$, parse $sk = (\{\mathbf{e}_i\}_{i\in S'})$ and $C = (c_0, \mathbf{c}', \{\mathbf{c}_i\}_{i\in W'})$. Since $|S \cap W| \ge t$, we have $|S' \cap W'| \ge d + 1$. Choose a subset $J \subseteq S' \cap W'$ with $|J| = d + 1$. For all $j \in J$, compute $b_j = \mathbf{e}_j^T(\mathbf{c}'; \mathbf{c}_j)$ and $b = \sum_{j\in J} L_j b_j$, where $L_j = \frac{\prod_{i\in J, i\neq j} -i}{\prod_{i\in J, i\neq j}(j-i)}$. Finally, compute $r = c_0 - b$. If $|r - \lfloor \frac{q}{2} \rfloor| \le \lfloor \frac{q}{4} \rfloor$ in $\mathbb{Z}$, return 1, otherwise return 0.

For correctness, we note that

$$(\mathbf{c}'; \mathbf{c}_j) = \begin{pmatrix} \mathbf{c}' \\ \mathbf{c}_j \end{pmatrix} = \begin{pmatrix} \mathbf{A}_0^T\mathbf{s} + D\mathbf{x} \\ (\mathbf{A}_j + \mathbf{B})^T\mathbf{s} + D\mathbf{R}_j^T\mathbf{x} \end{pmatrix} = (\mathbf{A}_0 \| \mathbf{A}_j + \mathbf{B})^T\mathbf{s} + D\begin{pmatrix} \mathbf{x} \\ \mathbf{R}_j^T\mathbf{x} \end{pmatrix}.$$

Denote $\mathbf{x}_j' = (\mathbf{x}; \mathbf{R}_j^T\mathbf{x})$, we have $(\mathbf{c}'; \mathbf{c}_j) = \mathbf{E}_j^T\mathbf{s} + D\mathbf{x}_j'$. In the decryption algorithm, we compute $b_j = \mathbf{e}_j^T(\mathbf{c}'; \mathbf{c}_j) = \mathbf{e}_j^T(\mathbf{E}_j^T\mathbf{s} + D\mathbf{x}_j') = \hat{\mathbf{u}}_j^T\mathbf{s} + Dy_j$, where $y_j = \mathbf{e}_j^T\mathbf{x}_j'$. Since $|J| = d + 1$, we have $\mathbf{u} = \sum_{j\in J} L_j\hat{\mathbf{u}}_j$, $b = \sum_{j\in J} L_j b_j = \mathbf{u}^T\mathbf{s} + y$, where $y = \sum_{j\in J} DL_j y_j$. Hence, $r = c_0 - b = \mathbf{u}^T\mathbf{s} + Dx_0 + M\lfloor \frac{q}{2} \rfloor - - - \mathbf{u}^T\mathbf{s} - - - y = Dx_0 - y + M\lfloor \frac{q}{2} \rfloor$. If $|Dx_0 - y| \le q/5$, then the decryption algorithm can decrypt the ciphertext correctly and output the right message M with overwhelming probability.

5.3.3 Security

Theorem 5.2 *For properly chosen m, s, q, α, let $\chi = \bar{\Psi}_\alpha$. Then if $LWE_{q,\chi}$ is hard, the CP-ABE scheme is secure against selective chosen-plaintext attack (sCPA).*

In particularly, if there exists an adversary $\mathcal{A}$ that breaks the sCPA security of the scheme, then there exists an algorithm $\mathcal{B}$ solves $LWE_{q,\chi}$.

Proof Suppose there exists a polynomial time adversary $\mathcal{A}$ that breaks the selective chosen-plaintext security of the CP-ABE scheme with non-negligible advantage. We construct an algorithm $\mathcal{B}$ that solves the (decisional) LWE problem with the same advantage. Note that $\mathcal{B}$ has an oracle $\mathcal{O}(\cdot)$, and he wants to decide whether the samples output by $\mathcal{O}(\cdot)$ is from $A_{\mathbf{s},\chi}$ or uniform. Let attribute set $\mathcal{R} = \{1, \ldots, \ell\}$, $\mathcal{D} = \{\ell+1, \ldots, \ell+d\}$, and denote $\mathcal{R}' = \mathcal{R} \cup \mathcal{D}$. $\mathcal{B}$ runs adversary $\mathcal{A}$ and simulates $\mathcal{A}$'s view in the selective CPA experiment as follows:

Init. Adversary $\mathcal{A}$ submits a challenge access structure (W^*, t^*) to $\mathcal{B}$, where $1 \le t^* \le \min\{|W^*|, d\}$. Let $W' = W^* \cup \{\ell+1, \ldots, \ell+d+1-t^*\}$.

Setup. After receiving W^*,

- $\mathcal{B}$ draws $(\mathbf{u}, v_u) \in \mathbb{Z}_q^n \times \mathbb{Z}_q$, $(\mathbf{A}_0, \mathbf{v}_0) \in \mathbb{Z}_q^{n\times m} \times \mathbb{Z}_q^m$ from $\mathcal{O}(\cdot)$.
- For each $i \in W'$, $\mathcal{B}$ randomly chooses $\mathbf{R}_i^* \in \{-1, 1\}^{m\times nk}$ and computes $\mathbf{A}_i = \mathbf{A}_0\mathbf{R}_i^* - \mathbf{G}$.
- . For each $i \in \mathcal{R}'\backslash W'$, $\mathcal{B}$ randomly chooses $\mathbf{R}_i^* \in \{-1, 1\}^{m\times nk}$ and computes $\mathbf{A}_i = \mathbf{A}_0\mathbf{R}_i^*$.

Finally, $\mathcal{B}$ sets $\mathsf{mpk} = (\mathbf{A}_0, \{\mathbf{A}_i\}_{i\in\mathcal{R}'}, \mathbf{u})$, and keeps $(\{\mathbf{R}_i^*\}_{i\in\mathcal{R}'}, v_u, \mathbf{v}_0)$ secret.

Key Extraction Query. After receiving a Key Extraction Query $S \subseteq \mathcal{R}$. If $S \vdash (W^*, t^*)$, $\mathcal{B}$ simply outputs $\perp$ (since it's an invalid query according to the security model). Otherwise, $|S \cap W^*| \le t^* - 1$. Let $S' = S \cup \{\ell+1, \ldots, \ell+d\}$, $|W' \cap S'| \le d$. Choose a subset $\hat{S}$, such that $W' \cap S' \subseteq \hat{S} \subset S'$ and $|\hat{S}| = d$. For each $i \in \hat{S}$, define $\mathbf{E}_i = (\mathbf{A}_0\|\mathbf{A}_i + \mathbf{G})$, choose $\mathbf{e}_i \leftarrow D_{\mathbb{Z}^{m+nk},s}$ and compute $\hat{\mathbf{u}}_i = \mathbf{E}_i\mathbf{e}_i$. Thus we have $d+1$ n-dimension vector $\{\mathbf{u}, \{\hat{\mathbf{u}}_i\}_{i\in\hat{S}}\}$. Let n degree d polynomials $p_1(x), \ldots, p_n(x) \in \mathbb{Z}_q[x]$ such that $\mathbf{u} = (p_1(0), \ldots, p_n(0))^T$, and for each $i \in \hat{S}$, $\hat{\mathbf{u}}_i = (p_1(i), \ldots, p_n(i))^T$. By the Lagrange interpolation formula, we can recover polynomials $p_1(x), \ldots, p_n(x) \in \mathbb{Z}_q[x]$. Note that, if $i \in S'\backslash\hat{S}$, we have $i \notin W'$. So $\mathbf{E}_i = (\mathbf{A}_0\|\mathbf{A}_i + \mathbf{G}) = (\mathbf{A}_0\|\mathbf{A}_0\mathbf{R}_i^* + \mathbf{G})$. Let $\hat{\mathbf{u}}_i = (p_1(i), \ldots, p_n(i))^T$, by using the trapdoor $\mathbf{T_B}$, $\mathcal{B}$ can compute $\mathbf{e}_i \leftarrow \mathsf{SampleD}(\mathbf{R}_i^*, \mathbf{E}_i, \mathbf{I}_n, \hat{\mathbf{u}}_i, s)$. Return secret key $sk = (\{\mathbf{e}_i\}_{i\in S'})$.

Challenge. When $\mathcal{A}$ submits $M_0, M_1 \in \{0, 1\}$, $\mathcal{B}$ randomly chooses $b \in \{0, 1\}$, and computes $c_0 = Dv_u + M_b\lfloor\frac{q}{2}\rfloor$ and $\mathbf{c}' = D\mathbf{v}_0$. For each $i \in W'$, computes $\mathbf{c}_i = D(\mathbf{R}_i^*)^T\mathbf{v}_0$. Return the challenge ciphertext $C^* = (c_0, \mathbf{c}', \{\mathbf{c}_i\}_{i\in W'})$.

$\mathcal{A}$ can make more key extraction queries on attribute set S that doesn't satisfy (W^*, t^*). Eventually, $\mathcal{A}$ outputs a bit b' as a guess for b. If $b' = b$, $\mathcal{B}$ outputs 1, else outputs 0.

Note that $\mathbf{A}_0$ and $\mathbf{u}$ are drawn from the LWE oracle, the tuple $(\mathbf{A}_0, \mathbf{u})$ is statistical to its distribution in the real attack according to Lemma 2.24. Since each $\mathbf{R}_i^*$ are chosen randomly from $\{-1, 1\}^{m \times m}$, applying Lemma 2.4, we have that $\{\mathbf{A}_i\}_{i \in \mathcal{R}'}$ is also statistical to uniform even given more information about $(\mathbf{R}_i^*)^T \mathbf{x}$. In one word, The master public key distribution in the simulation is statistical to that in the real attack, and the adversary gains little information about the matrices $\{\mathbf{R}_i^*\}_{i \in \mathcal{R}'}$ from the master public key.

Besides, the key extraction simulation uses algorithm SampleD, and its output distribution is statistically close to that in the real attack by Lemma 2.24. If $\mathcal{O}(\cdot)$ is a LWE oracle for some $\mathbf{s}^*$, we claim that C^* is a valid ciphertext for $\mathbf{s} = D\mathbf{s}^*$, and $\{\mathbf{R}_i^*\}_{i \in W'}$. In fact, for each $i \in W'$, we have $\mathbf{c}_i = D(\mathbf{R}_i^*)^T (\mathbf{A}_0^T \mathbf{s}^* + \mathbf{x}) = (\mathbf{A}_0 \mathbf{R}_i^*)^T (D\mathbf{s}^*) + D(\mathbf{R}_i^*)^T \mathbf{x} = (\mathbf{A}_i + \mathbf{G})^T \mathbf{s} + D(\mathbf{R}_i^*)^T \mathbf{x}$. Since $\mathcal{A}$ gains no information about $\{\mathbf{R}_i^*\}_{i \in W'}$ from The master public key, the ciphertext is the same as $\mathcal{A}$'s view in the real attack. Hence, if $\mathcal{A}$ can guess the right b with probability noticeably greater than $1/2$, then $\mathcal{B}$ can succeed in its game with the same probability. Else if $\mathcal{O}(\cdot)$ is chosen from uniform, then the ciphertext c_0 is uniform in $\mathbb{Z}_q$, thus the probability that $\mathcal{A}$ guesses the right b is exactly $1/2$. So if $\mathcal{A}$ can break the system, $\mathcal{B}$ can solve LWE problems, which yields the claim. □

5.3.4 Supporting Multi-valued Attributes

In this section, we present a CP-ABE scheme, which supports flexible threshold access policies on multi-valued attribute. Here, we need a new primitive called full-rank differences function.

Definition 5.3 ([1]) Let q be a prime and n be a positive integer. We say a function $H : \mathbf{Z}_q^n \to \mathbf{Z}_q^{n \times n}$ is an encoding with full-rank differences (FRD) if:

- for any $x \neq y$, the matrix $H(x) - H(y) \in \mathbb{Z}_q^{n \times n}$ is full rank;
- H is computable in polynomial time in $n \log q$.

Now we present the second construction by using FRD. As before, we assume the system has ℓ attributes, and attribute i is associated with a value space $\mathcal{R}_i \subseteq \mathbf{Z}_q^n \backslash \{\mathbf{0}\}$.[2] The attribute space is denoted by $\mathcal{R} = \mathcal{R}_1 \times \cdots \times \mathcal{R}_\ell$. Let $\mathcal{I}$ denote the index set $\{1, \ldots, \ell + d\}$. For convenience, we also denote $\mathcal{I}_1 = \{1, \ldots, \ell\}$, $\mathcal{I}_2 = \{\ell + 1, \ldots, \ell + d\}$ and $D = ((\ell + d)!)^2$.

As before, the scheme is parameterized by modulus q, dimension m, Gaussian parameter s and α that determines the error distribution $\chi = D_{\mathbb{Z},\alpha}$. Let $\mathbf{G}$ be the public

[2] We reserve $\mathbf{0}$ for the security proof, one can remove this restriction by assuming the FRD has a property that for any u, $H(u)$ is full rank.

primitive matrix with base $b = 2$ (i.e., $\mathbf{G} = \mathbf{G}_2 \in \mathbb{Z}_q^{n\times nk}$) as defined in Sect. 2.5. Let $H : \mathbb{Z}_q^n \to \mathbb{Z}_q^{n\times n}$. Usually, all these parameters are functions of security parameter n, and all of these will be instantiated later. All the additions here are performed in $\mathbb{Z}_q$.

$\mathsf{Setup}(n, m, q, d, \mathcal{R})$: Given positive integers n, m, q, d and an attribute set $\mathcal{R}$, the algorithm first computes $(\mathbf{A}_0, \mathbf{T}_{\mathbf{A}_0}) \leftarrow \mathsf{TrapGen}(1^n, 1^m, q, \mathbf{I}_n)$, and randomly chooses $\mathbf{u} = (u_1, \ldots, u_n)^T \xleftarrow{\$} \mathbb{Z}_q^n$. For each $i \in \mathcal{I}$, randomly choose $\mathbf{A}_i \xleftarrow{\$} \mathbb{Z}_q^{n\times nk}$. Return the master public key $\mathsf{mpk} = (\mathbf{A}_0, \{\mathbf{A}_i\}_{i\in\mathcal{I}}, \mathbf{u}, H)$, and the master secret key $\mathsf{msk} = \mathbf{T}_{\mathbf{A}_0}$.

$\mathsf{Extract}(\mathsf{msk}, S)$: Given the master secret key msk and a user's attribute set $S = \{att_i\}_{i\in I}$, where $I \subseteq \mathcal{I}_1$ and each $att_i \in \mathcal{R}_i$. For $j = 1, \ldots, n$, randomly choose degree d polynomial $p_j(x) \in \mathbb{Z}_q[x]$, such that $p_j(0) = u_j$ (i.e., $\mathbf{u} = (p_1(0), \ldots, p_n(0))^T$). For each $i \in I$, let $\hat{\mathbf{u}}_i = (p_1(i), \ldots, p_n(i))^T$ and $\mathbf{E}_i = (\mathbf{A}_0 \| \mathbf{A}_i + H(att_i)\mathbf{G})$, compute $\mathbf{e}_i \leftarrow \mathsf{SampleD}(\mathbf{T}_{\mathbf{A}_0}, \mathbf{E}_i, \hat{\mathbf{u}}_i, s)$, and for each $i \in \mathcal{I}_2$, let $\hat{\mathbf{u}}_i = (p_1(i), \ldots, p_n(i))^T$ and $\mathbf{E}_i = (\mathbf{A}_0 \| \mathbf{A}_i + \mathbf{G})$, compute $\mathbf{e}_i \leftarrow \mathsf{SampleD}(\mathbf{T}_{\mathbf{A}_0}, \mathbf{E}_i, \hat{\mathbf{u}}_i, s)$. Return secret key $sk = (\{\mathbf{e}_i\}_{i\in I\cup\mathcal{I}_2})$.
Note that $\mathbf{E}_i\mathbf{e}_i = \hat{\mathbf{u}}_i$, and for any subset $K \subseteq I \cup \mathcal{I}_2$, $|K| = d + 1$, we have $\mathbf{u} = \sum_{j\in K} L_j \cdot \hat{\mathbf{u}}_j$, where the Lagrangian coefficient $L_j = \frac{\prod_{i\in K, i\neq j} -i}{\prod_{i\in K, i\neq j}(j-i)}$.

$\mathsf{Enc}(\mathsf{mpk}, (W, t), M)$: Given the master public key $pk = (\mathbf{A}_0, \{\mathbf{A}_i\}_{i\in\mathcal{I}}, \mathbf{u}, H)$, an attribute set $W = \{att_i\}_{i\in J}$, an integer $t \leq \min(|W|, d)$, and a message bit $M \in \{0, 1\}$, the algorithm chooses $\mathbf{s} \xleftarrow{\$} \mathbb{Z}_q^n$ randomly, compute $c_0 = \mathbf{u}^T\mathbf{s} + Dx_0 + m\lfloor\frac{q}{2}\rfloor$, and $\mathbf{c}' = \mathbf{A}_0^T\mathbf{s} + D\mathbf{x}$, where $x_0 \leftarrow \chi, \mathbf{x} \leftarrow \chi^m$. For each $j \in J$, randomly choose $\mathbf{R}_j \in \{-1, 1\}^{m\times m}$, compute $\mathbf{c}_j = (\mathbf{A}_j + H(att_j)\mathbf{G})^T\mathbf{s} + D\mathbf{R}_j^T\mathbf{x}$. Let $J' = \{\ell + 1, \ldots, \ell + d + 1 - t\}$, and for each $j \in J'$, randomly choose $\mathbf{R}_j \in \{-1, 1\}^{m\times m}$, compute $\mathbf{c}_j = (\mathbf{A}_j + \mathbf{G})^T\mathbf{s} + D\mathbf{R}_j^T\mathbf{x}$. Return ciphertext $C = (c_0, \mathbf{c}', \{\mathbf{c}_j\}_{j\in J\cup J'})$.

$\mathsf{Dec}(C, sk)$: Given the ciphertext C and the secret key sk, let S be the attribute set associated to sk, (W, t) be the access structure associated to C. If $|S \cap W| < t$, then return $\perp$. Otherwise, parse $sk = (\{\mathbf{e}_i\}_{i\in I\cup\mathcal{I}_2})$ and $C = (c_0, \mathbf{c}', \{\mathbf{c}_i\}_{i\in J\cup J'})$. Let $S \cap W = \{att_i\}_{i\in K}$, where $K \subseteq I \cap J$. Since $|K| \geq t$, there exists a set $K' \subseteq K \cup J'$ with size $d + 1$. For all $j \in K'$, compute $b_j = \mathbf{e}_j^T(\mathbf{c}'; \mathbf{c}_j)$ and $b = \sum_{j\in K'} L_j b_j$, where $L_j = \frac{\prod_{i\in K', i\neq j} -i}{\prod_{i\in K', i\neq j}(j-i)}$. Finally, compute $r = c_0 - b$. If $|r - \lfloor\frac{q}{2}\rfloor| \leq \lfloor\frac{q}{4}\rfloor$ in $\mathbb{Z}$, return 1, otherwise return 0.

Obviously, this above scheme is an extension to the one in Sect. 5.3.2. By a similar analysis, one can show that the above scheme is correct and secure.

5.4 Functional Encryption

In 2011, Boneh, Sahai and Waters [5] introduced the notion of functional encryption as a generalization of many existing encryption schemes such as identity-based encryption (IBE) and attribute-based encryption (ABE). Unlike the IBE/ABE system, each user in a FE system holds a secret key sk_f corresponding to some function f, which can be used to obtain $f(m)$ from an encryption of m by running the decryption algorithm. A natural security goal is that any user who holding a secret key sk_f can only obtain the information of $f(m)$ (and the length of m) from an encryption of m, which can be formally captured by using a simulation-based security definition [5]. In this setting, a ciphertext policy ABE scheme can be seen as functional encryption with the function f is defined over a set of attributes A and a message m such that

$$f(A, m) = \begin{cases} (A, m), & \text{if } P(A) = 1 \text{ for some policy } P; \\ A, & \text{otherwise.} \end{cases}$$

In other words, the concatenation of the attributes set A and the message m that was used by the encryption algorithm of the CP-ABE scheme was treated as the message $m' = (A, m)$ of the functional encryption. The function f is defined to always output its first input A, since a CP-ABE scheme typically does not protect the privacy of the attribute set A used in the encryption algorithm.

Functional encryption is a very powerful primitive, which abstracts many other existing primitives, and has many promising applications. For example, if we encrypt a group photo which contains the faces of many persons, a user (e.g., the police) can be allowed to check whether a particular person is in the group without leaking the privacy of other persons in the group, by defining a function which searches the face of the targeted person in the group photo. However, a general function encryption seems to be too strong to be securely realized. Actually, Boneh et al. [5] showed that there exist some natural functionalities which cannot achieve the natural simulation-based security definition. For this reason, most functional encryption scheme only considered the game-based one which is a much weaker security notion but is achievable and useful in many applications. By relying on lattice assumptions, we already know how to construct FEs for a rich class of functions such as [9, 11], but most general FEs are only of theoretical interests.

5.5 Background and Further Reading

Sahai and Waters [23] introduced the notion of attribute-based encryption as an extension of identity-based encryption (IBE). Goyal et al. [13] further extended the idea of ABE and introduced two variants: key policy attribute-based encryption (KP-ABE) and ciphertext policy attribute-based encryption (CP-ABE). In a KP-ABE system, the ciphertext is associated with a set of descriptive attributes, while the private key of a party is associated with an access policy which is defined over a set of attributes and specifies which type of ciphertexts the key can decrypt. A CP-ABE system can be seen as a complementary form to KP-ABE system, where the private keys are associated with a set of attributes, while a policy defined over a set of attributes is attached to the ciphertext. A ciphertext can be decrypted by a party if the attributes associated with its private keys satisfy the ciphertext's policy.

Cheung and Newport [7] proposed the first CP-ABE system that supports and-gates, and proved its security under decision bilinear Diffie-Hellman (DBDH) assumption. Since then, there are many attribute-based encryptions that support various access structures. Such as and-gates schemes in [7, 8, 21], tree-based schemes in [12, 15, 19], and directly linear secret sharing scheme (LSSS) based constructions in [17, 18, 24]. Nishide et al. [21] first considered multi-valued attributes in their ABE scheme. Later, more works considered multi-valued attributes [8, 22]. Essentially, we can use a literal attribute-based system to obtain a multi-valued attribute-based system by simply defining each attributes' value as a new attribute (e.g., attribute "age=20") although this method might be very inefficient since for many known schemes, the number of system attributes determines the length of public key or ciphertext (e.g., [7, 12, 21, 24]).

In 2011, Zhang et al. [26] proposed a ciphertext policy attribute-based encryption from lattices, which supported somewhat limited access policies, say and-gates policies on positive and negative attributes. Their technique is basically adopted from [1]. Later, Agrawal et al. [3] constructed a functional encryption for inner product predicates from lattices. However, Katz et al. [16] pointed out that one can use inner products of length k vectors to evaluate degree d polynomials in t variables as long as $k > t^d$. Further, they also showed that how to use polynomial evaluation to evaluate conjunctions and disjunctions. Hence, essentially, one can obtain ciphertext policy attribute-based encryption that supports CNF or DNF formulae. Unfortunately, the inner product scheme with length k vectors has polynomial (in k) size public key [3]. So, if one wants to obtain a ciphertext policy attribute-based encryption by using the techniques in [16], the size of public key will be very large (as t and d become large), which might be unacceptable in practice. Since then, the past years have witnessed essential advances in constructing lattice-based ABEs [6, 10].

References

1. Agrawal, S., Boneh, D., Boyen, X.: Efficient lattice (H)IBE in the standard model. In: Gilbert, H. (ed.) Advances in Cryptology - EUROCRYPT 2010. Lecture Notes in Computer Science, vol. 6110, pp. 553–572. Springer, Heidelberg (2010)
2. Agrawal, S., Boyen, X., Vaikuntanathan, V., Voulgaris, P., Wee, H.: Functional encryption for threshold functions (or fuzzy IBE) from lattices. In: Fischlin, M., Buchmann, J., Manulis, M. (eds.) Public Key Cryptography - PKC 2012. Lecture Notes in Computer Science, vol. 7293, pp. 280–297. Springer, Heidelberg (2012)
3. Agrawal, S., Freeman, D., Vaikuntanathan, V.: Functional encryption for inner product predicates from learning with errors. In: Lee, D., Wang, X. (eds.) Advances in Cryptology - ASIACRYPT 2011. Lecture Notes in Computer Science, vol. 7073, pp. 21–40. Springer, Heidelberg (2011)
4. Ajtai, M.: Generating hard instances of lattice problems (extended abstract). In: Proceedings of the Twenty-Eighth Annual ACM Symposium on Theory of Computing, STOC '96, pp. 99–108. ACM (1996)
5. Boneh, D., Sahai, A., Waters, B.: Functional encryption: definitions and challenges. In: Ishai, Y. (ed.) Theory of Cryptography. Lecture Notes in Computer Science, vol. 6597, pp. 253–273. Springer, Berlin/Heidelberg (2011)

6. Boyen, X.: Attribute-based functional encryption on lattices. In: Sahai, A. (ed.) Theory of Cryptography. Lecture Notes in Computer Science, vol. 7785, pp. 122–142. Springer, Heidelberg (2013)
7. Cheung, L., Newport, C.: Provably secure ciphertext policy ABE. In: Proceedings of the 14th ACM Conference on Computer and Communications Security, CCS '07, pp. 456–465. ACM (2007)
8. Emura, K., Miyaji, A., Nomura, A., Omote, K., Soshi, M.: A ciphertext-policy attribute-based encryption scheme with constant ciphertext length. In: Bao, F., Li, H., Wang, G. (eds.) Information Security Practice and Experience. Lecture Notes in Computer Science, vol. 5451, pp. 13–23. Springer, Berlin/Heidelberg (2009)
9. Goldwasser, S., Kalai, Y., Popa, R.A., Vaikuntanathan, V., Zeldovich, N.: Reusable garbled circuits and succinct functional encryption. In: Proceedings of the Forty-fifth Annual ACM Symposium on Theory of Computing, STOC '13, pp. 555–564. ACM (2013)
10. Gorbunov, S., Vaikuntanathan, V., Wee, H.: Attribute-based encryption for circuits. J. ACM **62**(6), 45:1–45:33 (2015). https://doi.org/10.1145/2824233
11. Gorbunov, S., Vaikuntanathan, V., Wee, H.: Predicate encryption for circuits from LWE. In: Gennaro, R., Robshaw, M. (eds.) CRYPTO 2015, LNCS, vol. 9216, pp. 254–274. Springer (2015). https://doi.org/10.1007/978-3-662-48000-7_13
12. Goyal, V., Jain, A., Pandey, O., Sahai, A.: Bounded ciphertext policy attribute based encryption. In: Aceto, L., Damgård, I., Goldberg, L., Halldórsson, M., Ingólfsdóttir, A., Walukiewicz, I. (eds.) Automata, Languages and Programming, Lecture Notes in Computer Science, vol. 5126, pp. 579–591. Springer, Berlin/Heidelberg (2008)
13. Goyal, V., Pandey, O., Sahai, A., Waters, B.: Attribute-based encryption for fine-grained access control of encrypted data. In: Proceedings of the 13th ACM Conference on Computer and Communications Security, CCS '06, pp. 89–98. ACM (2006)
14. Herranz, J., Laguillaumie, F., Ràfols, C.: Constant size ciphertexts in threshold attribute-based encryption. In: Nguyen, P., Pointcheval, D. (eds.) Public Key Cryptography - PKC 2010. Lecture Notes in Computer Science, vol. 6056, pp. 19–34. Springer, Heidelberg (2010)
15. Ibraimi, L., Tang, Q., Hartel, P., Jonker, W.: Efficient and provable secure ciphertext-policy attribute-based encryption schemes. In: Bao, F., Li, H., Wang, G. (eds.) Information Security Practice and Experience. Lecture Notes in Computer Science, vol. 5451, pp. 1–12. Springer, Berlin/Heidelberg (2009)
16. Katz, J., Sahai, A., Waters, B.: Predicate encryption supporting disjunctions, polynomial equations, and inner products. In: Smart, N. (ed.) Advances in Cryptology - EUROCRYPT 2008. Lecture Notes in Computer Science, vol. 4965, pp. 146–162. Springer, Heidelberg (2008)
17. Lewko, A., Okamoto, T., Sahai, A., Takashima, K., Waters, B.: Fully secure functional encryption: attribute-based encryption and (hierarchical) inner product encryption. In: Gilbert, H. (ed.) Advances in Cryptology - EUROCRYPT 2010. Lecture Notes in Computer Science, vol. 6110, pp. 62–91. Springer, Berlin/Heidelberg (2010)
18. Lewko, A., Waters, B.: Decentralizing attribute-based encryption. In: Paterson, K. (ed.) Advances in Cryptology - EUROCRYPT 2011. Lecture Notes in Computer Science, vol. 6632, pp. 568–588. Springer, Berlin/Heidelberg (2011)
19. Liang, X., Cao, Z., Lin, H., Xing, D.: Provably secure and efficient bounded ciphertext policy attribute based encryption. In: Proceedings of the 4th International Symposium on Information, Computer, and Communications Security, ASIACCS '09, pp. 343–352. ACM (2009)
20. Micciancio, D., Regev, O.: Worst-case to average-case reductions based on gaussian measures. SIAM J. Comput. **37**, 267–302 (2007)
21. Nishide, T., Yoneyama, K., Ohta, K.: Attribute-based encryption with partially hidden encryptor-specified access structures. In: Bellovin, S., Gennaro, R., Keromytis, A., Yung, M. (eds.) Applied Cryptography and Network Security - ACNS 2008. Lecture Notes in Computer Science, vol. 5037, pp. 111–129. Springer, Berlin/Heidelberg (2008)
22. Okamoto, T., Takashima, K.: Fully secure functional encryption with general relations from the decisional linear assumption. In: Rabin, T. (ed.) Advances in Cryptology - CRYPTO 2010. Lecture Notes in Computer Science, vol. 6223, pp. 191–208. Springer, Heidelberg (2010)

23. Sahai, A., Waters, B.: Fuzzy identity-based encryption. In: Cramer, R. (ed.) Advances in Cryptology - EUROCRYPT 2005. Lecture Notes in Computer Science, vol. 3494, pp. 557–557. Springer, Berlin/Heidelberg (2005)
24. Waters, B.: Ciphertext-policy attribute-based encryption: an expressive, efficient, and provably secure realization. In: Catalano, D., Fazio, N., Gennaro, R., Nicolosi, A. (eds.) Public Key Cryptography - PKC 2011. Lecture Notes in Computer Science, vol. 6571, pp. 53–70. Springer, Berlin/Heidelberg (2011)
25. Zhang, J., Zhang, Z.: A ciphertext policy attribute-based encryption scheme without pairings. In: Wu, C.K., Yung, M., Lin, D. (eds.) Information Security and Cryptology, pp. 324–340. Springer, Berlin, Heidelberg (2012)
26. Zhang, J., Zhang, Z.: A ciphertext policy attribute-based encryption scheme without pairings. In: Wu, C.K., Yung, M., Lin, D. (eds.) Information Security and Cryptology - INSCRYPT 2011. Lecture Notes in Computer Science, pp. 324–340. Springer, Heidelberg (2012)
27. Zhang, J., Zhang, Z., Ge, A.: Ciphertext policy attribute-based encryption from lattices. In: Proceedings of the 7th ACM Symposium on Information, Computer and Communications Security, ASIACCS'12, pp. 16–17. Association for Computing Machinery, New York, NY, USA (2012)

Chapter 6
Key Exchanges

Abstract As discussed in Chap. 3, the use of public-key cryptography can avoid the problem of key distributions that is inherent in the symmetric-key cryptography. But most public-key primitives are typically much slower than symmetric-key ones, real applications with critical requirements on performance still heavily use symmetric-key primitives such AES even in more than 30 years after the introduction of public-key cryptography, and thus the key distribution problem may still survive in those applications. Key exchange (KE) is another fundamental cryptographic primitive for solving the key distribution problem, which directly allows users to securely generate a common secret key over an insecure network. In this chapter, we focus on constructing (two-party) KEs from lattices. Specifically, we will give a passively secure KE from the learning with errors problem. Then, we will show how to achieve authenticated key exchange (AKE) from lattices, assuming that each user has a certificate of a long-term public key. Finally, we will show how to construct AKE from lattices in the password-only setting, i.e., the two users who wants to establish a common session key only share a short low-entropy password.

6.1 Definition

Key exchange (KE) is a fundamental cryptographic primitive, allowing two parties to securely generate a common secret key over an insecure network. Because symmetric cryptographic tools (e.g., AES) are reliant on both parties having a shared key in order to securely transmit data, KE is one of the most used cryptographic tools in building secure communication protocols (e.g., SSL/TLS, IPSec, SSH). Following the introduction of the Diffie-Hellman (DH) protocol [41], cryptographers have devised a wide selection of KE protocols with various use-cases. One such class is authenticated key exchange (AKE), which enables each party to verify the other's identity so that an adversary cannot impersonate an honest party in the conversation.

AKEs with authenticated long-term public keys. For an AKE protocol, each party has a pair of *static keys*: a *static secret key* and a corresponding *static public key*. The

J. Zhang and Z. Zhang, *Lattice-Based Cryptosystems*,
https://doi.org/10.1007/978-981-15-8427-5_6

static public key is certified to belong to its owner using a public key or ID-based infrastructure. During an execution of the protocol, each party generates a pair of ephemeral keys—an *ephemeral secret key* and an *ephemeral public key*—and sends the *ephemeral public key* to the other party. Then, these keys are used along with the transcripts of the session to create a shared *session state*, which is then passed to a *key derivation function* to obtain a common session key. Intuitively, such a protocol is secure if no efficient adversary is able to extract any information about the session key from the publicly exchanged messages. More formally, Bellare and Rogaway [15] introduced an indistinguishability-based security model for AKE, the BR model, which captures key authentication such as *implicit mutual key authentication* and *confidentiality of agreed session keys*. The most prominent alternatives stem from Canetti and Krawczyk [31] and LaMacchia et al. [78], that also account for scenarios in which the adversary is able to obtain information about a static secret key or a session state other than the state of the target session. In practice, AKE protocols are usually required to have a property, perfect forward secrecy (PFS), that an adversary cannot compromise session keys after a completed session, even if it obtains the parties' static secret keys (e.g., via the Heartbleed attack[1]). As shown in [75], no two-pass *implicit* AKE protocol based on public key authentication can achieve PFS (but this may not be true for two-pass AKEs with *explicit* authentication [36]). Thus, the notion of weak PFS (wPFS) is usually considered for two-pass implicit AKE protocols, which states that the session key of an honestly run session remains private if the static keys are compromised after the session is finished [75].

We now recall the Bellare-Rogaway security model [15, 21], restricted to the case of two-pass AKE protocol.

Sessions. We fix a positive integer N to be the maximum number of honest parties that use the AKE protocol. Each party is uniquely identified by an integer i in $\{1, 2, \ldots, N\}$ and has a static key pair consisting of a static secret key sk_i and static public key pk_i, which is signed by a certificate authority (CA). A single run of the protocol is called a *session*. A session is activated at a party by an incoming message of the form (Π, I, i, j) or the form (Π, R, j, i, X_i), where Π is a protocol identifier; I and R are role identifiers; i and j are party identifiers. If user i receives a message of the form (Π, I, i, j), we say that i is the session initiator. User i then outputs the response X_i intended for user j. If user j receives a message of the form (Π, R, j, i, X_i), we say that j is the session responder; user j then outputs a response Y_j to user i. After exchanging these messages, both parties compute a session key.

If a session is activated at user i with i being the initiator, we associate with it a *session identifier* $\mathsf{sid} = (\Pi, I, i, j, X_i)$ or $\mathsf{sid} = (\Pi, I, i, j, X_i, Y_j)$. Similarly, if a session is activated at user j with j being the responder, the session identifier has the form $\mathsf{sid} = (\Pi, R, j, i, X_i, Y_j)$. For a session identifier $\mathsf{sid} = (\Pi, *, i, j, *[, *])$, the third coordinate—that is, the first party identifier—is called the owner of the session; the other party is called the peer of the session. A session is said to be *completed* when

---

[1] http://heartbleed.com/.

its owner computes a session key. The *matching session* of $\mathsf{sid} = (\Pi, I, i, j, X_i, Y_j)$ is the session with identifier $\widetilde{\mathsf{sid}} = (\Pi, R, j, i, X_i, Y_j)$ and vice versa.

Adversarial Capabilities. We model the adversary $\mathcal{A}$ as a probabilistic polynomial time (PPT) Turing machine with full control over all communication channels between parties, including control over session activations. In particular, $\mathcal{A}$ can intercept all messages, read them all and remove or modify any desired messages as well as inject its own messages. We also suppose $\mathcal{A}$ is capable of obtaining hidden information about the parties, including static secret keys and session keys to model potential leakage of them in genuine protocol executions. These abilities are formalized by providing $\mathcal{A}$ with the following oracles (we split the Send query as in [31] into Send_0, Send_1 and Send_2 queries for the case of two-pass protocols):

- $\mathsf{Send}_0(\Pi, I, i, j)$: $\mathcal{A}$ activates user i as an initiator. The oracle returns a message X_i intended for user j.
- $\mathsf{Send}_1(\Pi, R, j, i, X_i)$: $\mathcal{A}$ activates user j as a responder using message X_i. The oracle returns a message Y_j intended for user i.
- $\mathsf{Send}_2(\Pi, R, i, j, X_i, Y_j)$: $\mathcal{A}$ sends user i the message Y_j to complete a session previously activated with a $\mathsf{Send}_0(\Pi, I, i, j)$ query that returned X_i.
- $\mathsf{SessionKeyReveal}(\mathsf{sid})$: The oracle returns the session key associated with the session sid if it has been generated.
- $\mathsf{Corrupt}(i)$: The oracle returns the static secret key belonging to user i. A party whose key is given to $\mathcal{A}$ in this way is called *dishonest*; a party not compromised in this way is called *honest*.
- $\mathsf{Test}(\mathsf{sid}^*)$: The oracle chooses a bit $b \xleftarrow{\$} \{0, 1\}$. If $b = 0$, it returns a key chosen uniformly at random; if $b = 1$, it returns the session key associated with sid^*. Note that we impose some restrictions on this query. We only allow $\mathcal{A}$ to query this oracle once, and only on a fresh (see Definition 6.1) session sid^*.

Definition 6.1 (*Session Freshness for AKEs*) Let $\mathsf{sid}^* = (\Pi, I, i^*, j^*, X_i, Y_j)$ or $(\Pi, R, j^*, i^*, X_i, Y_j)$ be a completed session with initiator user i^* and responder user j^*. If the matching session exists, denote it $\widetilde{\mathsf{sid}}^*$. We say that sid^* is *fresh* if the following conditions hold:

- $\mathcal{A}$ has not made a $\mathsf{SessionKeyReveal}$ query on sid^*.
- $\mathcal{A}$ has not made a $\mathsf{SessionKeyReveal}$ query on $\widetilde{\mathsf{sid}}^*$ (if it exists).
- Neither user i^* nor j^* is dishonest if $\widetilde{\mathsf{sid}}^*$ does not exist. That is, $\mathcal{A}$ has not made a $\mathsf{Corrupt}$ query on either of them.

Recall that in the original BR model [15], no corruption query is allowed. In the above freshness definition, we allow the adversary to corrupt both parties of sid^* if the matching session exists, i.e., the adversary can obtain the parties' secret key in advance and then passively eavesdrops the session sid^* (and thus $\widetilde{\mathsf{sid}}^*$). We remark that this seems to be stronger than what is needed for capturing wPFS [75], where the adversary is only allowed to corrupt a party after an honest session sid^* (and thus $\widetilde{\mathsf{sid}}^*$) has been completed.

Security Game. The security of a two-pass AKE protocol is defined in terms of the following game. The adversary $\mathcal{A}$ makes any sequence of queries to the oracles above, so long as only one Test query is made on a fresh session, as mentioned above. The game ends when $\mathcal{A}$ outputs a guess b' for b. We say $\mathcal{A}$ wins the game if its guess is correct so that $b' = b$. The advantage of $\mathcal{A}$, $\mathbf{Adv}_{\Pi,\mathcal{A}}$, is defined as $|\Pr[b' = b] - 1/2|$.

Definition 6.2 (*Security of AKEs*) We say that an AKE protocol Π is secure if the following conditions hold:

- If two honest parties complete matching sessions then they compute the same session key with overwhelming probability.
- For any PPT adversary $\mathcal{A}$, the advantage $\mathbf{Adv}_{\Pi,\mathcal{A}}$ is negligible.

AKEs with low-entropy passwords. After decades of development, the community has witnessed great success in designing AKEs with long-term public keys, even in the setting of lattices [6, 92, 107]. However, people rarely make full use of the large character set in forming passwords and many tend to pick easily memorizable ones from a relatively small dictionary. AKEs based on high-entropy cryptographic keys usually do not apply to the case where only low-entropy passwords are available. Indeed, as shown in [58, 70], it can be trivially insecure to use a low-entropy password as a cryptographic key.

Informally, a secure password-based AKE (PAKE) should resist *off-line dictionary attacks* in which the adversary tries to determine the correct password using only information obtained during previous protocol executions, and limit the adversary to the trivial *on-line attacks* where the adversaries simply run the protocol with honest users using (a bounded number of) password trials. Formal security models for PAKE were developed in [13, 24]. Later, many provably secure PAKE protocols based on various hardness assumptions were proposed, where the research mainly falls into two lines: the first line starts from the work of Bellovin and Merritt [17], followed by plenty of excellent work focusing on PAKE in the random oracle/ideal cipher models and aiming at achieving the highest possible levels of performance [13, 24, 25, 84]; the second line dates back to the work of Katz, Ostrovsky and Yung [70], from which Gennaro and Lindell [53] abstracted out a generic PAKE framework (in the CRS model) based on smooth projective hash (SPH) functions [35]. This line of research devoted to seeking more efficient PAKE in the standard model [2, 3, 18, 30, 62, 72].

We recall the security model for password-based authenticated key exchange (PAKE) in [13, 70, 72]. Formally, the protocol relies on a setup assumption that a common reference string (CRS) and other public parameters are established (possibly by a trusted third party) before any execution of the protocol. Let $\mathcal{U}$ be the set of protocol users. For every distinct $A, B \in \mathcal{U}$, users A and B share a password $pw_{A,B}$. We assume that each $pw_{A,B}$ is chosen independently and uniformly from some dictionary set $\mathcal{D}$ for simplicity. Each user $A \in \mathcal{U}$ is allowed to execute the protocol multiple times with different partners, which is modeled by allowing A to have an unlimited number of *instances* with which to execute the protocol. Denote instance i of A as Π_A^i. An instance is for one-time use only and it is associated with the following variables that are initialized to $\perp$ or 0:

- sid_A^i, pid_A^i and sk_A^i denote the *session id, parter id*, and *session key*, for instance, Π_A^i. The session *id* consists of the (ordered) concatenation of all messages sent and received by Π_A^i; while the partner *id* specifies the user with whom Π_A^i believes it is interacting;
- acc_A^i and term_A^i are boolean variables denoting whether instance Π_A^i has accepted or terminated, respectively.

For any user $A, B \in \mathcal{U}$, instances Π_A^i and Π_B^j are *partnered* if $\mathsf{sid}_A^i = \mathsf{sid}_B^j \neq \perp$, $\mathsf{pid}_A^i = B$ and $\mathsf{pid}_B^j = A$. We say that a PAKE protocol is *correct* if instances Π_A^i and Π_B^j are partnered, then we have that $\mathrm{acc}_A^i = \mathrm{acc}_B^j = 1$ and $\mathsf{sk}_A^i = \mathsf{sk}_B^j \neq \perp$ hold (with overwhelming probability).
Adversarial abilities. The adversary $\mathcal{A}$ is a probabilistic polynomial time (PPT) algorithm with full control over all communication channels between users. In particular, $\mathcal{A}$ can intercept all messages, read them all, and remove or modify any desired messages as well as inject its own messages. $\mathcal{A}$ is also allowed to obtain the session key of an instance, which models possible leakage of session keys. These abilities are formalized by allowing the adversary to interact with the various instances via access to the following oracles:

- $\mathsf{Send}(A, i, \mathsf{msg})$: This sends message msg to instance Π_A^i. After receiving msg, instance Π_A^i runs according to the protocol specification and updates its states as appropriate. Finally, this oracle returns the message output by Π_A^i to the adversary. We stress that the adversary can prompt an unused instance Π_A^i to execute the protocol with partner B by querying $\mathsf{Send}(A, i, B)$, and obtain the first protocol message output by Π_A^i.
- $\mathsf{Execute}(A, i, B, j)$: If both instances Π_A^i and Π_B^j have not yet been used, this oracle executes the protocol between Π_A^i and Π_B^j, updates their states as appropriate and returns the transcript of this execution to the adversary.
- $\mathsf{Reveal}(A, i)$: This oracle returns the session key sk_A^i to the adversary if it has been generated (i.e., $\mathsf{sk}_A^i \neq \perp$).
- $\mathsf{Test}(A, i)$: This oracle chooses a random bit $b \xleftarrow{\$} \{0, 1\}$. If $b = 0$, it returns a key chosen uniformly at random; if $b = 1$, it returns the session key sk_A^i of instance Π_A^i. The adversary is only allowed to query this oracle once.

Definition 6.3 (*Session Freshness for PAKEs*) We say that an instance Π_A^i is fresh if the following conditions hold:

- the adversary $\mathcal{A}$ did not make a $\mathsf{Reveal}(A, i)$ query to instance Π_A^i;
- the adversary $\mathcal{A}$ did not make a $\mathsf{Reveal}(B, j)$ query to instance Π_B^j, where instances Π_A^i and Π_B^j are partnered;

Security Game. The security of a PAKE protocol is defined via the following game. The adversary $\mathcal{A}$ makes any sequence of queries to the oracles above, so long as only one $\mathsf{Test}(A, i)$ query is made to a fresh instance Π_A^i, with $\mathrm{acc}_A^i = 1$ at the time of this query. The game ends when $\mathcal{A}$ outputs a guess b' for b. We say $\mathcal{A}$ wins the game if its guess is correct, so that $b' = b$. The advantage $\mathbf{Adv}_{\Pi,\mathcal{A}}$ of adversary $\mathcal{A}$ in attacking a PAKE protocol Π is defined as $|2 \cdot \Pr[b' = b] - 1|$.

We say that an *on-line attack* happens when the adversary makes one of the following queries to some instance Π_A^i: $\mathsf{Send}(A, i, *)$, $\mathsf{Reveal}(A, i)$ or $\mathsf{Test}(A, i)$. In particular, the $\mathsf{Execute}$ queries are not counted as on-line attacks. Since the size of the password dictionary is small, a PPT adversary can always win by trying all password one-by-one in an on-line attack. The number $Q(\kappa)$ of on-line attacks represents a bound on the number of passwords the adversary could have tested in an on-line fashion. Informally, a PAKE protocol is secure if on-line password guessing attacks are already the best strategy (for all PPT adversaries).

Definition 6.4 (*Security of PAKEs*) We say that a PAKE protocol Π is secure if for all dictionary $\mathcal{D}$ and for all PPT adversaries $\mathcal{A}$ making at most $Q(\kappa)$ on-line attacks, it holds that $\mathbf{Adv}_{\Pi,\mathcal{A}}(\kappa) \leq Q(\kappa)/|\mathcal{D}| + \mathsf{negl}(\kappa)$.

6.2 Key Exchange Against Passive Attacks

Let κ be the security parameter, and let PRF be a secure pseudorandom function with key space Ω. Let $\Pi_{\text{pke}} = (\mathsf{KeyGen}, \mathsf{Enc}, \mathsf{Dec})$ be a CPA-secure PKE scheme with message space $\mathcal{P}$ of size $|\mathcal{P}| > 2^{\omega(\log \kappa)}$. Without loss of generality, we assume that $\Omega = \mathcal{P}$ (otherwise, one can use a hash function mapping from $\mathcal{P}$ to Ω). Then, two users i and j who want to establish an ephemeral session key over the Internet can execute the protocol illustrated in Fig. 6.1. Basically, user i will first generate a pair of ephemeral public key and secret key, and send the public key pk to user j. Upon receiving the public key from user i, user j randomly chooses a message $m \xleftarrow{\$} \mathcal{P}$, sends an encryption c of m to user i and computes his own session key $k_j = \mathsf{PRF}(m; \mathsf{pk}, c)$ by using m as the PRF key. Upon receiving the ciphertext c, user i first decrypts the ciphertext $m' \leftarrow \Pi_{\text{pke}}.\mathsf{Dec}(\mathsf{sk}, c)$ and computes the session key $k_i = \mathsf{PRF}(m'; \mathsf{pk}, c)$. Clearly, if the encryption algorithm is correct, then both users will compute the same session key (i.e., $k_i = k_j$). Moreover, given the transcript (pk, c) no probabilistic polynomial time adversary can distinguish the session key $k_i = k_j$ from random by the CPA-security of the encryption scheme and the indistinguishabilty of the PRF.

The protocol given in Fig. 6.1 is a generic construction of two-pass KE from any PKE, which can be further improved when instantiated with concrete PKE schemes. In the lattice setting, Ding et al. [42] present a simple construction of a two-pass KE from (ring-)LWE assumptions by directly using a mechanism to compute a shared session from two approximately equal values, which avoids the use of a full PKE scheme to encrypting a random message as in Fig. 6.1. We first define two functions Cha and Mod_2 which will be used in the following sections.

For an odd prime $q > 2$, take $\mathbb{Z}_q = \{-\frac{q-1}{2}, \ldots, \frac{q-1}{2}\}$ and define the subset $E := \{-\lfloor \frac{q}{4} \rfloor, \ldots, \lfloor \frac{q}{4} \rceil\}$ as the middle half of $\mathbb{Z}_q$. We also define Cha to be

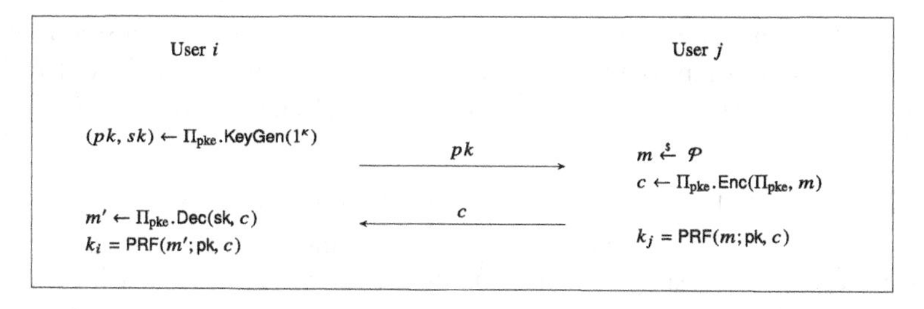

Fig. 6.1 A two-pass KE from PKE

the characteristic function of *the complement of* E, so $\mathsf{Cha}(v) = 0$ if $v \in E$ and 1 otherwise. Obviously, for any v in $\mathbb{Z}_q$, $v + \mathsf{Cha}(v) \cdot \frac{q-1}{2} \bmod q$ belongs to E. We define an auxiliary modular function, $\mathsf{Mod}_2 \colon \mathbb{Z}_q \times \{0, 1\} \to \{0, 1\}$ as $\mathsf{Mod}_2(v, b) = (v + b \cdot \frac{q-1}{2}) \bmod q \bmod 2$.

In the following lemma, we show that given the bit $b = \mathsf{Cha}(v)$, and a value $w = v + 2e$ with sufficiently small e, one can recover $\mathsf{Mod}_2(v, \mathsf{Cha}(v))$. In particular, we have $\mathsf{Mod}_2(v, b) = \mathsf{Mod}_2(w, b)$.

Lemma 6.1 *Let q be an odd prime, $v \in \mathbb{Z}_q$ and $e \in \mathbb{Z}_q$ such that $|e| < q/8$. Then, for $w = v + 2e$, we have $\mathsf{Mod}_2(v, \mathsf{Cha}(v)) = \mathsf{Mod}_2(w, \mathsf{Cha}(v))$.*

Proof Note that $w + \mathsf{Cha}(v)\frac{q-1}{2} \bmod q = v + \mathsf{Cha}(v)\frac{q-1}{2} + 2e \bmod q$. Now, $v + \mathsf{Cha}(v)\frac{q-1}{2} \bmod q$ is in E as we stated above; that is, $-\lfloor \frac{q}{4} \rfloor \leq v + \mathsf{Cha}(v)\frac{q-1}{2} \bmod q \leq \lfloor \frac{q}{4} \rceil$. Thus, since $-q/8 < e < q/8$, we have $-\lfloor \frac{q}{2} \rfloor \leq v + \mathsf{Cha}(v)\frac{q-1}{2} \bmod q + 2e \leq \lfloor \frac{q}{2} \rfloor$. Therefore, we have $v + \mathsf{Cha}(v)\frac{q-1}{2} \bmod q + 2e = v + \mathsf{Cha}(v)\frac{q-1}{2} + 2e \bmod q = w + \mathsf{Cha}(v)\frac{q-1}{2} \bmod q$. Thus, $\mathsf{Mod}_2(w, \mathsf{Cha}(v)) = \mathsf{Mod}_2(v, \mathsf{Cha}(v))$. □

Now, we extend the two functions Cha and Mod_2 to vectors over $\mathbb{Z}_q^n$ by applying them coefficient-wise to ring elements (Similarly, one can extend them to ring $R_q = \mathbb{Z}_q[X]/(X^n + 1)$ by identifying each element in R_q by its coefficient vector). Namely, for vector $\mathbf{v} = (v_0, \ldots, v_{n-1})^T \in \mathbb{Z}_q^n$ and binary-vector $\mathbf{b} = (b_0, \ldots, b_{n-1}) \in \{0, 1\}^n$, define $\widetilde{\mathsf{Cha}}(\mathbf{v}) = (\mathsf{Cha}(v_0), \ldots, \mathsf{Cha}(v_{n-1}))$ and $\widetilde{\mathsf{Mod}_2}(\mathbf{v}, \mathbf{b}) = (\mathsf{Mod}_2(v_0, b_0), \ldots, \mathsf{Mod}_2(v_{n-1}, b_{n-1}))$. For simplicity, we slightly abuse the notations and still use Cha (resp. Mod_2) to denote $\widetilde{\mathsf{Cha}}$ (resp. $\widetilde{\mathsf{Mod}_2}$). Clearly, the result in Lemma 6.1 still holds when extending to $\mathbb{Z}_q^n$. We also have the following lemma:

Lemma 6.2 *Let q be any odd prime and R_q be the ring defined above. Then, for any $\mathbf{b} \in \{0, 1\}^n$ and any $v' \in R_q$, the output distribution of $\mathsf{Mod}_2(v + v', \mathbf{b})$ given $\mathsf{Cha}(v)$ has min-entropy at least $-n \log(\frac{1}{2} + \frac{1}{|E|-1})$, where v is uniformly chosen from R_q at random. In particular, when $q > 203$, we have $-n \log(\frac{1}{2} + \frac{1}{|E|-1}) > 0.97n$.*

Proof Since each coefficient of v is independently and uniformly chosen from $\mathbb{Z}_q$ at random, we can simplify the proof by focusing on the first coefficient of v. Formally, letting $v = (v_0, \ldots, v_{n-1})$, $v' = (v'_0, \ldots, v'_{n-1})$ and $\mathbf{b} = (b_0, \ldots, b_{n-1})$, we condition on $\mathsf{Cha}(v_0)$:

- If $\mathsf{Cha}(v_0) = 0$, then $v_0 + v'_0 + b_0 \cdot \frac{q-1}{2}$ is uniformly distributed over $v'_0 + b_0 \cdot \frac{q-1}{2} + E \bmod q$. This shifted set has $(q + 1)/2$ elements, which are either consecutive integers—if the shift is small enough—or two sets of consecutive integers—if the shift is large enough to cause wrap-around. Thus, we must distinguish a few cases:
  - If $|E|$ is even and no wrap-around occurs, then the result of $\mathsf{Mod}_2(v_0 + v'_0, b_0)$ is clearly uniform on $\{0, 1\}$. Hence, the result of $\mathsf{Mod}_2(v_0 + v'_0, b_0)$ has no bias.
  - If $|E|$ is odd and no wrap-around occurs, then the result of $\mathsf{Mod}_2(v_0 + v'_0, b_0)$ has a bias $\frac{1}{2|E|}$ over $\{0, 1\}$. In other words, the $\mathsf{Mod}_2(v_0 + v'_0, b_0)$ will output either 0 or 1 with probability exactly $\frac{1}{2} + \frac{1}{2|E|}$.
  - If $|E|$ is odd and wrap-around does occur, then the set $v'_0 + b_0 \cdot \frac{q-1}{2} + E \bmod q$ splits into two parts, one with an even number of elements, and one with an odd number of elements. This leads to the same situation as with no wrap-around.
  - If $|E|$ is even and wrap-around occurs, then the sample space is split into either two even-sized sets, or two odd sized sets. If both are even, then once again the result of $\mathsf{Mod}_2(v_0 + v'_0, b_0)$ is uniform. If both are odd, it is easy to calculate that the result of $\mathsf{Mod}_2(v_0 + v'_0, b_0)$ has a bias with probability $\frac{1}{|E|}$ over $\{0, 1\}$.
- If $\mathsf{Cha}(v_0) = 1$, $v_0 + v'_0 + b_0 \cdot \frac{q-1}{2}$ is uniformly distributed over $v'_0 + b_0 \cdot \frac{q-1}{2} + \tilde{E}$, where $\tilde{E} = \mathbb{Z}_q \setminus E$. Now $|\tilde{E}| = |E| - 1$, so by splitting into the same cases as $\mathsf{Cha}(v_0) = 0$, the result of $\mathsf{Mod}_2(v_0 + v'_0, b)$ has a bias with probability $\frac{1}{|E|-1}$ over $\{0, 1\}$.

In all, we have that the result of $\mathsf{Mod}_2(v_0 + v'_0, b_0)$ conditioned on $\mathsf{Cha}(v_0)$ has min-entropy at least $-\log(\frac{1}{2} + \frac{1}{|E|-1})$. Since the bits in the result of $\mathsf{Mod}_2(v + v', \mathbf{b})$ are independent, we have that given $\mathsf{Cha}(v)$, the min-entropy $H_\infty(\mathsf{Mod}_2(v + v', \mathbf{b})) \geq -n \log(\frac{1}{2} + \frac{1}{|E|-1})$. This completes the first claim. The second claim directly follows from the fact that $-\log(\frac{1}{2} + \frac{1}{|E|-1}) > -\log(0.51) > 0.97$ when $q > 203$. □

6.2.1 Ding et al.'s KE Protocol

In this section, we describe (a variant of) the Ding et al.'s KE protocol, whose security is based on the hardness of the ring-LWE problem. Formally, let κ be the security parameter. Let n, q be positively integers, and let $\alpha > 0$ be a real. Let χ_α be the Gaussian distribution $D_{R,\alpha}$ (i.e., each coefficient of the ring element $x \xleftarrow{\$} D_{R,\alpha}$ is distributed as $D_{\mathbb{Z},\alpha}$). Let $H : \{0, 1\}^* \to \{0, 1\}^\kappa$ be a hash function. Let $R_q =$

$\mathbb{Z}_q[X]/(X^n + 1)$ be the ring of polynomials with degree at most $n - 1$. Let $a \in R_q$ be the common reference string (CRS) of the system, which is uniformly chosen at random and is publicly known to all users. Two users i and j who want to establish a session can do the following computations:

Initiation User i proceeds as follows:

1. Sample $r_i, e_i \xleftarrow{\$} \chi_\alpha$ and compute $x_i = ar_i + 2e_i$;
2. Send x_i to user j.

Response After receiving x_i from user i, user j proceeds as follows:

1′. Sample $r_j, e_j \xleftarrow{\$} \chi_\alpha$ and compute $y_j = ar_j + 2e_j$;
2′. Sample $f_j \xleftarrow{\$} \chi_\alpha$, compute $k_j = x_i r_j + 2f_j$;
3′. Compute $w_j = \mathsf{Cha}(k_j) \in \{0, 1\}^n$ and send (y_j, w_j) to user i;
4′. Compute $\sigma_j = \mathsf{Mod}_2(k_j, w_j)$ and derive the session key $\mathsf{ssk}_j = H(i, j, x_i, y_j, w_j, \sigma_j)$.

Finish User i receives the pair (y_j, w_j) from user j, and proceeds as follows:

3. Sample $f_i \xleftarrow{\$} \chi_\alpha$ and compute $k_i = y_j r_i + 2f_i$;
4. Compute $\sigma_i = \mathsf{Mod}_2(k_i, w_j)$ and derive the session key $\mathsf{ssk}_i = H(i, j, x_i, y_j, w_j, \sigma_i)$.

By definition, we have that

$$k_i = y_j r_i + 2f_i = (ar_j + 2e_j)r_i + 2f_i = ar_i r_j + 2(r_i e_j + f_i),$$
$$k_j = x_i r_j + 2f_j = (ar_i + 2e_i)r_j + 2f_j = ar_i r_j + 2(r_j e_i + f_j).$$

Thus, if we set the parameters such that $\|2(r_i e_j + f_i - r_j e_i - f_j)\|_\infty < q/8$, then we can have that $Mod(k_i, w_j) = \mathsf{Mod}_2(k_j, w_j)$, where $w_j = \mathsf{Cha}(k_j)$. This means that for appropriate choice of parameters, both users can compute the same session key, i.e., $\mathsf{ssk}_i = \mathsf{ssk}_j$ (Fig. 6.2).

For security, we need q is odd. In this setting, we have that σ_i (resp., σ_j) is not uniformly distributed over $\{0, 1\}^n$ even if k_i (resp., k_j) is uniformly distributed over R_q. In the original protocol of Ding et al. [42], they used a relatively complex method

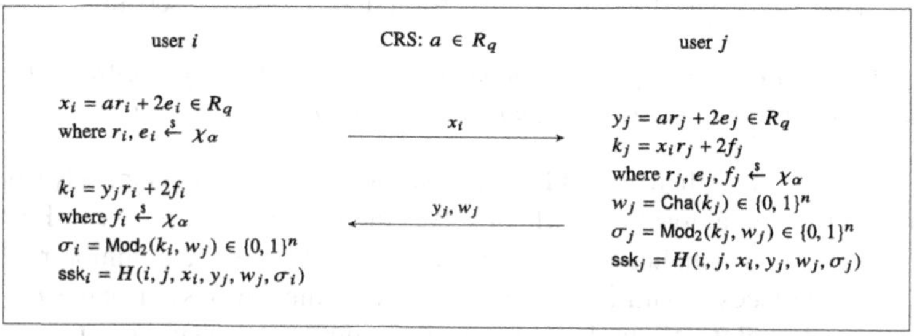

user i	CRS: $a \in R_q$	user j
$x_i = ar_i + 2e_i \in R_q$ where $r_i, e_i \xleftarrow{\$} \chi_\alpha$	x_i →	$y_j = ar_j + 2e_j \in R_q$ $k_j = x_i r_j + 2f_j$ where $r_j, e_j, f_j \xleftarrow{\$} \chi_\alpha$ $w_j = \mathsf{Cha}(k_j) \in \{0, 1\}^n$ $\sigma_j = \mathsf{Mod}_2(k_j, w_j) \in \{0, 1\}^n$ $\mathsf{ssk}_j = H(i, j, x_i, y_j, w_j, \sigma_j)$
$k_i = y_j r_i + 2f_i$ where $f_i \xleftarrow{\$} \chi_\alpha$ $\sigma_i = \mathsf{Mod}_2(k_i, w_j) \in \{0, 1\}^n$ $\mathsf{ssk}_i = H(i, j, x_i, y_j, w_j, \sigma_i)$	← y_j, w_j	

Fig. 6.2 Ding's KE from LWE

to remove the bias of σ_i (resp., σ_j) such that one can directly use σ_i and σ_j as the final session key. Here, we simply use a hash function as a randomness extractor to derive the session key. It is known that randomness extractors can be used to obtain an almost uniformly distributed key from a biased bit-string with high min-entropy [9, 34, 43, 100, 101]. In practice, as recommended by NIST [11], one can actually use the standard cryptographic hash functions such as SHA-2 to derive a uniformly distributed key if the source string has at least 2κ min-entropy, where κ is the length of the cryptographic hash function. Thus, it is enough to set n, q such that $q > 207$ and $0.97n > 2\kappa$ by Lemma 6.2.

6.2.2 Security

In this section, we show that the above KE protocol is passively secure under the ring-LWE assumption, namely, no PPT adversary given the transcript of the protocol can distinguish the session key from uniform with non-negligible advantage.

Theorem 6.1 *If the* $\text{RLWE}_{n,2,q,\alpha}$ *problem is hard, and* $H : \{0, 1\}^* \to \kappa$ *is a randomness extractor, then under approximate choice of parameters the above protocol is passively secure.*

Proof In the following, we prove the theorem by using a sequence of games from Game G_0 to Game G_3, where the session key in Game G_0 is honestly generated, while the session key in Game G_3 is uniformly random. The security established by each two consecutive games are computationally indistinguishable.
Game G_0. This game is essentially the same as the real protocol, which honestly executes the computations in the Initiation, Response and Finish phases.
Game G_1. This game is identical to Game G_0 except that we modify the Initiation and Finish phases as follows:

Initiation User i proceeds as follows:

1. Uniformly chosen $x_i \xleftarrow{\$} R_q$ at random;
2. Send x_i to user j.

Finish User i receives the pair (y_j, w_j) from user j, directly sets $\mathsf{ssk}_i = \mathsf{ssk}_j$.

Lemma 6.3 *If the* $\text{RLWE}_{n,1,q,\alpha}$ *problem is hard, then under appropriate choice of parameters Games* G_0 *and* G_1 *are computationally indistinguishable.*

Proof We show that if there is an adversary $\mathcal{A}$ that can distinguish Game G_1 from G_0, we can construct an algorithm $\mathcal{B}$ which solves the $\text{RLWE}_{n,1,q,\alpha}$ problem. Formally, given a $\text{RLWE}_{n,1,q,\alpha}$ challenge tuple $(\hat{a}, \hat{x})$, $\mathcal{B}$ directly sets the common reference string $a = \hat{a}$, replaces $\mathbf{x}_i$ with $\hat{x}$, and performs other computations as in game G_1. Note that if $(\hat{a}, \hat{x})$ is a real RLWE tuple, then the game simulated by $\mathcal{B}$ is almost identical to game G_0 (note that by the correctness we have that $\sigma_i = \sigma_j$ holds under appropriate choice of parameters by Lemma 6.1, which implies that $\mathsf{ssk}_i = \mathsf{ssk}_j$). Otherwise, if $(\hat{a}, \hat{x})$ is uniformly random over $R_q \times R_q$, then game simulated by $\mathcal{B}$ is identical to game G_1. This means that if the adversary $\mathcal{A}$ given $(x_i, y_j, w_j, \mathsf{ssk}_i, \mathsf{ssk}_j)$ as inputs can distinguish Game G_1 from G_0, then $\mathcal{B}$ can solve the $\mathrm{RLWE}_{n,1,q,\alpha}$ problem, which completes the proof. □

Game G_2. This game is identical to Game G_1 except that we modify the Response phase as follows:

Response After receiving x_i from user i, user j proceeds as follows:

1′. Randomly chooses $y_j, k_j \xleftarrow{\$} R_q$;
2′. Compute $w_j = \mathsf{Cha}(k_j) \in \{0, 1\}^n$ and send (y_j, w_j) to user i;
3′. Compute $\sigma_j = \mathsf{Mod}_2(k_j, w_j)$ and derive the session key $\mathsf{ssk}_j = H(i, j, x_i, y_j, w_j, \sigma_j)$.

Lemma 6.4 *If the* $\mathrm{RLWE}_{n,2,q,\alpha}$ *problem is hard, then under appropriate choice of parameters Games* G_1 *and* G_2 *are computationally indistinguishable.*

Proof We show that if there is an adversary $\mathcal{A}$ that can distinguish Game G_2 from G_1, we can construct an algorithm $\mathcal{B}$ which solves the $\mathrm{RLWE}_{n,2,q,\alpha}$ problem. Formally, given a $\mathrm{RLWE}_{n,2,q,\alpha}$ challenge tuples $(\hat{a}, \hat{y})$ and $(\hat{x}, \hat{k})$, $\mathcal{B}$ directly sets the common reference string $a = \hat{a}$, replaces (x_i, y_j, k_j) with $(\hat{x}, \hat{y}, \hat{k})$ and performs other operations as in game G_2. Note that if $(\hat{a}, \hat{y})$ and $(\hat{x}, \hat{k})$ are two real RLWE tuples, then the game simulated by $\mathcal{B}$ is identical to game G_1. Otherwise, if $(\hat{a}, \hat{x})$ is uniformly random over $R_q \times R_q$, then game simulated by $\mathcal{B}$ is identical to game G_2. This means that if the adversary $\mathcal{A}$ given $(x_i, y_j, w_j, \mathsf{ssk}_i, \mathsf{ssk}_j)$ as inputs can distinguish Game G_2 from G_1, then $\mathcal{B}$ can solve the $\mathrm{RLWE}_{n,2,q,\alpha}$ problem, which completes the proof. □

Game G_3. This game is identical to Game G_2 except that we modify the Response phase as follows:

Response After receiving x_i from user i, user j directly chooses the session key $\mathsf{ssk}_j \xleftarrow{\$} \{0, 1\}^\kappa$ at random.

Lemma 6.5 *If* $H : \{0, 1\}^* \to \{0, 1\}^n$ *is a randomness extractor, then under appropriate choice of parameters games* G_2 *and* G_3 *are statistically indistinguishable.*

Proof Note that the session key in game G_2 derived by using σ_j as input. If we set the parameters such that $q > 203$ and $0.97n > 2\kappa$, we can have that in game G_2 σ_j (conditioned on x_i, y_j, w_j) is statistically close to uniform by Lemma 6.2. This shows that games G_3 and G_2 are statistically indistinguishable. □

Since the session key in game G_3 is essentially uniformly chosen at random, by Lemmas 6.3–6.5, we have that given the transcript (x_i, y_j, w_j), the adversary cannot distinguish the session key $\mathsf{ssk}_i = \mathsf{ssk}_j$ from uniform. □

6.3 Implicit Authenticated Key Exchange

In this section, we present an implicit authenticated key exchange from the ring-LWE problem. As before, the techniques used in this section can be easily extended to the standard LWE problem.

6.3.1 Design Rational

At a high level, the AKE protocol in this section is inspired by HMQV [75], which makes the protocol share some similarities to HMQV. However, there are also many differences between the protocol and HMQV due to the different underlying algebraic structures. To better illustrate the similarities and differences between the AKE protocol and HMQV, we first briefly recall the HMQV protocol [75]. Let $\mathbb{G}$ be a cyclic group with generator $g \in \mathbb{G}$. Let $(P_i = g^{s_i}, s_i)$ and $(P_j = g^{s_j}, s_j)$ be the static public/secret key pairs of party i and user j, respectively. During the protocol, both parties exchange ephemeral public keys, i.e., user i sends $X_i = g^{r_i}$ to user j, and user j sends $Y_j = g^{r_j}$ to user i. Then, both parties compute the same key material $k_i = (P_j^d Y_j)^{s_i c + r_i} = g^{(s_i c + r_i)(s_j d + r_j)} = (P_i^c X_i)^{s_j d + r_j} = k_j$ where $c = H_1(j, X)$ and $d = H_1(i, Y)$ are computed by using a function H_1 and use it as input of a key derivation function H_2 to generate a common session key, i.e., $\mathsf{ssk}_i = H_2(k_i) = H_2(k_j) = \mathsf{ssk}_j$.

As mentioned above, HMQV has many nice properties such as only two-pass messages, implicit key authentication, high efficiency, and without using any explicit entity authentication techniques (e.g., signatures). The main goal is to construct a lattice-based counterpart such that it not only enjoys all those nice properties of HMQV, but also belongs to post-quantum cryptography, i.e., the underlying hardness assumption is believed to hold even against quantum computer. However, such a task is highly non-trivial since the success of HMQV greatly relies on the nice properties of cyclic groups such as commutativity (i.e., $(g^a)^b = (g^b)^a$) and perfect (and public) randomization (i.e. g^a can be perfectly randomized by computing $g^a g^r$ with a uniformly chosen r at random).

Fortunately, as noticed in [23, 42, 92], the Ring-LWE problem supports some kind of "approximate" commutativity and can be used to build a passive-secure key exchange protocol. Specifically, let R_q be a ring, and χ be a Gaussian distribution over R_q. Then, given two Ring-LWE tuples with both secret and errors chosen from χ, e.g., $(a, b_1 = as_1 + e_1)$ and $(a, b_2 = as_2 + e_2)$ for randomly chosen $a \xleftarrow{\$} R_q, s_1, s_2, e_1, e_2 \xleftarrow{\$} \chi$, the approximate equation $s_1 b_2 \approx s_1 a s_2 \approx s_2 b_1$ holds with overwhelming probability for proper parameters. By the same observation, we construct an AKE protocol (as illustrated in Fig. 6.3), where both the static and ephemeral public keys are actually Ring-LWE elements corresponding to a globally public element $a \in R_q$. In order to overcome the inability of "approximate" commutativity, the protocol has to send a signal information w_j computed by using a

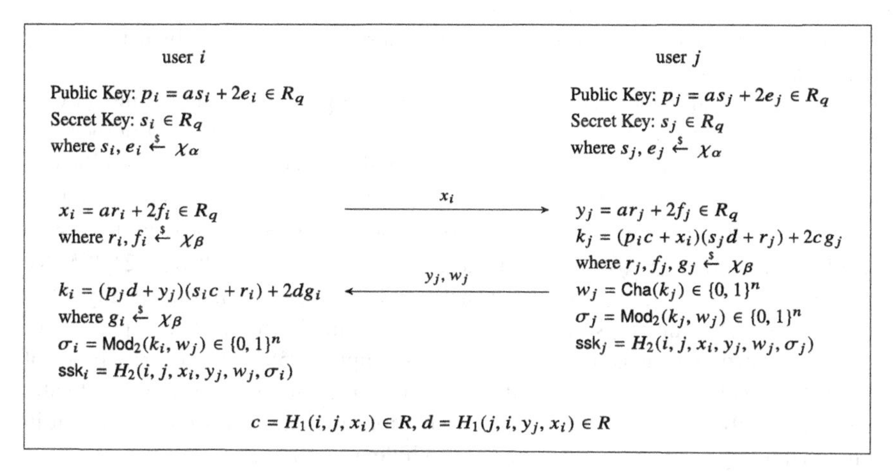

Fig. 6.3 An AKE protocol from Ring-LWE

function Cha [42]. Combining this with another useful function Mod_2, both parties are able to compute the same key material $\sigma_i = \sigma_j$ (from the approximately equal values k_i and k_j) with a guarantee that $\sigma_j = \mathsf{Mod}_2(k_j, w_j)$ has high min-entropy even conditioned on the partial information $w_j = \mathsf{Cha}(k_j)$ of k_j (thus it can be used to derive a uniform session key ssk_j).

However, the strategy of sending out the information $w_j = \mathsf{Cha}(k_j)$ inherently brings an undesired byproduct. Specifically, unlike HMQV, the security of the AKE protocol cannot be proven in the CK model which allows the adversaries to obtain the session state (e.g., k_i at user i or k_j at user j) via *session state reveal queries*. This is because in a traditional definition of session identifier that consists of all the exchanged messages, the two "different" sessions with identifiers $\mathsf{sid} = (i, j, x_i, y_j, w_j)$ and $\mathsf{sid}' = (i, j, x_i, y_j, w'_j)$ have the same session state, i.e., k_i at user i.[2] This also means that we cannot directly use $\sigma_i = \sigma_j$ as the session key, because the binding between the value of σ_i and the session identifier (especially for the signal part w_j) is too loose. In particular, the fact that $\sigma_i = \mathsf{Mod}_2(k_i, w_j)$ corresponding to sid is simply a shift of $\sigma'_i = \mathsf{Mod}_2(k_i, w'_j)$ corresponding to sid'(by the definition of the Mod_2 function), may potentially help the adversary distinguish σ_i with the knowledge of σ'_i. We prevent the adversary from utilizing this weakness by setting the session key as the output of the hash function H_2 (modeled as a random oracle) which tightly binds the session identifier sid and the key material σ_i (i.e., $\mathsf{ssk}_i = H_2(\mathsf{sid}, \sigma_i)$). This technique works due to another useful property of Mod_2, which guarantees that $\sigma_i = \mathsf{Mod}_2(k_i, w_j)$ preserves the high min-entropy property

[2]This problem might not exist if one consider a different definition of session identifier, e.g., the one that was uniquely determined at the beginning of the protocol execution.

of k_i for any w_j (and thus is enough to generate a secure session key by using a good randomness extractor H_2, e.g., a random oracle).[3]

In order to finally get a security proof of the AKE protocol in the BR model with weakly perfect forward secrecy, we have to make use of the following property of Gaussian distributions, namely, some kind of "public randomization". Specifically, let χ_α and χ_β be two Gaussian distributions with standard deviation α and β, respectively. Then, the sum of the two distributions is still a Gaussian distribution χ_γ with standard deviation $\gamma = \sqrt{\alpha^2 + \beta^2}$. In particular, if $\beta \gg \alpha$ (e.g., $\beta/\alpha = 2^{\omega(\log \kappa)}$ for some security parameter κ), we have that the distribution χ_γ is statistically close to χ_β. This technique is also known as "noise flooding" and has been applied, for instance, in proving robustness of the LWE assumption [59]. The security proof of the protocol is based on the observation that such a technique allows to statistically hide the distribution of χ_α in a bigger distribution χ_β, and for now let us keep it in mind that a large distribution will be used to hide a small one.

To better illustrate the technique, we take user j as an example, i.e., the one who combines his static and ephemeral secret keys by computing $\hat{r}_j = s_j d + r_j$ where $d = H_1(j, i, y_j, x_i)$. We notice that the value $\hat{r}_j$ actually behaves like a "signature" on the messages that user j knows so far. In other words, it should be difficult to compute $\hat{r}_j$ if we do not know the corresponding "signing key" s_j. Indeed, this combination is necessary to provide the implicit entity authentication. However, it also poses an obstacle to get a security proof since the simulator may also be unaware of s_j. Fortunately, if the randomness r_j is chosen from a big enough Gaussian distribution, then the value $\hat{r}_j$ almost obliterates all information of s_j. More specifically, the simulator can directly choose $\hat{r}_j$ such that $\hat{r}_j = s_j d + r_j$ for some unknown r_j by computing $y_j = (a\hat{r}_j + 2\hat{f}_j) - p_j d$, and programming the random oracle $d = H_1(j, i, y_j, x_i)$ correspondingly. The properties of Gaussian distributions and the random oracle H_1 imply that y_j has almost identical distribution as in the real run of the protocol. Now, we check the randomness of $k_j = (p_i c + x_i)\hat{r}_j + 2cg_j$. Note that for the test session, we can always guarantee that at least one of the pair (p_i, x_i) is honestly generated (and thus is computationally indistinguishable from uniformly distributed element under the Ring-LWE assumption), or else there is no "secrecy" to protect if both p_i and x_i are chosen by the adversary. That is, $p_i c + x_i$ is always random if c is invertible in R_q. Again, by programming $c = H_1(i, j, x_i)$, the simulator can actually replace $p_i c + x_i$ with $\hat{x}_i = cu_i$ for a uniformly distributed ring element u_i. In this case, we have that $k_j = \hat{x}_i \hat{r}_j + 2cg_j = c(u_i \hat{r}_j + 2g_j)$ should be computationally indistinguishable from a uniformly distributed element under the Ring-LWE assumption. In other words, when proving the security one can replace k_j with a uniformly distributed element to derive a high min-entropy key material σ_j by using the Mod_2 function as required.

---

[3] We remark that this is also the reason why the nice reconciliation mechanism in [92] cannot be used in the protocol. Specifically, it is unclear whether the reconciliation function $\mathsf{rec}(\cdot, \cdot)$ in [92] could also preserve the high min-entropy property of the first input (i.e., which might not be uniformly random) for any (maliciously chosen) second input.

Unfortunately, directly using "noise flooding" has a significant drawback, i.e., the requirement of a super polynomially large standard deviation β, which may lead to a nightmare for practical performance due to a super polynomially large modulus q for correctness and a very large ring dimension n for the hardness of the underlying Ring-LWE problems. Fortunately, we can reduce the big cost by further employing the rejection sampling technique [82]. Rejection sampling is a crucial technique in signature schemes to make the distribution of signatures independent of the signing key and has been applied in many other lattice-based signature schemes [7, 44, 63, 66].

In the case the combination of the static and ephemeral secret keys, $\hat{r}_j = s_j d + r_j$, at user j is essentially a signature on all the public messages under user j's public key (we again take party j as an example, but note that similar analysis also holds for user i). Thus, we can freely use the rejection sampling technique to relax the requirement on a super polynomially large β. Namely, we can use a much smaller β, but require user j to use r_j if $\hat{r}_j = s_j d + r_j$ follows the distribution χ_β, and to resample a new r_j otherwise. We note that by deploying rejection sampling in the AKE it is the first time that rejection sampling is used beyond signature schemes in lattice-based cryptography. As for signatures, rejection sampling is done locally, and thus will not affect the interaction between the two parties, i.e., two-pass messages. Even though the computational performance of each execution might become worse with certain (small) probability (due to rejection sampling), the average computational cost is much better than the setting of using a super polynomially large β.

6.3.2 The Protocol

We now describe the protocol in detail. Let n be a power of 2, and q be an odd prime such that $q \bmod 2n = 1$. Take $R = \mathbb{Z}[x]/(x^n + 1)$ and $R_q = \mathbb{Z}_q[x]/(x^n + 1)$ as above. For any positive $\gamma \in \mathbb{R}$, let $H_1 : \{0, 1\}^* \to \chi_\gamma = D_{R,\gamma}$ be a hash function that always outputs invertible elements in R_q.[4] Let $H_2 : \{0, 1\}^* \to \{0, 1\}^\kappa$ be the key derivation function, where κ is the bit-length of the final shared key. We model both functions as random oracles [14]. Let χ_α, χ_β be two discrete Gaussian distributions with parameters $\alpha, \beta \in \mathbb{R}^+$. Let $a \in R_q$ be the global public parameter uniformly chosen from R_q at random, and M be a constant determined by Theorem 2.1. Let $p_i = as_i + 2e_i \in R_q$ be user i's static public key, where (s_i, e_i) is the corresponding static secret key; both s_i and e_i are taken from the distribution χ_α. Similarly, user j has static public key $p_j = as_j + 2e_j$ and static secret key (s_j, e_j).

[4] In practice, one can first use a hash function (e.g., SHA-2) to obtain a uniformly random string, and then use it to sample from $D_{\mathbb{Z}^n,\gamma}$. The algorithm outputs a sample only if it is invertible in R_q, otherwise, it tries another sample and repeats. By Lemma 10 in [98], we can have a good probability to sample an invertible element in each trial for an appropriate choice of γ.

Initiation user i proceeds as follows:

1. Sample $r_i, f_i \stackrel{\$}{\leftarrow} \chi_\beta$ and compute $x_i = ar_i + 2f_i$;
2. Compute $c = H_1(i, j, x_i)$, $\hat{r}_i = s_i c + r_i$ and $\hat{f}_i = e_i c + f_i$;
3. Go to Step 4 with probability $\min\left(\frac{D_{\mathbb{Z}^{2n},\beta}(\mathbf{z})}{MD_{\mathbb{Z}^{2n},\beta,\mathbf{z}_1}(\mathbf{z})}, 1\right)$, where $\mathbf{z} \in \mathbb{Z}^{2n}$ is the coefficient vector of $\hat{r}_i$ concatenated with the coefficient vector of $\hat{f}_i$, and $\mathbf{z}_1 \in \mathbb{Z}^{2n}$ is the coefficient vector of $s_i c$ concatenated with the coefficient vector of $e_i c$; otherwise go back to Step 1;
4. Send x_i to user j.

Response After receiving x_i from user i, user j proceeds as follows:

1′. Sample $r_j, f_j \stackrel{\$}{\leftarrow} \chi_\beta$ and compute $y_j = ar_j + 2f_j$;
2′. Compute $d = H_1(j, i, y_j, x_i)$, $\hat{r}_j = s_j d + r_j$ and $\hat{f}_j = e_j d + f_j$;
3′. Go to Step 4′ with probability $\min\left(\frac{D_{\mathbb{Z}^{2n},\beta}(\mathbf{z})}{MD_{\mathbb{Z}^{2n},\beta,\mathbf{z}_1}(\mathbf{z})}, 1\right)$, where $\mathbf{z} \in \mathbb{Z}^{2n}$ is the coefficient vector of $\hat{r}_j$ concatenated with the coefficient vector of $\hat{f}_j$, and $\mathbf{z}_1 \in \mathbb{Z}^{2n}$ is the coefficient vector of $s_j d$ concatenated with the coefficient vector of $e_j d$; otherwise go back to Step 1′;
4′. Sample $g_j \stackrel{\$}{\leftarrow} \chi_\beta$, compute $k_j = (p_i c + x_i)\hat{r}_j + 2cg_j$, where $c = H_1(i, j, x_i)$;
5′. Compute $w_j = \mathsf{Cha}(k_j) \in \{0, 1\}^n$ and send (y_j, w_j) to user i;
6′. Compute $\sigma_j = \mathsf{Mod}_2(k_j, w_j)$ and derive the session key $\mathsf{ssk}_j = H_2(i, j, x_i, y_j, w_j, \sigma_j)$.

Finish User i receives the pair (y_j, w_j) from user j, and proceeds as follows:

5. Sample $g_i \stackrel{\$}{\leftarrow} \chi_\beta$ and compute $k_i = (p_j d + y_j)\hat{r}_i + 2dg_i$ with $d = H_1(j, i, y_j, x_i)$;
6. Compute $\sigma_i = \mathsf{Mod}_2(k_i, w_j)$ and derive the session key $\mathsf{ssk}_i = H_2(i, j, x_i, y_j, w_j, \sigma_i)$.

Remark 6.1 Deploying the protocol practically in a large scale requires the support of a PKI with a trusted certificate authority (CA). In this setting, all the system parameters (such as a) will be generated by the CA like other PKI-based protocols.

In the above protocol, both parties will make use of rejection sampling, i.e., they will repeat the first three steps with certain probability. By Theorem 2.1, the probability that each party will repeat the steps is about $1 - \frac{1}{M}$ for some constant M and appropriately chosen β. Thus, one can hope that both parties will send something to each other after an averaged M times repetitions of the first three steps. Next, we will show that once they send something to each other, both parties will finally compute a shared session key.

6.3.3 Correctness

To show the correctness of the AKE protocol, i.e., that both parties compute the same session key $\mathsf{ssk}_i = \mathsf{ssk}_j$, it suffices to show that $\sigma_i = \sigma_j$. Since σ_i and σ_j are both the output of Mod_2 with $\mathsf{Cha}(k_j)$ as the second argument, we need only to show that k_i and k_j are sufficiently close by Lemma 6.1. Note that the two parties will compute k_i and k_j as follows:

$$\begin{aligned} k_i &= (p_j d + y_j)\hat{r}_i + 2dg_i \\ &= a(s_j d + r_j)\hat{r}_i + 2(e_j d + f_j)\hat{r}_i \\ &\quad +2dg_i \\ &= a\hat{r}_i\hat{r}_j + 2\widetilde{g}_i \end{aligned} \qquad \begin{aligned} k_j &= (p_i c + x_i)\hat{r}_j + 2cg_j \\ &= a(s_i c + r_i)\hat{r}_j + 2(e_i c + f_i)\hat{r}_j \\ &\quad +2cg_j \\ &= a\hat{r}_i\hat{r}_j + 2\widetilde{g}_j \end{aligned},$$

where $\widetilde{g}_i = \hat{f}_j\hat{r}_i + dg_i$, and $\widetilde{g}_j = \hat{f}_i\hat{r}_j + cg_j$. Then $k_i = k_j + 2(\widetilde{g}_i - \widetilde{g}_j)$, and we have $\sigma_i = \sigma_j$ if $\|\widetilde{g}_i - \widetilde{g}_j\|_\infty < q/8$ by Lemma 6.1.

6.3.4 Security

Theorem 6.2 *Let n be a power of 2 satisfying $0.97n \geq 2\kappa$, prime $q > 203$ satisfying $q = 1 \bmod 2n$, real $\beta = \omega(\alpha\gamma n\sqrt{n \log n})$ and let H_1, H_2 be random oracles. Then, if $\mathrm{RLWE}_{q,\alpha}$ is hard, the proposed AKE is secure with respect to Definition 6.2.*

The intuition behind the proof is quite simple. Since the public element a and the public key of each party (e.g., $p_i = as_i + 2e_i$) actually consist of a $\mathrm{RLWE}_{q,\alpha}$ tuple with Gaussian parameter α (scaled by 2), the parties' static public keys are computationally indistinguishable from uniformly distributed elements in R_q under the Ring-LWE assumption. Similarly, both the exchanged elements x_i and y_j are also computationally indistinguishable from uniformly distributed elements in R_q under the $\mathrm{RLWE}_{q,\beta}$ assumption.

Without loss of generality, we take user j as an example to check the distribution of the session key. Note that if k_j is uniformly distributed over R_q, we have $\sigma_j \in \{0,1\}^n$ has high min-entropy (i.e., $0.97n > 2\kappa$) even conditioned on w_j by Lemma 6.2. Since H_2 is a random oracle, we have that ssk_j is uniformly distributed over $\{0,1\}^\kappa$ as expected. Now, let us check the distribution of $k_j = (p_i c + x_i)(s_j d + r_j) + 2cg_j$. As one can imagine, we want to establish the randomness of k_j based on pseudorandomness of "Ring-LWE samples" with public element $\hat{a}_j = c^{-1}(p_i c + x_i) = p_i + c^{-1}x_i$, the secret $\hat{s}_j = s_j d + r_j$, as well as the error term $2g_j$ (thus we have $k_j = c(\hat{a}_j\hat{s}_j + 2g_j)$). Actually, k_j is pseudorandom due to the following fact: (1) c is invertible in R_q; (2) $\hat{a}_j$ is uniformly distributed over R_q whenever p_i or x_i is uniform, and (3) $\hat{s}_j$ has distribution statistically close to χ_β by the strategy of rejection sampling in Theorem 2.1. In other words, $\hat{a}_j\hat{s}_j + 2g_j$ is statistically close to a $\mathrm{RLWE}_{q,\beta}$ sample, and thus is pseudorandom.

Formally, let N be the maximum number of parties, and m be maximum number of sessions for each party. We distinguish the following five types of adversaries:

Type I: $\mathsf{sid}^* = (\Pi, I, i^*, j^*, x_{i^*}, (y_{j^*}, w_{j^*}))$ is the test session, and y_{j^*} is output by a session activated at user j by a $\mathsf{Send}_1(\Pi, R, j^*, i^*, x_{i^*})$ query.

Type II: $\mathsf{sid}^* = (\Pi, I, i^*, j^*, x_{i^*}, (y_{j^*}, w_{j^*}))$ is the test session, and y_{j^*} is **not** output by a session activated at user j^* by a $\mathsf{Send}_1(\Pi, R, j^*, i^*, x_{i^*})$ query.

Type III: $\mathsf{sid}^* = (\Pi, R, j^*, i^*, x_{i^*}, (y_{j^*}, w_{j^*}))$ is the test session, and x_{i^*} is **not** output by a session activated at user i^* by a $\mathsf{Send}_0(\Pi, I, i^*, j^*)$ query.

Type IV: $\mathsf{sid}^* = (\Pi, R, j^*, i^*, x_{i^*}, (y_{j^*}, w_{j^*}))$ is the test session, and x_{i^*} is output by a session activated at user i^* by a $\mathsf{Send}_0(\Pi, I, i^*, j^*)$ query, but i^* either never completes the session, or i^* completes it with exact y_{j^*}.

Type V: $\mathsf{sid}^* = (\Pi, R, j^*, i^*, x_{i^*}, (y_{j^*}, w_{j^*}))$ is the test session, and x_{i^*} is output by a session activated at user i^* by a $\mathsf{Send}_0(\Pi, I, i^*, j^*)$ query, but i^* completes the session with another $y'_j \neq y_{j^*}$.

The five types of adversaries give a complete partition of all the adversaries. The weak perfect forward secrecy (wPFS) is captured by allowing **Type I** and **Type IV** adversaries to obtain the static secret keys of both user i^* and j^* by using $\mathsf{Corrupt}$ queries. Since sid^* definitely has no matching session for **Type II**, **Type III**, and **Type V** adversaries, no corruption to either user i^* or user j^* is allowed by Definition 6.1. The security proofs for the five types of adversaries are similar, except the forking lemma [12] is involved for **Type II**, **Type III**, and **Type V** adversaries by using the assumption that H_1 is a random oracle. Informally, the adversary must first "commit" x_i (y_j, resp.) before seeing c (d, resp.), thus it cannot determine the value $p_i c + x_i$ or $p_j d + y_i$ in advance (but the simulator can set the values by programming H_1 when it tries to embed Ring-LWE instances with respect to either $p_i c + x_i$ or $p_j d + y_i$ as discussed before).

In the following, we only give the security proof for **Type I** adversaries in Lemma 6.6. In the security proof, we will use a matrix form of the ring-LWE problem. Formally, let B_{β,ℓ_1,ℓ_2} be the distribution of $(\mathbf{a}, \mathbf{B} = (b_{i,j})) \in R_q^{\ell_1} \times R_q^{\ell_1 \times \ell_2}$, where $\mathbf{a} = (a_0, \ldots, a_{\ell_1 - 1}) \xleftarrow{\$} R_q^{\ell_1}$, $\mathbf{s} = (s_0, \ldots, s_{\ell_2 - 1}) \xleftarrow{\$} R_q^{\ell_2}$, $e_{i,j} \xleftarrow{\$} \chi_\beta$, and $b_{i,j} = a_i s_j + 2e_{i,j}$ for $i \in \{0, \ldots, \ell_1 - 1\}$ and $j \in \{0, \ldots, \ell_2 - 1\}$. For polynomially bounded ℓ_1 and ℓ_2, one can show that the distribution of $B_{\chi_\beta,\ell_1,\ell_2}$ is pseudorandom based on the $\mathrm{RLWE}_{q,\beta}$ assumption [94].

Lemma 6.6 *Let n be a power of 2 satisfying $0.97n \geq 2\kappa$, prime $q > 203$ satisfying $q = 1 \bmod 2n$, real $\beta = \omega(\alpha\gamma n\sqrt{n \log n})$. Then, if $\mathrm{RLWE}_{q,\alpha}$ is hard, the proposed AKE is secure against any PPT **Type I** adversary $\mathcal{A}$ in the random oracle model.*

*In particular, if there is a PPT **Type I** adversary $\mathcal{A}$ breaking the protocol with non-negligible advantage ϵ, then there is a PPT algorithm $\mathcal{B}$ solving $\mathrm{RLWE}_{q,\alpha}$ with advantage at least $\frac{\epsilon}{m^2N^2} - \mathsf{negl}(\kappa)$.*

Proof We prove this lemma via a sequence of games $G_{1,l}$ for $0 \leq l \leq 7$, where the first game $G_{1,0}$ is almost the same as the real one except that the simulator randomly guesses the test session at the beginning of the game and aborts the simulation if the guess is wrong, while the last game $G_{1,7}$ is a fake one with randomly and independently chosen session key for the test session (thus the adversary can only win the game with negligible advantage). The security is established by showing that any two consecutive games are computationally indistinguishable.

Game $G_{1,0}$. $\mathcal{S}$ chooses $i^*, j^* \xleftarrow{\$} \{1, \ldots, N\}$, $s_{i^*}, s_{j^*} \xleftarrow{\$} \{1, \ldots, m\}$ and hopes that the adversary will use $\mathsf{sid}^* = (\Pi, I, i^*, j^*, x_{i^*}, (y_{j^*}, w_{j^*}))$ as the test session, where x_{i^*} is output by the s_{i^*}-th session of user i^*, and y_{j^*} is output by the s_j^*-th session of user j^* activated by a $\mathsf{Send}_1(\Pi, R, j^*, i^*, x_{i^*})$ query. Then, $\mathcal{S}$ chooses $a \xleftarrow{\$} R_q$, generates static public keys for all parties (by choosing $s_i, e_i \xleftarrow{\$} \chi_\alpha$), and simulates the security game for $\mathcal{A}$. Specifically, $\mathcal{S}$ maintains two tables L_1, L_2 for the random oracles H_1, H_2, respectively, and answers the queries from $\mathcal{A}$ as follows:

- $H_1(in)$: If there does not exist a tuple (in, out) in L_1, choose an invertible element $out \in \chi_\gamma$ at random, and add (in, out) into L_1. Then, return out to $\mathcal{A}$.
- $H_2(in)$ queries: If there does not exist a tuple (in, out) in L_2, choose a vector $out \xleftarrow{\$} \{0, 1\}^\kappa$, and add (in, out) into L_2. Then, return out to $\mathcal{A}$.
- $\mathsf{Send}_0(\Pi, I, i, j)$: $\mathcal{A}$ activates a new session of i with intended user j, $\mathcal{S}$ proceeds as follows:

  1. Sample $r_i, f_i \xleftarrow{\$} \chi_\beta$ and compute $x_i = ar_i + 2f_i$;
  2. Compute $c = H_1(i, j, x_i)$, $\hat{r}_i = s_i c + r_i$ and $\hat{f}_i = e_i c + f_i$;
  3. Go to Step 4 with probability $\min\left(\frac{D_{\mathbb{Z}^{2n},\beta}(\mathbf{z})}{MD_{\mathbb{Z}^{2n},\beta,\mathbf{z}_1}(\mathbf{z})}, 1\right)$, where $\mathbf{z} \in \mathbb{Z}^{2n}$ is the coefficient vector of $\hat{r}_i$ concatenated with the coefficient vector of $\hat{f}_i$, and $\mathbf{z}_1 \in \mathbb{Z}^{2n}$ is the coefficient vector of $s_i c$ concatenated with the coefficient vector of $e_i c$; otherwise go back to Step 1;
  4. Return x_i to $\mathcal{A}$;

- $\mathsf{Send}_1(\Pi, R, j, i, x_i)$: $\mathcal{S}$ proceeds as follows:

  1′. Sample $r_j, f_j \xleftarrow{\$} \chi_\beta$ and compute $y_j = ar_j + 2f_j$;
  2′. Compute $d = H_1(j, i, y_j, x_i)$, $\hat{r}_j = s_j d + r_j$ and $\hat{f}_j = e_j d + f_j$;
  3′. Go to Step 4′ with probability $\min\left(\frac{D_{\mathbb{Z}^{2n},\beta}(\mathbf{z})}{MD_{\mathbb{Z}^{2n},\beta,\mathbf{z}_1}(\mathbf{z})}, 1\right)$, where $\mathbf{z} \in \mathbb{Z}^{2n}$ is the coefficient vector of $\hat{r}_j$ concatenated with the coefficient vector of $\hat{f}_j$, and $\mathbf{z}_1 \in \mathbb{Z}^{2n}$ is the coefficient vector of $s_j d$ concatenated with the coefficient vector of $e_j d$; otherwise go back to Step 1′;
  4′. Sample $g_j \xleftarrow{\$} \chi_\beta$, compute $k_j = (p_i c + x_i)\hat{r}_j + 2cg_j$, where $c = H_1(i, j, x_i)$;
  5′. Compute $w_j = \mathsf{Cha}(k_j) \in \{0, 1\}^n$ and return (y_j, w_j) to $\mathcal{A}$;
  6′. Compute $\sigma_j = \mathsf{Mod}_2(k_j, w_j)$ and derive the session key $\mathsf{ssk}_j = H_2(i, j, x_i, y_j, w_j, \sigma_j)$.

- $\mathsf{Send}_2(\Pi, I, i, j, x_i, (y_j, w_j))$: $\mathcal{S}$ computes k_i and ssk_i as follows:

  5. Sample $g_i \xleftarrow{\$} \chi_\beta$ and compute $k_i = (p_j d + y_j)\hat{r}_i + 2dg_i$ where $d = H_1(j, i, y_j, x_i)$;

6. Compute $\sigma_i = \mathsf{Mod}_2(k_i, w_j)$ and derive the session key $\mathsf{ssk}_i = H_2(i, j, x_i, y_j, w_j, \sigma_i)$.

- $\mathsf{SessionKeyReveal}(\mathsf{sid})$: Let $\mathsf{sid} = (\Pi, *, i, *, *, *, *)$, $\mathcal{S}$ returns ssk_i if the session key of sid has been generated.
- $\mathsf{Corrupt}(i)$: Return the static secret key s_i of i to $\mathcal{A}$.
- $\mathsf{Test}(\mathsf{sid})$: Let $\mathsf{sid} = (\Pi, I, i, j, x_i, (y_j, w_j))$, $\mathcal{S}$ aborts if $(i, j) \neq (i^*, j^*)$, or x_i and y_j are not output by the s_{i^*}-th session of user i^* and the s_j^*-th session of user j^*, respectively. Else, $\mathcal{S}$ chooses $b \xleftarrow{\$} \{0, 1\}$, returns $sk_i' \xleftarrow{\$} \{0, 1\}^\kappa$ if $b = 0$. Otherwise, return the session key ssk_i of sid.

Claim *The probability that $\mathcal{S}$ will not abort in $G_{1,0}$ is at least $\frac{1}{m^2N^2}$.* □

Proof This claim directly follows from the fact that $\mathcal{S}$ randomly chooses $i^*, j^* \xleftarrow{\$} \{1, \ldots, N\}$ and $s_{i^*}, s_j^* \xleftarrow{\$} \{1, \ldots, m\}$ independently from the view of $\mathcal{A}$. □

Game $G_{1,1}$. $\mathcal{S}$ behaves almost the same as in $G_{1,0}$ except in the following case:

- $\mathsf{Send}_1(\Pi, R, j, i, x_i)$: If $(i, j) \neq (i^*, j^*)$, or it is not the s_j^*-th session of j^*, $\mathcal{S}$ answers the query as in Game $G_{1,0}$. Otherwise, it proceeds as follows:

  1′. Sample $r_j, f_j \xleftarrow{\$} \chi_\beta$ and compute $y_j = ar_j + 2f_j$;
  2′. Sample an invertible element $d \xleftarrow{\$} \chi_\gamma$, and compute $\hat{r}_j = s_j d + r_j$, $\hat{f}_j = e_j d + f_j$;
  3′. Go to Step 4′ with probability $\min\left(\frac{D_{\mathbb{Z}^{2n},\beta}(\mathbf{z})}{MD_{\mathbb{Z}^{2n},\beta,\mathbf{z}_1}(\mathbf{z})}, 1\right)$, where $\mathbf{z} \in \mathbb{Z}^{2n}$ is the coefficient vector of $\hat{r}_j$ concatenated with the coefficient vector of $\hat{f}_j$, and $\mathbf{z}_1 \in \mathbb{Z}^{2n}$ is the coefficient vector of $s_j d$ concatenated with the coefficient vector of $e_j d$; otherwise go back to Step 1′;
  4′. Abort if there is a tuple $((j, i, y_j, x_i), *)$ in L_1. Else, add $((j, i, y_j, x_i), d)$ into L_1. Then, sample $g_j \xleftarrow{\$} \chi_\beta$ and compute $k_j = (p_i c + x_i)\hat{r}_j + 2cg_j$, where $c = H_1(i, j, x_i)$;
  5′. Compute $w_j = \mathsf{Cha}(k_j) \in \{0, 1\}^n$ and return (y_j, w_j) to $\mathcal{A}$;
  6′. Compute $\sigma_j = \mathsf{Mod}_2(k_j, w_j)$ and derive the session key $\mathsf{ssk}_j = H_2(i, j, x_i, y_j, w_j, \sigma_j)$.

  Let $F_{1,\ell}$ be the event that $\mathcal{A}$ outputs a guess b' that equals to b in Game $G_{1,\ell}$.

Claim *If RLWE$_{q,\beta}$ is hard, then* $\Pr[F_{1,1}] = \Pr[F_{1,0}] - \mathsf{negl}(\kappa)$. □

Proof Since H_1 is a random oracle, Game $G_{1,0}$ and Game $G_{1,1}$ are identical if the adversary $\mathcal{A}$ does not make a H_1 query $((j, i, y_j, x_i), *)$ before $\mathcal{S}$ generates y_j. Thus, the claim follows if the probability that $\mathcal{A}$ makes such a query in both Games is negligible. Actually, if $\mathcal{A}$ can make the query before seeing y_j with non-negligible probability, we can construct an algorithm $\mathcal{B}$ that breaks the RLWE$_{q,\beta}$ assumption.

Formally, after given a ring-LWE challenge tuple $(u, \mathbf{b}) \in R_q \times R_q^\ell$ in matrix form for some polynomially bounded ℓ, $\mathcal{B}$ sets $a = u$ and behaves like in Game $G_{1,0}$ until $\mathcal{B}$ has to generate y_j for the s_j^*-th session of j^* intended for user i^*.

Instead of generating a fresh y_j, $\mathcal{B}$ simply sets y_j as the first unused elements in $\mathbf{b} = (b_0, \ldots, b_{\ell-1})$, and checks if there is a tuple $((j, i, y_j, x_i), *)$ in L_1. If yes, it returns 1 and aborts, else it returns 0 and aborts.

It is easy to check that $\mathcal{A}$ has the same view as in $G_{1,0}$ and $G_{1,1}$ until the point that $\mathcal{B}$ has to compute y_j. Moreover, if $\mathbf{b} = (b_0 = ur_0 + 2f_0, \ldots, b_{\ell-1} = ur_{\ell-1} + 2f_{\ell-1})$ for some randomly choosing $r_{\ell'}, f_{\ell'} \stackrel{\$}{\leftarrow} \chi_\beta$, where $\ell' \in \{0, 1, \ldots, \ell - 1\}$, we have the probability that $\mathcal{A}$ will make the H_1 query with (j, i, y_j, x_i) is non-negligible by assumption. While if $\mathbf{b}$ is uniformly distributed over $\mathbb{R}_q^\ell$, we have the probability that $\mathcal{A}$ will make the H_1 query with (j, i, y_j, x_i) is negligible. This shows that $\mathcal{B}$ can be used to solve Ring-LWE assumption by interacting with $\mathcal{A}$. □

Game $G_{1,2}$. S behaves almost the same as in $G_{1,1}$ except in the following case:

- $\mathsf{Send}_1(\Pi, R, j, i, x_i)$: If $(i, j) \neq (i^*, j^*)$, or it is not the s_j^*-th session of j^*, S answers the query as in Game $G_{1,1}$. Otherwise, it proceeds as follows:

1′. Sample an invertible element $d \stackrel{\$}{\leftarrow} \chi_\gamma$, and choose $\mathbf{z} \stackrel{\$}{\leftarrow} D_{\mathbb{Z}^{2n},\beta}$;
2′. Parse $\mathbf{z}$ as two ring elements $\hat{r}_j, \hat{f}_j \in R_q$, and define $y_j = a\hat{r}_j + 2\hat{f}_j - p_j d$;
3′. Go to Step 4′ with probability $1/M$; otherwise go back to Step 1′;
4′. Abort if there is a tuple $((j, i, y_j, x_i), *)$ in L_1. Else, add $((j, i, y_j, x_i), d)$ into L_1. Then, sample $g_j \stackrel{\$}{\leftarrow} \chi_\beta$ and compute $k_j = (p_i c + x_i)\hat{r}_j + 2cg_j$, where $c = H_1(i, j, x_i)$;
5′. Compute $w_j = \mathsf{Cha}(k_j) \in \{0, 1\}^n$ and return (y_j, w_j) to $\mathcal{A}$;
6′. Compute $\sigma_j = \mathsf{Mod}_2(k_j, w_j)$ and derive the session key $\mathsf{ssk}_j = H_2(i, j, x_i, y_j, w_j, \sigma_j)$.

Claim *If $\beta = \omega(\alpha\gamma n\sqrt{n \log n})$, then* $\Pr[F_{1,2}] = \Pr[F_{1,1}] - \mathsf{negl}(\kappa)$. □

Proof By Lemmas 2.3 and 2.17, we have that both $\|s_j d\| \leq \alpha\gamma n\sqrt{n}$ and $\|e_j d\| \leq \alpha\gamma n\sqrt{n}$ (in Game $G_{1,1}$) hold with overwhelming probability. This means that $\beta = \omega(\alpha\gamma n\sqrt{n \log n})$ satisfies the requirement in Theorem 2.1, and thus the distribution of $(d, \mathbf{z})$ in Game $G_{1,2}$ is statistically close to that in $G_{1,1}$. The claim follows from the fact that the equation $y_j = a\hat{r}_j + 2\hat{f}_j - p_j d$ holds in both Game $G_{1,1}$ and $G_{1,2}$. □

Game $G_{1,3}$. S behaves almost the same as in $G_{1,2}$, except for the following case:

- $\mathsf{Send}_0(\Pi, I, i, j)$: If $(i, j) \neq (i^*, j^*)$, or it is not the s_{i^*}-th session of i^*, S answers as in Game $G_{1,2}$. Otherwise, it proceeds as follows:

1. Sample $r_i, f_i \stackrel{\$}{\leftarrow} \chi_\beta$ and compute $x_i = ar_i + 2f_i$;
2. Sample an invertible element $c \stackrel{\$}{\leftarrow} \chi_\gamma$, and compute $\hat{r}_i = s_i c + r_i$, $\hat{f}_i = e_i c + f_i$;
3. Go to Step 4 with probability $\min\left(\frac{D_{\mathbb{Z}^{2n},\beta}(\mathbf{z})}{M D_{\mathbb{Z}^{2n},\beta,\mathbf{z}_1}(\mathbf{z})}, 1\right)$, where $\mathbf{z} \in \mathbb{Z}^{2n}$ is the coefficient vector of $\hat{r}_i$ concatenated with the coefficient vector of $\hat{f}_i$, and $\mathbf{z}_1 \in \mathbb{Z}^{2n}$ is the coefficient vector of $s_i c$ concatenated with the coefficient vector of $e_i c$; otherwise go back to Step 1;

4. Abort if there is a tuple $((i, j, x_i), *)$ in L_1. Else, add $((i, j, x_i), c)$ into L_1. Return x_i to $\mathcal{A}$.

Claim *If $RLWE_{q,\beta}$ is hard, then* $\Pr[F_{1,3}] = \Pr[F_{1,2}] - \mathsf{negl}(\kappa)$. □

Proof The proof is similar to the proof of Claim 2, we omit the details. □

Game $G_{1,4}$. $\mathcal{S}$ behaves almost the same as in $G_{1,3}$ except for the following case:

- $\mathsf{Send}_0(\Pi, I, i, j)$: If $(i, j) \neq (i^*, j^*)$, or it is not the s_{i^*}-th session of i^*, $\mathcal{S}$ answers as in Game $G_{1,3}$. Otherwise, it proceeds as follows:

1. Sample an invertible element $c \xleftarrow{\$} \chi_\gamma$, and choose $\mathbf{z} \xleftarrow{\$} D_{\mathbb{Z}^{2n},\beta}$;
2. Parse $\mathbf{z}$ as two ring elements $\hat{r}_i, \hat{f}_i \in R_q$, and define $x_i = a\hat{r}_i + 2\hat{f}_i - p_i c$;
3. Go to Step 4 with probability $1/M$; otherwise go back to Step 1;
4. Abort if there is a tuple $((i, j, x_i), *)$ in L_1. Else, add $((i, j, x_i), c)$ into L_1. Return x_i to $\mathcal{A}$.

Claim *If $\beta = \omega(\alpha\gamma n\sqrt{n \log n})$, then* $\Pr[F_{1,4}] = \Pr[F_{1,3}] - \mathsf{negl}(\kappa)$. □

Proof The proof is similar to the proof of Claim 3, we omit the details. □

Game $G_{1,5}$. $\mathcal{S}$ behaves almost the same as in $G_{1,4}$ except for the following case:

- $\mathsf{Send}_2(\Pi, I, i, j, x_i, (y_j, w_j))$: If $(i, j) \neq (i^*, j^*)$, or it is not the s_{i^*}-th session of i^*, $\mathcal{S}$ behaves as in Game $G_{1,4}$. Otherwise, if (y_j, w_j) is output by the s_j^*-th session of user j^*, $\mathcal{S}$ sets $\mathsf{ssk}_i = \mathsf{ssk}_j$, where ssk_j is the session key of $\mathsf{sid} = (\Pi, R, j, i, x_i, (y_j, w_j))$. Else, $\mathcal{S}$ samples $g_i \xleftarrow{\$} \chi_\beta$ and computes $k_i = (p_j d + y_j)\hat{r}_i + 2dg_i$ where $d = H_1(j, i, y_j, x_i)$. Finally, it computes $\sigma_i = \mathsf{Mod}_2(k_i, w_j)$ and derives the session key $\mathsf{ssk}_i = H_2(i, j, x_i, y_j, w_j, \sigma_i)$.

Claim $\Pr[F_{1,5}] = \Pr[F_{1,4}] - \mathsf{negl}(\kappa)$. □

Proof This claim follows since $G_{1,5}$ is just a conceptual change of $G_{1,4}$ by the correctness of the protocol. □

Game $G_{1,6}$. $\mathcal{S}$ behaves almost the same as in $G_{1,5}$ except in the following case:

- $\mathsf{Send}_0(\Pi, I, i, j)$: If $(i, j) \neq (i^*, j^*)$, or it is not the s_{i^*}-th session of i^*, $\mathcal{S}$ answers as in Game $G_{1,5}$. Otherwise, it proceeds as follows:

1. Sample an invertible element $c \xleftarrow{\$} \chi_\gamma$, and choose $\hat{x}_i \xleftarrow{\$} R_q$;
2. Define $x_i = \hat{x}_i - p_i c$;
3. Go to Step 4 with probability $1/M$; otherwise go back to Step 1;
4. Abort if there is a tuple $((i, j, x_i), *)$ in L_1. Else, add $((i, j, x_i), c)$ into L_1. Return x_i to $\mathcal{A}$.

- $\mathsf{Send}_2(\Pi, I, i, j, x_i, (y_j, w_j))$: If $(i, j) \neq (i^*, j^*)$, or it is not the s_{i^*}-th session of i^*, or (y_j, w_j) is output by the s_j^*-th session of user j^*, $\mathcal{S}$ behaves the same as in $G_{1,5}$. Otherwise, it proceeds as follows:

5. Randomly choose $k_i \xleftarrow{\$} R_q$;
6. Compute $\sigma_i = \mathsf{Mod}_2(k_i, w_j)$ and derive the session key $\mathsf{ssk}_i = H_2(i, j, x_i, y_j, w_j, \sigma_i)$.

Note that in Game $G_{1,6}$, we have made two changes: (1) The term $a\hat{r}_i + 2\hat{f}_i$ in Game $G_{1,5}$ is replaced by a uniformly chosen element $\hat{x} \in R_q$ at random; (2) The value $k_i = (p_j d + y_j)\hat{r}_i + 2dg_i$ in Game $G_{1,5}$ is replaced by a uniformly chosen string $k_i \xleftarrow{\$} R_q$, when (y_j, w'_j) is output by the s_j^*-th session of user j^* but $w_j \neq w'_j$. In the following, we will employ the "deferred analysis" proof technique in [54], which informally allows us to proceed the security games by patiently postponing some tough probability analysis to a later game. Specially, for $\ell = 5, 6, 7$, denote $Q_{1,l}$ as the event in Game $G_{1,\ell}$ that (1) (y_j, w'_j) is output by the s_j^*-th session of user j^* but $w_j \neq w'_j$; and (2) $\mathcal{A}$ makes a query to H_2 that is exactly used to generate the session key ssk_i for the s_{i^*}-th session of user i^*, i.e., $\mathsf{ssk}_i = H_2(i, j, x_i, y_j, w_j, \sigma_i)$ for $\sigma_i = \mathsf{Mod}_2(k_i, w_j)$. Ideally, if $Q_{1,5}$ does not happen, then the adversary cannot distinguish whether a correctly computed k_i or a randomly chosen one is used (since H_2 is a random oracle, and the adversary gains no information about k_i even if it obtains the session key ssk_i). However, we cannot prove the claim immediately due to technical reason. Instead, we will show that $\Pr[Q_{1,5}] \approx \Pr[Q_{1,6}] \approx \Pr[Q_{1,7}]$ and $\Pr[Q_{1,7}]$ is negligible in κ.

Claim *If $RLWE_{q,\beta}$ is hard,* $\Pr[Q_{1,6}] = \Pr[Q_{1,5}] - \mathsf{negl}(\kappa)$, *and* $\Pr[F_{1,6}|\neg Q_{1,6}] = \Pr[F_{1,5}|\neg Q_{1,5}] - \mathsf{negl}(\kappa)$. □

Proof Note that H_2 is a random oracle, the event $Q_{1,5}$ is independent from the distribution of the corresponding ssk_i. Namely, no matter whether or not $\mathcal{A}$ obtains ssk_i, $\Pr[Q_{1,5}]$ is the same, which also holds for $\Pr[Q_{1,6}]$. In addition, under the $\mathrm{RLWE}_{q,\beta}$ assumption, we have $\hat{x}_i = a\hat{r}_i + 2\hat{f}_i$ in $G_{1,5}$ is computationally indistinguishable from uniform distribution over R_q, and thus the public information (i.e., static public keys and public transcripts) in $G_{1,5}$ and $G_{1,6}$ is computationally indistinguishable. In particular, the view of the adversary $\mathcal{A}$ before $Q_{1,\ell}$ happens for $\ell = 5, 6$ is computationally indistinguishable, which implies that $\Pr[Q_{1,6}] = \Pr[Q_{1,5}] - \mathsf{negl}(\kappa)$. Besides, if $Q_{1,l}$ for $l = 5, 6$ does not happen, the distribution of ssk_i is the same in both games. In other words, $\Pr[F_{1,6}|\neg Q_{1,6}] = \Pr[F_{1,5}|\neg Q_{1,5}] - \mathsf{negl}(\kappa)$. □

Game $G_{1,7}$. S behaves almost the same as in $G_{1,6}$ except in the following case:

- $\mathsf{Send}_1(\Pi, R, j, i, x_i)$: If $(i, j) \neq (i^*, j^*)$, or it is not the s_j^*-th session of j^*, S answers the query as in Game $G_{1,6}$. Otherwise, it proceeds as follows:

1′. Sample an invertible element $d \xleftarrow{\$} \chi_\gamma$, and choose $\hat{y}_j \xleftarrow{\$} R_q$;
2′. Define $y_j = \hat{y}_j - p_j d$;
3′. Go to Step 4′ with probability $1/M$; otherwise go back to Step 1′;
4′. Abort if there is a tuple $((j, i, y_j, x_i), *)$ in L_1. Else, add $((j, i, y_j, x_i), d)$ into L_1. Then, the simulator S uniformly chooses $k_j \xleftarrow{\$} R_q$ at random;
5′. Compute $w_j = \mathsf{Cha}(k_j) \in \{0, 1\}^n$ and return (y_j, w_j) to $\mathcal{A}$;
6′. Compute $\sigma_j = \mathsf{Mod}_2(k_j, w_j)$ and derive the session key $\mathsf{ssk}_j = H_2(i, j, x_i, y_j, w_j, \sigma_j)$.

Claim *Let n be a power of 2, prime $q > 203$ satisfying $q = 1 \bmod 2n$, $\beta = \omega(\alpha\gamma n\sqrt{n \log n})$. Then, if $RLWE_{q,\beta}$ is hard, Game $G_{1,6}$ and $G_{1,7}$ are computationally indistinguishable. In particular, we have* $\Pr[Q_{1,7}] = \Pr[Q_{1,6}] - \mathsf{negl}(\kappa)$, *and* $\Pr[F_{1,7}|\neg Q_{1,7}] = \Pr[F_{1,6}|\neg Q_{1,6}] - \mathsf{negl}(\kappa)$. □

Proof Assume there is an adversary that distinguishes Game $G_{1,6}$ and $G_{1,7}$, we now construct a distinguisher $\mathcal{D}$ that solves the Ring-LWE problem. Specifically, let $(\mathbf{u} = (u_0, \ldots, u_{\ell-1}), \mathbf{B}) \in R_q^{\ell} \times R_q^{\ell\times\ell}$ be a challenge Ring-LWE tuple in matrix form for some polynomially bounded ℓ, $\mathcal{D}$ first sets public parameter $a = u_0$. Then, it randomly chooses invertible elements $\mathbf{v} = (v_1, \ldots, v_{\ell-1}) \leftarrow \chi_\gamma^{\ell-1}$, and compute $\hat{\mathbf{u}} = (v_1 \cdot u_1, \ldots, v_{\ell-1}u_{\ell-1})$. Finally, $\mathcal{D}$ behaves the same as $\mathcal{S}$ in Game $G_{1,6}$, except for the following cases:

- $\mathsf{Send}_0(\Pi, I, i, j)$: If $(i, j) \neq (i^*, j^*)$, or it is not the s_{i^*}-th session of i^*, $\mathcal{S}$ answers as in Game $G_{1,6}$. Otherwise, it proceeds as follows:

  1. Set c and $\hat{x}_i$ be the first unused element in $\mathbf{v}$ and $\hat{\mathbf{u}}$, respectively;
  2. Define $x_i = \hat{x}_i - p_i c$;
  3. Go to Step 4 with probability $1/M$; otherwise go back to Step 1;
  4. Abort if there is a tuple $((i, j, x_i), *)$ in L_1. Else, add $((i, j, x_i), c)$ into L_1. Return x_i to $\mathcal{A}$.

- $\mathsf{Send}_1(\Pi, R, j, i, x_i)$: If $(i, j) \neq (i^*, j^*)$, or it is not the s_j^*-th session of j^*, $\mathcal{S}$ answers the query as in Game $G_{1,6}$. Otherwise, it proceeds as follows:

  1′. Sample an invertible element $d \xleftarrow{\$} \chi_\gamma$, and set $\hat{y}_j$ be the first unused element in $\mathbf{b}_0 = (b_{0,0}, \ldots, b_{0,\ell-1})$;
  2′. Define $y_j = \hat{y}_j - p_j d$;
  3′. Go to Step 4′ with probability $1/M$; otherwise go back to Step 1′;
  4′. Abort if there is a tuple $((j, i, y_j, x_i), *)$ in L_1. Else, add $((j, i, y_j, x_i), d)$ into L_1. Then, let $\ell_1 \geq 1$ be the index that $\hat{x}_i$ appears in $\hat{\mathbf{u}}$, and $\ell_2 \geq 0$ be the index that $\hat{y}_j$ appears in $\mathbf{b}_0$, the simulator $\mathcal{S}$ sets $k_j = cb_{\ell_1,\ell_2}$;
  5′. Compute $w_j = \mathsf{Cha}(k_j) \in \{0, 1\}^n$ and return (y_j, w_j) to $\mathcal{A}$;
  6′. Compute $\sigma_j = \mathsf{Mod}_2(k_j, w_j)$ and derive the session key $\mathsf{ssk}_j = H_2(i, j, x_i, y_j, w_j, \sigma_j)$.

Since $\mathbf{v}$ is randomly and independently chosen from $\chi_\gamma^{\ell-1}$, the distribution of c is identical to that in Game $G_{1,6}$ and Game $G_{1,7}$. Besides, since each v_i is invertible in R_q, we have $\hat{\mathbf{u}}$ is uniformly distributed over $R_q^{\ell-1}$, which shows that the distribution of $\hat{x}_i$ is identical to that in Game $G_{1,6}$ and Game $G_{1,7}$. Moreover, if $(\mathbf{u}, \mathbf{B}) \in R_q^{\ell} \times R_q^{\ell\times\ell}$ is a Ring-LWE challenge tuple in matrix form, we have $\hat{y}_j = u_0 s_{\ell_2} + 2e_{0,\ell_2}$ and $k_j = cb_{\ell_1,\ell_2} = cu_{\ell_1}s_{\ell_2} + 2ce_{\ell_1,\ell_2} = \hat{x}_i s_{\ell_2} + 2ce_{\ell_1,\ell_2} = (x_i + p_i c)s_{\ell_2} + 2ce_{\ell_1,\ell_2}$ for some randomly chosen $s_{\ell_2}, e_{0,\ell_2}, e_{\ell_1,\ell_2} \xleftarrow{\$} \chi_\beta$. This shows that the view of $\mathcal{A}$ is the same as in Game $G_{1,6}$. While if $(\mathbf{u}, \mathbf{B}) \in R_q^{\ell} \times R_q^{\ell\times\ell}$ is uniformly distributed over $R_q^{\ell} \times R_q^{\ell\times\ell}$, we have both $\hat{y}_j$ and $k_j = cb_{\ell_1,\ell_2}$ are uniformly distributed over R_q (since c is invertible). Thus, the view of $\mathcal{A}$ is the same as in $G_{1,7}$. In all, we have shown that $\mathcal{D}$ can be used to break Ring-LWE assumption if $\mathcal{A}$ can distinguish Game $G_{1,6}$ and $G_{1,7}$. □

Claim *If* $0.97n > 2\kappa$, *we have* $\Pr[Q_{1,7}] = \mathsf{negl}(\kappa)$ □

Proof Let $k_{i,\ell}$ be the element "computed" by $\mathcal{S}$ for the s_i^*-th session at user i^* in Games $G_{1,\ell}$, and $k_{j,\ell}$ be the element "computed" by $\mathcal{S}$ for the s_j^*-th session at user j^*. By the correctness of the protocol, we have that $k_{i,5} = k_{j,5} + \hat{g}$ for some $\hat{g}$ with small coefficients in $G_{1,5}$. Since we have proven that the view of the adversary before $Q_{1,\ell}$ happens in Game $G_{1,5}$, $G_{1,6}$ and $G_{1,7}$ is computationally indistinguishable, the equation $k_{i,7} = k_{j,7} + \hat{g}'$ should still hold for some $\hat{g}'$ with small coefficients in the adversary's view until $Q_{1,7}$ happens in $G_{1,7}$. Let (y_j, w_j) be output by the s_j^*-th session of user $j = j^*$, and (y_j, w_j') be the message that is used to complete the test session (i.e., the s_{i^*}-th session of user $i = i^*$). Note that $k_{j,7}$ is randomly chosen from R_q, and the adversary can only obtain the information of $k_{j,7}$ from the public w_j, the dependence of $\hat{g}$ on k_j should be totally determined by the information of w_j. Thus, we have that $\sigma_i' = \mathsf{Mod}_2(k_i, w_j') = \mathsf{Mod}_2(k_j + \hat{g}', w_j')$ conditioned on w_j has high min-entropy by Lemma 6.2. In other words, the probability that the adversary makes a query $H_2(i, j, x_i, y_j, w_j', \sigma_i')$ is at most $2^{-0.97n} + \mathsf{negl}(\kappa)$, which is negligible in κ. □

Claim $\Pr[F_{1,7}|\neg Q_{1,7}] = 1/2 + \mathsf{negl}(\kappa)$ □

Proof Let (y_j, w_j) be output by the s_j^*-th session of user $j = j^*$, (y_j, w_j') be the message that is used to complete the test session (i.e., the s_{i^*}-th session of user $i = i^*$). We distinguish the following two cases:

- $w_j = w_j'$: In this case, we have $\mathsf{ssk}_i = \mathsf{ssk}_j = H_2(i, j, x_i, y_j, w_j, \sigma_i)$, where $\sigma_i = \sigma_j = \mathsf{Mod}_2(k_j, w_j)$. Note that in $G_{1,7}$, k_j is randomly chosen from the uniform distribution over R_q, we have that $\sigma_j \in \{0, 1\}^n$ (conditioned on w_j) has min-entropy at least $0.97n$ by Lemma 6.2. Thus, the probability that $\mathcal{A}$ has made a H_2 query with σ_i is less than $2^{-0.97n} + \mathsf{negl}(\kappa)$.
- $w_j \neq w_j'$: By assumption that $Q_{1,7}$ does not happen, we have that $\mathcal{A}$ will never make a H_2 query with σ_i.

The probability that $\mathcal{A}$ has made a H_2 query with σ_i is negligible. This claim follows from the fact that if the adversary does not make a query with σ_i exactly, the distribution of ssk_i is uniform over $\{0, 1\}^\kappa$ due to the random oracle property of H_2, i.e., $\Pr[F_{1,7}|\neg Q_{1,7}] = 1/2 + \mathsf{negl}(\kappa)$. □

Combining the claims 1–10, we have that Lemma 6.6 follows.

6.4 Password-Based Authenticated Key Exchange

In this section, we give a PAKE framework (in the CRS model) from PKE with associated ASPH, which uses only *two-round* messages. The construction mainly benefit from two useful features of the underlying primitives: (1) the PKE is *splittable*, which informally requires that each ciphertext of the PKE scheme consists of two relatively independent parts, where the first part is designed for realizing the "functionality" of encryption, while the second part helps to achieve CCA-security; and (2) the ASPH is *non-adaptive* [72], i.e., the projection function only depends on the hash key, and the smoothness property holds even when the ciphertext depends on the projection key. We also give a concrete construction of splittable PKE with associated non-adaptive ASPH from learning with errors (LWE). As in [72], the PKE construction relies on simulation-sound non-interactive zero-knowledge (SS-NIZK) proofs, and thus, in general, is computationally inefficient. Fortunately, one can construct an efficient SS-NIZK from lattices in the random oracle model, and finally obtains an efficient two-round lattice-based PAKE in the random oracle model, which is at least $O(\log n)$ times more efficient in the communication overhead than the three-round lattice-based PAKE (in the standard model) [71].

6.4.1 Design Rational

We begin with the GL-framework [53] from CCA-secure public-key encryption (PKE) with associated smooth projective hash (SPH) functions. Informally, the SPH for a PKE scheme is a keyed hash function which maps a ciphertext-plaintext pair into a hash value, and can be computed in two ways: either using the hash key hk or using a projection key hp (which can be efficiently determined from hk and a targeted ciphertext c). The GL-framework for PAKE roughly relies on the following two properties of SPH:

Correctness: if c is an encryption of the password pw using randomness r, then the hash value $\mathrm{H}_{\mathsf{hk}}(c, pw) = \mathsf{Hash}(\mathsf{hp}, (c, pw), r)$, where both functions H and Hash can be efficiently computed from the respective inputs.

Smoothness: if c is not an encryption of pw, the value $\mathrm{H}_{\mathsf{hk}}(c, pw)$ is statistically close to uniform given hp, c and pw (over the random choice of hk).

Specifically, the GL-framework for PAKE has three-round messages: (1) the client computes an encryption c_1 of the password pw using randomness r_1 and sends c_1 to the server; (2) the server randomly chooses a hash key hk_2, computes a projection key hp_2 (from hk_2 and c_1) together with an encryption c_2 of the password pw using randomness r_2, and sends (hp_2, c_2) to the client; (3) the client sends a projection key hp_1 corresponding to a randomly chosen hash key hk_1 and c_2. After exchanging the above three messages, both users can compute the same session key $sk = \mathrm{H}_{\mathsf{hk}_1}(c_2, pw) \oplus \mathsf{Hash}(\mathsf{hp}_2, (c_1, pw), r_1) = \mathsf{Hash}(\mathsf{hp}_1, (c_2, pw), r_2) \oplus \mathrm{H}_{\mathsf{hk}_2}(c_1, pw)$ by the correctness of the SPH. Note that if the PKE scheme is CCA-secure, no user can obtain useful information about the password held by the other user from the received ciphertext. Thus, if the client (resp., the server) does not hold the correct password pw, his view is independent from the "session key" computed by the server (resp., the client) by the smoothness of the SPH. We stress that the above discussion is very informal and omits many details. For example, a verification key vk should be sent in the first message such that the client can generate a signature σ on the protocol transcripts in the third message (and thus the total communication cost is determined by $|\mathsf{hp}| + |c| + |vk| + |\sigma|$).

Clearly, a lattice-based PAKE is immediate if a PKE with associated SPH could be obtained from lattice assumptions. However, the literature [93] suggests that it is highly non-trivial, if not impossible, to instantiate SPH from lattices. Instead, Katz and Vaikuntanathan [71] provided a solution from a weaker notion of SPH—Approximate SPH (ASPH), which weakens both the correctness and smoothness properties of the SPH notion in [53]. First, ASPH only provides "approximate correctness" in the sense that $H_{\mathsf{hk}}(c, pw)$ and $\mathsf{Hash}(\mathsf{hp}, (c, pw), r)$ may differ at a few positions when parsed as bit-strings. Second, the smoothness property of ASPH only holds for some (c, pw) that pw is not equal to the decryption of c, and hence leaves a gap that there exists (c, pw) for which ASPH provides neither correctness nor smoothness guarantee. This relaxation is necessary for instantiating ASPH on lattices, since in the lattice setting there is no clear boundary between "c is an encryption of pw" and "c is not an encryption of pw", which is actually one of the main difficulties for realizing SPH from lattices.

Thus, if one directly plugs ASPH into the GL-framework [53], neither the correctness nor the security of the resulting PAKE is guaranteed. Because both users may not compute the same session key, and the adversary may break the protocol by exploiting the (inherent) gap introduced by ASPH. The authors [71] fixed the issues by relying on error-correcting codes (ECC) and the robustness of the GL-framework [53]. Specifically, in addition to sending a projection key hp_1, the client also randomly chooses a session key sk, computes $tk = H_{\mathsf{hk}_1}(c_2, pw) \oplus \mathsf{Hash}(\mathsf{hp}_2, (c_1, pw), r_1)$, and appends $\Delta = tk \oplus \mathsf{ECC}(sk)$ to the third message (i.e., tk is used as a masking key to deliver sk to the server), where ECC and ECC^{-1} are the corresponding encoding and decoding algorithms. After receiving the third message, the server can compute the session key $sk' = \mathsf{ECC}^{-1}(tk' \oplus \Delta)$, where $tk' = \mathsf{Hash}(\mathsf{hp}_1, (c_2, pw), r_2) \oplus H_{\mathsf{hk}_2}(c_1, pw)$. By the "approximate correctness" of the ASPH, we know that $tk' \oplus \Delta$ is not far from the codeword $\mathsf{ECC}(sk)$. Thus, both users can obtain the same session key $sk = sk'$ by the correctness of an appropriately chosen ECC, which finally allows [71] to obtain a three-round PAKE from PKE with associated ASPH.

However, the techniques of [71] are not enough to obtain a two-round PAKE (in particular, they cannot be applied into the PAKE framework [72]) due to the following two main reasons. First, the ASPH in [71] is *adaptive* (i.e., the projection key hp depends on the ciphertext c, and the smoothness only holds when c is independent of hp), which seems to inherently require at least three-round messages [53, 72]. Second, the strategy of delivering a random session key to deal with the "approximate correctness" of ASPH can only be applied when one user (e.g., the client) obtained the masking key tk and may be vulnerable to active attacks (e.g., modifications) because of the loose relation between the marking part (namely, Δ) and other protocol messages. This is not a problem for the GL-framework [53], since it had three-round messages and used one-time signatures, which allows the authors of [71] to simply send Δ in the third message and tightly bind it with other protocol messages by incorporating it into the one-time signature. Nevertheless, the above useful features are not available in the more efficient PAKE framework [72].

In order to get a two-round PAKE from PKE with associated ASPH, we strengthen the underlying primitive with several reasonable properties. First, we require that the ASPH is non-adaptive, i.e., the projection function only depends on the hash key, and the smoothness property holds even when the ciphertext c depends on hp. Second, we require that the underlying PKE is splittable. Informally, this property says that a ciphertext $c = (u, v)$ of the PKE scheme can be "independently" computed by two functions (f, g), where $u = f(pk, pw, \cdots)$ mainly takes a plaintext pw as input and plays the role of "encrypting" pw, while $v = g(pk, \mathsf{label}, \cdots)$ mainly takes a label as input and plays the role of providing non-malleability for CCA-security.[5] Third, we require that the hash value of the ASPH is determined by the hash key hk, the first part u of the ciphertext $c = (u, v)$, as well as the password pw. At a high level, the first enhancement allows us to safely compute the masking key tk after receiving the first message, while the second and third enhancements enable us to leverage the non-malleability of the underlying CCA-secure PKE scheme to tightly bind the masking part Δ with other protocol messages. Concretely, we let the client to send the projection hash key hp_1 together with the ciphertext c_1 in a single message, and let the server compute the masking key tk immediately after it has obtained the first part $u_2 = f(pk, pw, \cdots)$ of the ciphertext $c_2 = (u_2, v_2)$, and compute the second part $v_2 = g(pk, \mathsf{label}, \cdots)$ with a label consisting of $\mathsf{hp}_1, c_1, \mathsf{hp}_2, u_2$ and $\Delta = tk \oplus sk$ for some randomly chosen session key sk. The protocol ends with a message $(\mathsf{hp}_2, c_2, \Delta)$ sent by the server to the client. A high-level overview of the two-round PAKE framework is given in Fig. 6.4.

Note that the PKEs with associated SPH in [72] can be used to instantiate the two-round PAKE framework, but the only known lattice-based PKE with associated ASPH [71] does not satisfy the requirements. Actually, it is highly non-trivial to realize non-adaptive ASPH from lattices. One of the main reason is that the smoothness should hold even when the ciphertext c is adversarially chosen and dependent on the projection key hp (and thus is stronger than that in [71]), which gives the adversary an ability to obtain non-trivial information about the secret hash key hk and makes the above (inherent) gap introduced by the ASPH notion more problematic. In order to ensure the stronger smoothness property, we first develop an adaptive smoothing lemma for q-ary lattices, which may be of independent interest. Then, we combine it with several other techniques [55, 71, 89, 90, 96] to achieve the goal. As in [72], the PKE is computationally inefficient due to the use of simulation-sound non-interactive zero-knowledge (SS-NIZK) proofs. However, we can obtain an efficient SS-NIZK from lattices in the random oracle model, and finally get an efficient lattice-based PAKE. Despite the less message rounds, the PAKE (in the random oracle model) is also at least $O(\log n)$ times more efficient in the communication overhead than the one in [71], because they used the correlated products technique [95] and signatures. Specifically, the communication cost of [71] is determined by $|vk| + |c| + |\mathsf{hp}|$, where vk is the verification key of signatures (which usually consists of matrices on lattices [105]), c is the ciphertext of the underlying PKE scheme and hp is the projective hashing key. Since [71] used the correlated

[5] Similar properties were also considered for identity-based encryptions [4, 108].

products technique [95] (which introduces an expansion factor n w.r.t. the basic CPA-secure PKE scheme) to achieve CCA-secure PKE, their communication cost is dominated by $|c|$ (which is at least $O(\log n)$ times larger than $|\mathsf{hp}|$ when setting $k = O(n)$ or $\ell = O(n)$ in the notation). Since the framework does not use signatures, the communication cost is mainly determined by $|c| + |\mathsf{hp}|$. Although the use of Stern-like ZK introduces an $\omega(\log n)$ expansion factor, the ciphertext c of the PKE scheme is still $n/\omega(\log n)$ times shorter than that of [71]. Thus, the communication cost of the PAKE is now dominated by $|\mathsf{hp}|$, which is asymptotically the same as that in [71]. This is why we can (only) save a factor of $O(\log n)$ in the total communication cost. Note that one can also use the PKE with ASPH to instantiate the three-round PAKE framework in [71] with improved efficiency, but currently there seems no other way to do it significantly better even in the random oracle model.

6.4.2 PAKE from Splittable PKE with Associated ASPH

In this section, we give the definition of splittable PKE with associated ASPH. Informally, the splittable property of a PKE scheme Π_{pke} requires that the encryption algorithm can be split into two functions.

Definition 6.5 (*Splittable PKE*) A labeled CCA-secure PKE scheme $\Pi_{\text{pke}} = (\mathsf{KeyGen}, \mathsf{Enc}, \mathsf{Dec})$ is splittable if there exists a pair of two efficiently computable functions (f, g) such that the followings hold:

1. for any $(\mathsf{pk}, \mathsf{sk}) \leftarrow \mathsf{KeyGen}(1^\kappa)$, string $\mathsf{label} \in \{0,1\}^*$, plaintext $pw \in \mathcal{P}$ and randomness $r \in \{0,1\}^*$, we have $c = (u, v) = \mathsf{Enc}(\mathsf{pk}, \mathsf{label}, pw; r)$, where $u = f(\mathsf{pk}, pw, r)$ and $v = g(\mathsf{pk}, \mathsf{label}, pw, r)$. Moreover, the first part u of the ciphertext $c = (u, v)$ fixes the plaintext pw in the sense that for any v' and $\mathsf{label}' \in \{0,1\}^*$, the probability that $\mathsf{Dec}(\mathsf{sk}, \mathsf{label}', (u, v')) \notin \{\perp, pw\}$ is negligible in κ over the random choices of sk and r;
2. the security of Π_{pke} still holds in a CCA game with modified challenge phase: the adversary $\mathcal{A}$ first submits two equal-length plaintexts $pw_0, pw_1 \in \mathcal{P}$. Then, the challenger C chooses a random bit $b^* \xleftarrow{\$} \{0,1\}$, randomness $r^* \xleftarrow{\$} \{0,1\}^*$, and returns $u^* = f(\mathsf{pk}, pw_{b^*}, r^*)$ to $\mathcal{A}$. Upon receiving u^*, $\mathcal{A}$ outputs a string $\mathsf{label} \in \{0,1\}^*$. Finally, C computes $v^* = g(\mathsf{pk}, \mathsf{label}, pw_{b^*}, r^*)$, and returns the challenge ciphertext $c^* = (u^*, v^*)$ to $\mathcal{A}$;

Definition 6.5 captures the "splittable" property in both the functionality and the security of the PKE scheme. In particular, the modified CCA game allows the adversary to see the first part u^* of c^* and then adaptively determine label to form the complete challenge ciphertext $c^* = (u^*, v^*)$. We note that similar properties had been used in the context of identity-based encryption (IBE) [4, 108], where one part of the ciphertext is defined as a function of the plaintext, and the other part is a function of the user identity. By applying generic transformations such as the CHK technique [29] from IBE (with certain property) to PKE, it is promising to get a splittable PKE such that the g function simply outputs a tag or a signature which can be used to publicly verify the validity of the whole ciphertext. Finally, we stress that the notion of splittable PKE is not the main goal, but rather a crucial intermediate step to reaching two-round PAKE.

6.4.2.1 Approximate Smooth Projective Hash Functions

Smooth projective hash (SPH) functions were first introduced by Cramer and Shoup [35] for achieving CCA-secure PKEs. Later, several works [53, 72] extended the notion for PAKE. Here, we tailor the definition of approximate SPH (ASPH) in [71] to the application. Formally, let $\Pi_{\text{pke}} = (\mathsf{KeyGen}, \mathsf{Enc}, \mathsf{Dec})$ be a splittable PKE scheme with respect to functions (f, g), and let $\mathcal{P}$ be an efficiently recognizable plaintext space of Π_{pke}. As in [71], we require that Π_{pke} defines a notion of ciphertext validity in the sense that the validity of a label-ciphertext pair (label, c) with respect to any public key pk can be efficiently determined using pk alone, and all honestly generated ciphertexts are valid. We also assume that given a valid ciphertext c, one can easily parse $c = (u, v)$ as the outputs of (f, g). Now, fix a key pair $(\mathsf{pk}, \mathsf{sk}) \leftarrow \mathsf{KeyGen}(1^\kappa)$, and let C_{pk} denote the set of valid label-ciphertexts with respect to pk. Define sets X, L and $\bar{L}$ as follows:

$$X = \{(\mathsf{label}, c, pw) \mid (\mathsf{label}, c) \in C_{\mathsf{pk}};\ pw \in \mathcal{P}\}$$
$$L = \{(\mathsf{label}, c, pw) \in X \mid \mathsf{label} \in \{0,1\}^*;\ c = \mathsf{Enc}(\mathsf{pk}, \mathsf{label}, pw)\}$$
$$\bar{L} = \{(\mathsf{label}, c, pw) \in X \mid \mathsf{label} \in \{0,1\}^*;\ pw = \mathsf{Dec}(\mathsf{sk}, \mathsf{label}, c)\}.$$

By the definitions, for any ciphertext c and $\mathsf{label} \in \{0,1\}^*$, there is at most a single plaintext $pw \in \mathcal{P}$ such that $(\mathsf{label}, c, pw) \in \bar{L}$.

Definition 6.6 (*ϵ-approximate SPH*) An ϵ-approximate SPH function is defined by a sampling algorithm that, given a public key pk of Π_{pke}, outputs $(K, \ell, \{\mathrm{H}_{\mathsf{hk}} : X \to \{0,1\}^\ell\}_{\mathsf{hk} \in K}, S, \mathsf{Proj} : K \to S)$ such that

- There are efficient algorithms for (1) sampling a hash key $\mathsf{hk} \xleftarrow{\$} K$, (2) computing $\mathrm{H}_{\mathsf{hk}}(x) = \mathrm{H}_{\mathsf{hk}}(u, pw)$ for all $\mathsf{hk} \in K$ and $x = (\mathsf{label}, (u, v), pw) \in X$,[6] and (3) computing $\mathsf{hp} = \mathsf{Proj}(\mathsf{hk})$ for all $\mathsf{hk} \in K$.
- For all $x = (\mathsf{label}, (u, v), pw) \in L$ and randomness r such that $u = f(\mathsf{pk}, pw, r)$ and $v = g(\mathsf{pk}, \mathsf{label}, pw, r)$, there exists an efficient algorithm computing the value $\mathsf{Hash}(\mathsf{hp}, x, r) = \mathsf{Hash}(\mathsf{hp}, (u, pw), r)$, and satisfies $\Pr[\mathsf{Ham}(\mathrm{H}_{\mathsf{hk}}(u, pw), \mathsf{Hash}(\mathsf{hp}, (u, pw), r)) \ge \epsilon \cdot \ell] = \mathsf{negl}(\kappa)$ over the choice of $\mathsf{hk} \xleftarrow{\$} K$.
- For any (even unbounded) function $h : S \to X \backslash \bar{L}$, $\mathsf{hk} \xleftarrow{\$} K$, $\mathsf{hp} = \mathsf{Proj}(\mathsf{hk})$, $x = h(\mathsf{hp})$ and $\rho \xleftarrow{\$} \{0,1\}^\ell$, the statistical distance between $(\mathsf{hp}, \mathrm{H}_{\mathsf{hk}}(x))$ and (hp, ρ) is negligible in the security parameter κ.

---

[6]For all $x = (\mathsf{label}, (u, v), pw) \in X$, we slightly abuse the notation $\mathrm{H}_{\mathsf{hk}}(x) = \mathrm{H}_{\mathsf{hk}}(u, pw)$ by omitting (label, v) from its inputs. Similarly, the notation $\mathsf{Hash}(\mathsf{hp}, x, r) = \mathsf{Hash}(\mathsf{hp}, (u, pw), r)$ will be used later.

Common reference string: pk		
Client A (pw)		Server B (pw)
$r_1 \xleftarrow{\$} \{0,1\}^*$ $\mathsf{hk}_1 \xleftarrow{\$} K$ $\mathsf{hp}_1 = \mathsf{Proj}(\mathsf{hk}_1)$ $\mathsf{label}_1 := A\|B\|\mathsf{hp}_1$ $u_1 = f(\mathsf{pk}, pw, r_1)$ $v_1 = g(\mathsf{pk}, \mathsf{label}_1, pw, r_1)$	$A, \mathsf{hp}_1, c_1 = (u_1, v_1)$ $\longrightarrow$ $\longleftarrow$ $\mathsf{hp}_2, c_2 = (u_2, v_2), \Delta$	$r_2 \xleftarrow{\$} \{0,1\}^*$ $\mathsf{hk}_2 \xleftarrow{\$} K$ $sk \xleftarrow{\$} \{0,1\}^\kappa$ $\mathsf{hp}_2 = \mathsf{Proj}(\mathsf{hk}_2)$ $u_2 = f(\mathsf{pk}, pw, r_2)$ $tk = \mathsf{Hash}(\mathsf{hp}_1, (u_2, pw), r_2)$ $\oplus \mathrm{H}_{\mathsf{hk}_2}(u_1, pw)$
$tk' = \mathrm{H}_{\mathsf{hk}_1}(u_2, pw)$ $\oplus \mathsf{Hash}(\mathsf{hp}_2, (u_1, pw), r_1)$ $sk = \mathsf{ECC}^{-1}(tk' \oplus \Delta)$		$\Delta = tk \oplus \mathsf{ECC}(sk)$ $\mathsf{label}_2 := A\|B\|\mathsf{hp}_1\|c_1\|\mathsf{hp}_2\|\Delta$ $v_2 = g(\mathsf{pk}, \mathsf{label}_2, pw, r_2)$

Fig. 6.4 PAKE from splittable PKE with ASPH

Compared to the ASPH notion in [71], the ASPH notion in Definition 6.6 mainly has three modifications: (1) the projection function only depends on the hash key; (2) the value $\mathrm{H}_{\mathsf{hk}}(x) = \mathrm{H}_{\mathsf{hk}}(u, pw)$ is determined by the hash key hk, the first part u of the ciphertext $c = (u, v)$, as well as the plaintext pw (i.e., it is independent from the pair (label, v)); and (3) the smoothness property holds even for adaptive choice of $x = h(\mathsf{hp}) \notin \bar{L}$. Looking ahead, the first modification allows us to achieve PAKE with two-round messages, whereas the last two are needed for proving the security of the resulting PAKE. One can check that the PKEs with associated SPH (based on either DDH or decisional linear assumptions) in [72] satisfy Definition 6.6 with $\epsilon = 0$ (under certain choices of f and g). We will construct a splittable PKE with associated ASPH from lattices in Sect. 6.4.4.

6.4.2.2 A Framework for Two-Round PAKE

Let $\Pi_{\mathrm{pke}} = (\mathsf{KeyGen}, \mathsf{Enc}, \mathsf{Dec})$ be a splittable PKE scheme with respect to functions (f, g). Let $(K, \ell, \{\mathrm{H}_{\mathsf{hk}} : X \to \{0,1\}^\ell\}_{\mathsf{hk} \in K}, S, \mathsf{Proj} : K \to S)$ be the associated ϵ-approximate SPH for some $\epsilon \in (0, 1/2)$. Let the session key space be $\{0,1\}^\kappa$, where κ is the security parameter. Let $\mathsf{ECC} : \{0,1\}^\kappa \to \{0,1\}^\ell$ be an error-correcting code which can correct 2ϵ-fraction of errors, and let $\mathsf{ECC}^{-1} : \{0,1\}^\ell \to \{0,1\}^\kappa$ be the decoding algorithm. We assume that for uniformly distributed $\rho \in \{0,1\}^\ell$, the distribution of $w = \mathsf{ECC}^{-1}(\rho)$ conditioned on $w \neq \perp$ is uniform over $\{0,1\}^\kappa$. A high-level overview of the PAKE is given in Fig. 6.4.

Public parameters. The public parameter consists of a public key pk of the scheme Π_{pke}, which can be generated by a trusted third party using $\mathsf{KeyGen}(1^\kappa)$. No users in the system need to know the secret key corresponding to pk.

Protocol Execution. Consider an execution of the protocol between a client A and a server B holding a shared password $pw \in \mathcal{D} \subset \mathcal{P}$, where $\mathcal{D}$ is the set of valid passwords in the system. First, A chooses random coins $r_1 \xleftarrow{\$} \{0,1\}^*$ for encryption, a hash key $\mathsf{hk}_1 \xleftarrow{\$} K$ for the ASPH, and computes the projection key $\mathsf{hp}_1 = \mathsf{Proj}(\mathsf{hk}_1)$. Then, it defines $\mathsf{label}_1 := A\|B\|\mathsf{hp}_1$ and computes $(u_1, v_1) = \mathsf{Enc}(\mathsf{pk}, \mathsf{label}_1, pw; r_1)$, where $u_1 = f(\mathsf{pk}, pw, r_1)$ and $v_1 = g(\mathsf{pk}, \mathsf{label}_1, pw, r_1)$. Finally, A sends $(A, \mathsf{hp}_1, c_1 = (u_1, v_1))$ to the server B.

Upon receiving $(A, \mathsf{hp}_1, c_1 = (u_1, v_1))$ from the client A, the server B checks if c_1 is a valid ciphertext with respect to pk and $\mathsf{label}_1 := A\|B\|\mathsf{hp}_1$.[7] If not, B rejects and aborts. Otherwise, B chooses random coins $r_2 \xleftarrow{\$} \{0,1\}^*$ for encryption, a hash key $\mathsf{hk}_2 \xleftarrow{\$} K$ for the ASPH, and a random session key $sk \xleftarrow{\$} \{0,1\}^\kappa$. Then, it computes $\mathsf{hp}_2 = \mathsf{Proj}(\mathsf{hk}_2)$, $u_2 = f(\mathsf{pk}, pw, r_2)$, $tk = \mathsf{Hash}(\mathsf{hp}_1, (u_2, pw), r_2) \oplus \mathrm{H}_{\mathsf{hk}_2}(u_1, pw)$, and $\Delta = tk \oplus \mathsf{ECC}(sk)$. Finally, let $\mathsf{label}_2 := A\|B\|\mathsf{hp}_1\|c_1\|\mathsf{hp}_2\|\Delta$, the server B computes $v_2 = g(\mathsf{pk}, \mathsf{label}_2, pw, r_2)$ and sends the message $(\mathsf{hp}_2, c_2 = (u_2, v_2), \Delta)$ to the client A.

After receiving $(\mathsf{hp}_2, c_2 = (u_2, v_2), \Delta)$ from the server B, the client A checks if c_2 is a valid ciphertext with respect to pk and $\mathsf{label}_2 := A\|B\|\mathsf{hp}_1\|c_1\|\mathsf{hp}_2\|\Delta$. If not, A rejects and aborts. Otherwise, A computes $tk' = \mathrm{H}_{\mathsf{hk}_1}(u_2, pw) \oplus \mathsf{Hash}(\mathsf{hp}_2, (u_1, pw), r_1)$, and decodes to obtain $sk = \mathsf{ECC}^{-1}(tk' \oplus \Delta)$. If $sk = \perp$ (i.e., an error occurs during decoding), A rejects and aborts. Otherwise, A accepts $sk \in \{0,1\}^\kappa$ as the shared session key. This completes the description of the protocol.

In the following, we say that a user (or an instance of a user) accepts an incoming message msg as *a valid protocol message* if no abort happens during the computations after receiving msg. Note that a client/server will only obtain a session key when he accepts a received message as a valid protocol message.

Correctness. It suffices to show that honestly users can obtain the same session key $sk \in \{0,1\}^\kappa$ with overwhelming probability. First, all honestly generated ciphertexts are valid. Second, $\mathrm{H}_{\mathsf{hk}_1}(u_2, pw) \oplus \mathsf{Hash}(\mathsf{hp}_1, (u_2, pw), r_2) \in \{0,1\}^\ell$ has at most ϵ-fraction non-zeros by the ϵ-approximate correctness of the ASPH. Similarly, $\mathsf{Hash}(\mathsf{hp}_2, (u_1, pw), r_1) \oplus \mathrm{H}_{\mathsf{hk}_2}(u_1, pw) \in \{0,1\}^\ell$ has at most ϵ-fraction non-zeros. Thus, $tk' \oplus tk$ has at most 2ϵ-fraction non-zeros. Since ECC can correct 2ϵ-fraction of errors by assumption, we have that $sk = \mathsf{ECC}^{-1}(tk' \oplus tk \oplus \mathsf{ECC}(sk))$ holds. This completes the correctness argument.

6.4.3 Security

We now show that the above PAKE is secure. Formally,

Theorem 6.3 *If* $\Pi_{\text{pke}} = (\mathsf{KeyGen}, \mathsf{Enc}, \mathsf{Dec})$ *is a splittable CCA-secure PKE scheme associated with an* ϵ*-approximate SPH* $(K, \ell, \{\mathrm{H}_{\mathsf{hk}} : X \to \{0,1\}^\ell\}_{\mathsf{hk}\in K}, S, \mathsf{Proj} : K \to S)$, *and* $\mathsf{ECC} : \{0,1\}^\kappa \to \{0,1\}^\ell$ *is an error-correcting code which can correct* 2ϵ*-fraction of errors, then the above protocol is a secure PAKE.*

[7]Recall that the validity of a ciphertext can be efficiently determined using pk alone.

Before giving the proof, we first give some intuitions. Without loss of generality we assume $0 \in \mathcal{P} \backslash \mathcal{D}$ (i.e., 0 is not a valid password in the system). First, by the CCA-security of the PKE scheme Π_{pke}, the adversary cannot obtain any useful information of the real password pw via the Execute query (i.e., by eavesdropping on a protocol execution). In particular, it is computationally indistinguishable for the adversary if the encryption of pw is replaced by an encryption of 0 in answering the Execute queries. Since $0 \notin \mathcal{D}$, by the smoothness of the ASPH we have that the session keys corresponding to the instances used in the Execute queries are indistinguishable from uniform in the adversary's view.

Second, if the adversary simply relays the messages between honest instances, the proof is the same for the Execute queries. In case that the adversary modifies the message (i.e., the label-ciphertext pair) output by some instance, then one can use the decryption oracle provided by the CCA-security to decrypt the modified ciphertext, and check if the decrypted result pw' is equal to the real password pw. For $pw' = pw$ the attack is immediately considered successful (note that this will only increase the advantage of the adversary). By the CCA-security of Π_{pke} and the fact that pw is uniformly chosen from $\mathcal{D}$ at random, we have $\Pr[pw' = pw]$ is at most $1/|\mathcal{D}|$. Thus, for $Q(\kappa)$ times on-line attacks, this will only increase the adversary's advantage by at most $Q(\kappa)/|\mathcal{D}|$. Otherwise (i.e., $pw' \neq pw$) we again have that the corresponding session key is indistinguishable from uniform in the adversary's view by the smoothness of the ASPH.

Proof We now formally prove Theorem 6.3 via a sequence of games from G_0 to G_{10}, where G_0 is the real security game, and G_{10} is a random game with uniformly chosen session keys. The security is established by showing that the adversary's advantage in game G_0 and G_{10} will differ at most $Q(\kappa)/|\mathcal{D}| + \mathsf{negl}(\kappa)$. Let $\mathbf{Adv}_{\mathcal{A},i}(\kappa)$ be the adversary $\mathcal{A}$'s advantage in game G_i.

Game G_0. This game is the real security game, where all the oracle queries are honestly answered following the protocol specification.

Game G_1. This game is similar to game G_0 except that in answering each Execute query the value tk' is directly computed using the corresponding hash keys hk_1 and hk_2, i.e., $tk' = \mathrm{H}_{\mathsf{hk}_1}(u_2, pw) \oplus \mathrm{H}_{\mathsf{hk}_2}(u_1, pw)$.

Lemma 6.7 *Let* $(K, \ell, \{\mathrm{H}_{\mathsf{hk}} : X \to \{0,1\}^\ell\}_{\mathsf{hk} \in K}, S, \mathsf{Proj} : K \to S)$ *be an* ϵ*-approximate SPH, and* $\mathsf{ECC} : \{0,1\}^\kappa \to \{0,1\}^\ell$ *be an error-correcting code which can correct* 2ϵ*-fraction of errors, then* $|\mathbf{Adv}_{\mathcal{A},1}(\kappa) - \mathbf{Adv}_{\mathcal{A},0}(\kappa)| \le \mathsf{negl}(\kappa)$.

Proof Since the simulator knows both hk_1 and hk_2, this lemma follows from the approximate correctness of the ASPH and the correctness of the ECC. □

Game G_2. This game is similar to game G_1 except that the ciphertext c_1 is replaced with an encryption of $0 \notin \mathcal{D}$ in answering each Execute query.

Lemma 6.8 *If* $\Pi_{\text{pke}} = (\mathsf{KeyGen}, \mathsf{Enc}, \mathsf{Dec})$ *is a CCA-secure scheme, then we have that* $|\mathbf{Adv}_{\mathcal{A},2}(\kappa) - \mathbf{Adv}_{\mathcal{A},1}(\kappa)| \le \mathsf{negl}(\kappa)$.

Proof Since the adversary $\mathcal{A}$ can only make polynomial times Execute queries, it is enough to consider that $\mathcal{A}$ only makes a single Execute query by a standard hybrid argument. In this case, the only difference between game G_1 and G_2 is that the encryption of pw is replaced by an encryption of $0 \notin \mathcal{D}$. We now show that any PPT adversary $\mathcal{A}$ that distinguishes the two games with non-negligible advantage can be directly transformed into an algorithm $\mathcal{B}$ that breaks the CCA-security of the underlying Π_{pke} scheme with the same advantage.

Formally, given a challenge public key pk, the algorithm $\mathcal{B}$ sets pk as the CRS of the protocol, and interacts with $\mathcal{A}$ as in game G_1. When $\mathcal{B}$ has to answer the adversary's $\mathsf{Execute}(A, i, B, j)$ query, it first randomly chooses a hash key $\mathsf{hk}_1 \xleftarrow{\$} K$ for the ASPH and computes the projection key $\mathsf{hp}_1 = \mathsf{Proj}(\mathsf{hk}_1)$. Then, $\mathcal{B}$ submits two plaintexts $(pw, 0)$ and $\mathsf{label}_1 := A\|B\|\mathsf{hp}_1$ to its own challenger, and obtains a challenge ciphertext c_1^*. Finally, $\mathcal{B}$ uses c_1^* to form the answer of the $\mathsf{Execute}(A, i, B, j)$ query, and returns whatever $\mathcal{A}$ outputs as its own guess.

Note that if c_1^* is an encryption of pw, then $\mathcal{B}$ exactly simulates the attack environment of game G_1 for adversary $\mathcal{A}$, else it simulates the attack environment of G_2 for $\mathcal{A}$. Thus, if $\mathcal{A}$ can distinguish G_1 and G_2 with non-negligible advantage, then $\mathcal{B}$ can break the CCA-security of Π_{pke} with the same advantage. □

Game G_3. This game is similar to game G_2 except that in answering each Execute query: (1) the value tk is directly computed by using the corresponding hash keys hk_1 and hk_2, i.e., $tk = \mathrm{H}_{\mathsf{hk}_1}(u_2, pw) \oplus \mathrm{H}_{\mathsf{hk}_2}(u_1, pw)$; (2) the ciphertext c_2 is replaced with an encryption of $0 \notin \mathcal{D}$.

Lemma 6.9 *If* $\Pi_{\mathsf{pke}} = (\mathsf{KeyGen}, \mathsf{Enc}, \mathsf{Dec})$ *is a splittable CCA-secure scheme,* $(K, \ell, \{\mathrm{H}_{\mathsf{hk}} : X \to \{0,1\}^{\ell}\}_{\mathsf{hk}\in K}, S, \mathsf{Proj} : K \to S)$ *is an* ϵ*-approximate SPH, and* $\mathsf{ECC} : \{0,1\}^{\kappa} \to \{0,1\}^{\ell}$ *is an error-correcting code which can correct* 2ϵ*-fraction of errors, then we have that* $|\mathbf{Adv}_{\mathcal{A},3}(\kappa) - \mathbf{Adv}_{\mathcal{A},2}(\kappa)| \le \mathsf{negl}(\kappa)$.

Proof This lemma can be shown by using a sequence of games similar to that from G_0 to G_2 except the modified CCA-security game considered in Definition ?? is used instead of the standard CCA-security game, we omit the details. □

Game G_4. This game is similar to game G_3 except that a random session key $\mathsf{sk}_A^i = \mathsf{sk}_B^j$ is set for both Π_A^i and Π_B^j in answering each $\mathsf{Execute}(A, i, B, j)$ query.

Lemma 6.10 *If* $(K, \ell, \{\mathrm{H}_{\mathsf{hk}} : X \to \{0,1\}^{\ell}\}_{\mathsf{hk}\in K}, S, \mathsf{Proj} : K \to S)$ *is an* ϵ*-approximate SPH, then we have that* $|\mathbf{Adv}_{\mathcal{A},4}(\kappa) - \mathbf{Adv}_{\mathcal{A},3}(\kappa)| \le \mathsf{negl}(\kappa)$.

Proof Since both ciphertexts $c_1 = (u_1, v_1)$ and $c_2 = (u_2, v_2)$ in answering each $\mathsf{Execute}(A, i, B, j)$ query are encryptions of $0 \notin \mathcal{D}$, the value $tk' = tk = \mathrm{H}_{\mathsf{hk}_1}(u_2, pw) \oplus \mathrm{H}_{\mathsf{hk}_2}(u_1, pw)$ is statistically close to uniform by the smoothness of the ASPH. Thus, the masking part $\Delta = tk \oplus \mathsf{ECC}(sk)$ in answering each $\mathsf{Execute}(A, i, B, j)$ query statistically hides $sk \in \{0,1\}^{\kappa}$ from the adversary $\mathcal{A}$. Since $sk \in \{0,1\}^{\kappa}$ is uniformly random, the modification in game G_4 can only introduce a negligible statistical difference. Since $\mathcal{A}$ can only make polynomial times Execute queries, this lemma follows by a standard hybrid argument. □

Game G_5. This game is similar to game G_4 except that the simulator generates the CRS pk by running $(\mathsf{pk}, \mathsf{sk}) \leftarrow \mathsf{KeyGen}(1^\kappa)$, and keeps sk private.

Lemma 6.11 $\mathbf{Adv}_{\mathcal{A},5}(\kappa) = \mathbf{Adv}_{\mathcal{A},4}(\kappa)$.

Proof This lemma follows from the fact that the modification from game G_4 to G_5 is just conceptual. □

Before continuing, we divide the adversary's Send query into three types depending on the message which may be sent as part of the protocol:

- $\mathsf{Send}_0(A, i, B)$: the adversary prompts an unused instance Π_A^i to execute the protocol with partner B. This oracle updates $\mathsf{pid}_A^i = B$ and returns the message $\mathsf{msg}_1 = (A, \mathsf{hp}_1, c_1)$ output by Π_A^i to the adversary.
- $\mathsf{Send}_1(B, j, (A, \mathsf{hp}_1, c_1))$: the adversary sends message $\mathsf{msg}_1 = (A, \mathsf{hp}_1, c_1)$ to an unused instance Π_B^j. This oracle updates $(\mathsf{pid}_B^j, \mathsf{sk}_B^j, \mathrm{acc}_B^j, \mathrm{term}_B^j)$ as appropriate and returns the message $\mathsf{msg}_2 = (\mathsf{hp}_2, c_2, \Delta)$ output by Π_B^j to the adversary (only if Π_B^j accepts msg_1 as a valid protocol message).
- $\mathsf{Send}_2(A, i, (\mathsf{hp}_2, c_2, \Delta))$: the adversary sends message $\mathsf{msg}_2 = (\mathsf{hp}_2, c_2, \Delta)$ to instance Π_A^i. This oracle updates $(\mathsf{sk}_B^j, \mathrm{acc}_B^j, \mathrm{term}_B^j)$ as appropriate.

Game G_6. This game is similar to game G_5 except that each $\mathsf{Send}_1(B, j, \mathsf{msg}_1' = (A', \mathsf{hp}_1', c_1'))$ query is handled as follows:

- If msg_1' was output by a previous $\mathsf{Send}_0(A', *, B)$ query, the simulator C performs exactly as in game G_5;
- Otherwise, let $\mathsf{label}_1' := A' \| B \| \mathsf{hp}_1'$, and distinguish the following two cases:
  - If c_1' is not a valid ciphertext with respect to pk and label_1', the simulator C rejects this query;
  - Else, C decrypts $(\mathsf{label}_1', c_1')$ using the secret key sk corresponding to pk, and let pw' be the decryption result. If pw' is equal to the real password pw shared by A and B (i.e., $pw' = pw$), the simulator C declares that $\mathcal{A}$ succeeds and terminates the experiment. Otherwise, C answers this query as in game G_5 but sets the session key sk_B^j, for instance, Π_B^j by using an independently and uniformly chosen element from $\{0, 1\}^\kappa$.

Lemma 6.12 *If* $(K, \ell, \{\mathsf{H}_{\mathsf{hk}} : X \to \{0, 1\}^\ell\}_{\mathsf{hk} \in K}, S, \mathsf{Proj} : K \to S)$ *is an* ϵ-*approximate SPH, then we have that* $\mathbf{Adv}_{\mathcal{A},5}(\kappa) \le \mathbf{Adv}_{\mathcal{A},6}(\kappa) + \mathsf{negl}(\kappa)$.

Proof We only have to consider the case that $\mathsf{msg}_1' = (A', \mathsf{hp}_1', c_1')$ was not output by any previous $\mathsf{Send}_0(A', *, B)$ query and c_1' is a valid cipertext with respect to pk and label_1' (note that B will always reject invalid ciphertexts in the real run of the protocol). Since C knows the secret key sk corresponding to pk in both game G_5 and G_6, it can always decrypt $(\mathsf{label}_1', c_1')$ to obtain the decryption result pw'. Obviously, the modification for the case $pw' = pw$ can only increase the advantage of the adversary $\mathcal{A}$. As for the case $pw' \neq pw$, we have $(\mathsf{label}_1', c_1', pw) \notin \bar{L}$. By the smoothness of the underlying ASPH (in Definition 6.6), the masking part

$\Delta = tk \oplus \mathsf{ECC}(sk)$ output by Π_B^j statistically hides $sk \in \{0,1\}^\kappa$ from the adversary $\mathcal{A}$ with knowledge of $\mathsf{hp}_2 = \mathsf{Proj}(\mathsf{hk}_2)$ (because tk has a term $\mathrm{H}_{\mathsf{hk}_2}(u_1', pw)$ for $c_1' = (u_1', v_1')$ and $\mathsf{hk}_2 \xleftarrow{\$} K$). Using the fact that sk is essentially uniformly chosen from $\{0,1\}^\kappa$, we have that the modification for the case $pw' \neq pw$ in game G_6 can only introduce a negligible statistical difference. In all, we have that $\mathbf{Adv}_{\mathcal{A},5}(\kappa) \leq \mathbf{Adv}_{\mathcal{A},6}(\kappa) + \mathsf{negl}(\kappa)$. □

Game G_7. This game is similar to game G_6 except that each $\mathsf{Send}_2(A, i, \mathsf{msg}_2' = (\mathsf{hp}_2', c_2', \Delta'))$ query is handled as follows: let $\mathsf{msg}_1 = (A, \mathsf{hp}_1, c_1)$ be the message output by a previous $\mathsf{Send}_0(A, i, B)$ query (note that such a query must exist),

- If msg_2' was output by a previous $\mathsf{Send}_1(B, j, \mathsf{msg}_1)$ query, the simulator C performs as in game G_6 except that C computes tk' directly using the corresponding hash keys hk_1 and hk_2, and sets the session key $\mathsf{sk}_A^i = \mathsf{sk}_B^j$;
- Otherwise, let $\mathsf{label}_2' := A\|B\|\mathsf{hp}_1\|c_1\|\mathsf{hp}_2'\|\Delta'$, and distinguish the following two cases:
  - If c_2' is not a valid ciphertext with respect to pk and label_2', the simulator C rejects this query;
  - Else, C decrypts $(\mathsf{label}_2', c_2')$ using the secret key sk corresponding to pk, and let pw' be the decryption result. If $pw' = pw$, the simulator C declares that $\mathcal{A}$ succeeds and terminates the experiment. Otherwise, C performs the computations on behalf of Π_A^i as in game G_6. If Π_A^i accepts msg_2' as a valid protocol message, C sets the session key sk_A^i, for instance, Π_A^i by using an independently and uniformly chosen element from $\{0,1\}^\kappa$ (note that Π_A^i might reject msg_2' if the decoding algorithm returns $\perp$, and thus no session key is generated in this case, i.e., $\mathrm{acc}_A^i = 0$ and $\mathsf{sk}_A^i = \perp$).

Lemma 6.13 *If* $(K, \ell, \{\mathrm{H}_{\mathsf{hk}} : X \to \{0,1\}^\ell\}_{\mathsf{hk} \in K}, S, \mathsf{Proj} : K \to S)$ *is an* ϵ*-approximate SPH, and* $\mathsf{ECC} : \{0,1\}^\kappa \to \{0,1\}^\ell$ *is an error-correcting code which can correct* 2ϵ*-fraction of errors, then* $\mathbf{Adv}_{\mathcal{A},6}(\kappa) \leq \mathbf{Adv}_{\mathcal{A},7}(\kappa) + \mathsf{negl}(\kappa)$.

Proof First, if both msg_1 and msg_2' were output by previous oracle queries, then the simulator C knows the corresponding hash keys hk_1 and hk_2 needed for computing tk', and it is just a conceptual modification to compute tk' using $(\mathsf{hk}_1, \mathsf{hk}_2)$ and set $\mathsf{sk}_A^i = \mathsf{sk}_B^j$. Second, as discussed in the proof of Lemma 3.11, C knows the secret key sk corresponding to pk in both game G_6 and G_7, it can always decrypt $(\mathsf{label}_2', c_2')$ to obtain the decryption result pw'. Obviously, the modification for the case $pw' = pw$ can only increase the advantage of the adversary $\mathcal{A}$. Moreover, if $pw' \neq pw$, we have $(\mathsf{label}_2', c_2', pw) \notin \bar{L}$. By the smoothness of the ASPH, the value $tk' \in \{0,1\}^\ell$ computed by Π_A^i is statistically close to uniform over $\{0,1\}^\ell$ (because tk' has a term $\mathrm{H}_{\mathsf{hk}_1}(u_2', pw)$ for $c_2' = (u_2', v_2')$). By the assumption on ECC^{-1}, if $sk = \mathsf{ECC}^{-1}(tk' \oplus \Delta') \neq \perp$, then it is statistically close to uniform over $\{0,1\}^\kappa$. Thus, the modification for the case $pw' \neq pw$ in game G_6 can only introduce a negligible statistical difference. In all, we can have that $\mathbf{Adv}_{\mathcal{A},6}(\kappa) \leq \mathbf{Adv}_{\mathcal{A},7}(\kappa) + \mathsf{negl}(\kappa)$ holds. □

Game G_8. This game is similar to game G_7 except that the ciphertext c_1 is replaced with an encryption of $0 \notin \mathcal{D}$ in answering each $\mathsf{Send}_0(A, i, B)$ query.

Lemma 6.14 *If* $\Pi_{\text{pke}} = (\mathsf{KeyGen}, \mathsf{Enc}, \mathsf{Dec})$ *is a CCA-secure scheme, we have that* $|\,\mathbf{Adv}_{\mathcal{A},8}(\kappa) - \mathbf{Adv}_{\mathcal{A},7}(\kappa)| \leq \mathsf{negl}(\kappa)$.

Proof By a standard hybrid argument, it is enough to consider that $\mathcal{A}$ only makes a single $\mathsf{Send}_0(A, i, B)$ query. In this case, the only difference between game G_8 and G_7 is that the encryption of pw is replaced with an encryption of $0 \notin \mathcal{D}$. We now show that any PPT adversary $\mathcal{A}$ that distinguishes the two games with non-negligible advantage can be directly transformed into an algorithm $\mathcal{B}$ that breaks the CCA-security of the underlying Π_{pke} scheme.

Formally, given a challenge public key pk, the algorithm $\mathcal{B}$ sets pk as the CRS of the protocol and simulates the attack environment for $\mathcal{A}$ as in game G_7. When $\mathcal{B}$ has to answer the adversary's $\mathsf{Send}_0(A, i, B)$ query, it first randomly chooses a hash key $\mathsf{hk}_1 \stackrel{\$}{\leftarrow} K$ for the ASPH and computes the projection key $\mathsf{hp}_1 = \mathsf{Proj}(\mathsf{hk}_1)$. Then, $\mathcal{B}$ submits two plaintexts $(pw, 0)$ and $\mathsf{label}_1 := A \| B \| \mathsf{hp}_1$ to its own challenger and obtains a challenge ciphertext c_1^*. Finally, $\mathcal{B}$ sends $(A, \mathsf{hp}_1, c_1^*)$ to the adversary $\mathcal{A}$. When $\mathcal{B}$ has to decrypt some valid label-ciphertext pair $(\mathsf{label}_1', c_1') \neq (\mathsf{label}_1, c_1^*)$, it submits $(\mathsf{label}_1', c_1')$ to its own CCA-security challenger for decryption. At some time, the adversary $\mathcal{A}$ outputs a bit $b \in \{0, 1\}$, $\mathcal{B}$ outputs b as its own guess.

Note that if c_1^* is an encryption of pw, then $\mathcal{B}$ exactly simulates the attack environment of game G_7 for adversary $\mathcal{A}$, else it simulates the attack environment of game G_8 for $\mathcal{A}$. Thus, if $\mathcal{A}$ can distinguish game G_7 and G_8 with non-negligible advantage, then $\mathcal{B}$ can break the CCA-security of the PKE scheme Π_{pke} with the same advantage, which completes the proof. □

Game G_9. This game is similar to game G_8 except that each $\mathsf{Send}_1(B, j, \mathsf{msg}_1' = (A', \mathsf{hp}_1', c_1'))$ query is handled as follows:

- If msg_1' was output by a previous $\mathsf{Send}_0(A', *, B)$ query, the simulator C performs as in game G_8 except that it computes tk directly using the corresponding hash keys $(\mathsf{hk}_1, \mathsf{hk}_2)$, and sets the session key sk_B^j, for instance, Π_B^j by using an independently and uniformly chosen element from $\{0, 1\}^\kappa$;
- Otherwise, C performs exactly as in game G_8.

Lemma 6.15 *If* $(K, \ell, \{\mathrm{H}_{\mathsf{hk}} : X \to \{0, 1\}^\ell\}_{\mathsf{hk} \in K}, S, \mathsf{Proj} : K \to S)$ *is an* ϵ*-approximate SPH, and* $\mathsf{ECC} : \{0, 1\}^\kappa \to \{0, 1\}^\ell$ *is an error-correcting code which can correct* 2ϵ*-fraction of errors, then* $|\,\mathbf{Adv}_{\mathcal{A},9}(\kappa) - \mathbf{Adv}_{\mathcal{A},8}(\kappa)| \leq \mathsf{negl}(\kappa)$.

Proof Note that if msg_1' was output by a previous $\mathsf{Send}_0(A', *, B)$ query, then we have that (1) the simulator C knows the corresponding hash keys $(\mathsf{hk}_1, \mathsf{hk}_2)$ and (2) $c_1' = (u_1', v_1')$ is an encryption of $0 \notin \mathcal{D}$. In other words, C can directly compute tk using $(\mathsf{hk}_1, \mathsf{hk}_2)$ and tk is statistically close to uniform (because $pw \neq 0$, and tk has a term $\mathrm{H}_{\mathsf{hk}_2}(u_1', pw)$ that is statistically close to uniform by the smoothness of the ASPH). Thus, the masking part $\Delta = tk \oplus \mathsf{ECC}(sk)$ output by Π_B^j statistically hides $sk \in \{0, 1\}^\kappa$ from the adversary $\mathcal{A}$. Since sk is essentially uniformly chosen from

$\{0, 1\}^\kappa$, we have that the modification in game G_9 can only introduce a negligible statistical difference, which means that $|\,\mathbf{Adv}_{\mathcal{A},9}(\kappa) - \mathbf{Adv}_{\mathcal{A},8}(\kappa)| \leq \mathsf{negl}(\kappa)$. □

Game G_{10}. This game is similar to game G_9 except that each $\mathsf{Send}_1(B, j, \mathsf{msg}_1' = (A', \mathsf{hp}_1', c_1'))$ query is handled as follows:

- If msg_1' was output by a previous $\mathsf{Send}_0(A', *, B)$ query, the simulator C performs as in game G_9 except that the ciphertext c_2 is replaced with an encryption of $0 \notin \mathcal{D}$;
- Otherwise, C performs exactly as in game G_9.

Lemma 6.16 *If* $\Pi_{\text{pke}} = (\mathsf{KeyGen}, \mathsf{Enc}, \mathsf{Dec})$ *is a splittable CCA-secure scheme, then* $|\,\mathbf{Adv}_{\mathcal{A},10}(\kappa) - \mathbf{Adv}_{\mathcal{A},9}(\kappa)| \leq \mathsf{negl}(\kappa)$.

Proof As before, it is enough to consider that $\mathcal{A}$ only makes a single $\mathsf{Send}_1(B, j, \mathsf{msg}_1' = (A', \mathsf{hp}_1', c_1'))$ query with msg_1' output by some $\Pi_{A'}^i$. We now show that any PPT adversary $\mathcal{A}$ that distinguishes the two games with non-negligible advantage can be directly transformed into an algorithm $\mathcal{B}$ that breaks the modified CCA-security game of the underlying Π_{pke} scheme with the same advantage.

Formally, given a challenge public key pk, the algorithm $\mathcal{B}$ sets pk as the CRS of the protocol and interacts with $\mathcal{A}$ as in game G_9. When $\mathcal{B}$ has to answer a $\mathsf{Send}_1(B, j, \mathsf{msg}_1' = (A', \mathsf{hp}_1', c_1'))$ query for some $c_1' = (u_1', v_1')$, it first randomly chooses a hash key $\mathsf{hk}_2 \xleftarrow{\$} K$ for the ASPH, a random session key $sk \xleftarrow{\$} \{0, 1\}^\kappa$, and computes $\mathsf{hp}_2 = \mathsf{Proj}(\mathsf{hk}_2)$. Then, $\mathcal{B}$ submits two plaintexts $(pw, 0)$ to its own challenger. After obtaining u_2^*, $\mathcal{B}$ computes $tk = \mathrm{H}_{\mathsf{hk}_1}(u_2^*, pw) \oplus \mathrm{H}_{\mathsf{hk}_2}(u_1', pw)$, $\Delta = tk \oplus \mathsf{ECC}(sk)$, and submits $\mathsf{label}_2 := A' \| B \| \mathsf{hp}_1' \| c_1' \| \mathsf{hp}_2 \| \Delta$ to its own modified CCA-security challenger to obtain the challenge ciphertext $c_2^* = (u_2^*, v_2^*)$. Finally, $\mathcal{B}$ sends $(\mathsf{hp}_2, c_2^*, \Delta)$ to the adversary $\mathcal{A}$. When $\mathcal{B}$ has to decrypt some valid label-ciphertext pair $(\mathsf{label}_2', c_2') \neq (\mathsf{label}_2, c_2^*)$, it submits $(\mathsf{label}_2', c_2')$ to its own challenger for decryption. At some time, the adversary $\mathcal{A}$ outputs a bit $b \in \{0, 1\}$, $\mathcal{B}$ outputs b as its own guess.

Note that if c_2^* is an encryption of pw, then $\mathcal{B}$ perfectly simulates the attack environment of game G_9 for adversary $\mathcal{A}$, else it simulates the attack environment of G_{10} for $\mathcal{A}$. Thus, if $\mathcal{A}$ can distinguish game G_9 and G_{10} with non-negligible advantage, then algorithm $\mathcal{B}$ can break the modified CCA-security of the PKE scheme Π_{pke} with the same advantage, which completes the proof. □

Lemma 6.17 *If the adversary* $\mathcal{A}$ *only makes at most* $Q(\kappa)$ *times on-line attacks, then we have that* $\mathbf{Adv}_{\mathcal{A},10}(\kappa) \leq Q(\kappa)/|\mathcal{D}| + \mathsf{negl}(\kappa)$.

Proof Let $\mathcal{E}$ be the event that $\mathcal{A}$ submits a ciphertext that decrypts to the real password pw. If $\mathcal{E}$ does not happen, we have that the advantage of $\mathcal{A}$ is negligible in κ (because all the session keys are uniformly chosen at random). Now, we estimate the probability that $\mathcal{E}$ happens. Since in game G_{10}, all the ciphertexts output by oracle queries are encryptions of $0 \notin \mathcal{D}$, the adversary cannot obtain useful information of the real password pw via the oracle queries. Thus, for any adversary $\mathcal{A}$ that makes at most $Q(\kappa)$ times on-line attacks, the probability that $\mathcal{E}$ happens is at most $Q(\kappa)/|\mathcal{D}|$, i.e., $\Pr[E] \leq Q(\kappa)/|\mathcal{D}|$. By a simple calculation, we have $\mathbf{Adv}_{\mathcal{A},10}(\kappa) \leq Q(\kappa)/|\mathcal{D}| + \mathsf{negl}(\kappa)$. □

In all, we have that $\mathbf{Adv}_{\mathcal{A},0}(\kappa) \leq Q(\kappa)/|\mathcal{D}| + \mathsf{negl}(\kappa)$ by Lemma 6.7–6.17. This completes the proof of Theorem 6.3. □

6.4.4 Instantiations

In order to construct a splittable PKE with associated ASPH from lattices, the basic idea is to incorporate the specific algebraic properties of lattices into the Naor-Yung paradigm [90, 96], which is a generic construction of CCA-secure PKE scheme from any CPA-secure PKE scheme and simulation-sound non-interactive zero-knowledge (NIZK) proof [96], and was used to achieve the first one-round PAKEs from DDH and decisional linear assumptions [72].

Looking ahead, we will use a CPA-secure PKE scheme from lattices and a simulation-sound NIZK proof for specific statements, so that we can freely apply Lemmas 2.1 and 2.13 to construct a non-adaptive approximate SPH and achieve the stronger smoothness property. Formally, we need a simulation-sound NIZK proof for the following relation:

$$R_{pke} := \left\{ \begin{array}{l} ((\mathbf{A}_0, \mathbf{A}_1, \mathbf{c}_0, \mathbf{c}_1, \beta), (\mathbf{s}_0, \mathbf{s}_1, \mathbf{w})) : \\ \qquad \left\| \mathbf{c}_0 - \mathbf{A}_0^t \begin{pmatrix} \mathbf{s}_0 \\ 1 \\ \mathbf{w} \end{pmatrix} \right\| \leq \beta \wedge \left\| \mathbf{c}_1 - \mathbf{A}_1^t \begin{pmatrix} \mathbf{s}_1 \\ 1 \\ \mathbf{w} \end{pmatrix} \right\| \leq \beta \\ , \end{array} \right\}$$

where $\mathbf{A}_0, \mathbf{A}_1 \in \mathbb{Z}_q^{n\times m}$, $\mathbf{c}_0, \mathbf{c}_1 \in \mathbb{Z}_q^m$, $\beta \in \mathbb{R}$, $\mathbf{s}_0, \mathbf{s}_1 \in \mathbb{Z}_q^{n_1}$, $\mathbf{w} \in \mathbb{Z}_q^{n_2}$ for some integers $n = n_1 + n_2 + 1, m, q \in \mathbb{Z}$. Note that under the existence of (enhanced) trapdoor permutations, there exist NIZK proofs with efficient prover for any NP relation [16, 47, 57]. Moreover, Sahai [96] showed that one can transform any general NIZK proof into a simulation-sound one. Thus, there exists a simulation-sound NIZK proof with efficient prover for the relation R_{pke}.

For the purpose, we require that the NIZK proof supports labels [1], which can be obtained from a normal NIZK proof by a standard way (e.g., appending the label to the statement [46, 72]). Let (CRSGen, Prove, Verify) be a labeled NIZK proof for relation R_{pke}. The algorithm $\mathsf{CRSGen}(1^\kappa)$ takes a security parameter κ as input, outputs a common reference string crs, i.e., $crs \leftarrow \mathsf{CRSGen}(1^\kappa)$. The algorithm Prove takes a pair $(x, wit) = ((\mathbf{A}_0, \mathbf{A}_1, \mathbf{c}_0, \mathbf{c}_1, \beta), (\mathbf{s}_0, \mathbf{s}_1, \mathbf{w})) \in R_{pke}$ and a label $\in \{0, 1\}^*$ as inputs, outputs a proof π, i.e., $\pi \leftarrow \mathsf{Prove}(crs, x, wit, \mathsf{label})$. The algorithm Verify takes as inputs x, a proof π and a label $\in \{0, 1\}^*$, outputs a bit $b \in \{0, 1\}$ indicating whether π is valid or not, i.e., $b \leftarrow \mathsf{Verify}(crs, x, \pi, \mathsf{label})$. For completeness, we require that for any $(x, wit) \in R_{pke}$ and any label $\in \{0, 1\}^*$, $\mathsf{Verify}(crs, x, \mathsf{Prove}(crs, x, wit, \mathsf{label}), \mathsf{label}) = 1$.

6.4.5 A Splittable PKE from Lattices

Let $n_1, n_2 \in \mathbb{Z}$ and prime q be polynomials in the security parameter κ. Let $n = n_1 + n_2 + 1$, $m = O(n \log q) \in \mathbb{Z}$, and $\alpha, \beta \in \mathbb{R}$ be the system parameters. Let $\mathcal{P} = \{-\alpha q + 1, \ldots, \alpha q - 1\}^{n_2}$ be the plaintext space. Let (CRSGen, Prove, Verify) be a simulation-sound NIZK proof for R_{pke}. The PKE scheme $\Pi_{\text{pke}} =$ (KeyGen, Enc, Dec) is defined as follows:

KeyGen(1^κ): Given the security parameter κ, compute $(\mathbf{A}_0, \mathbf{R}_0) \leftarrow \mathsf{TrapGen}(1^n, 1^m, q)$, $(\mathbf{A}_1, \mathbf{R}_1) \leftarrow \mathsf{TrapGen}(1^n, 1^m, q)$ and $crs \leftarrow \mathsf{CRSGen}(1^\kappa)$. Return the public and secret key pair $(\mathsf{pk}, \mathsf{sk}) = ((\mathbf{A}_0, \mathbf{A}_1, crs), \mathbf{R}_0)$.

Enc(pk, label, $\mathbf{w} \in \mathcal{P}$): Given $\mathsf{pk} = (\mathbf{A}_0, \mathbf{A}_1, crs)$, label $\in \{0, 1\}^*$ and plaintext $\mathbf{w}$, randomly choose $\mathbf{s}_0, \mathbf{s}_1 \xleftarrow{\$} \mathbb{Z}_q^{n_1}$, $\mathbf{e}_0, \mathbf{e}_1 \xleftarrow{\$} D_{\mathbb{Z}^m, \alpha q}$. Finally, return the ciphertext $C = (\mathbf{c}_0, \mathbf{c}_1, \pi)$, where

$$\mathbf{c}_0 = \mathbf{A}_0^t \begin{pmatrix} \mathbf{s}_0 \\ 1 \\ \mathbf{w} \end{pmatrix} + \mathbf{e}_0, \quad \mathbf{c}_1 = \mathbf{A}_1^t \begin{pmatrix} \mathbf{s}_1 \\ 1 \\ \mathbf{w} \end{pmatrix} + \mathbf{e}_1,$$

and $\pi \leftarrow \mathsf{Prove}(crs, (\mathbf{A}_0, \mathbf{A}_1, \mathbf{c}_0, \mathbf{c}_1, \beta), (\mathbf{s}_0, \mathbf{s}_1, \mathbf{w}), \mathsf{label})$.

Dec(sk, label, C): Given $\mathsf{sk} = \mathbf{R}_0$, label $\in \{0, 1\}^*$ and ciphertext $C = (\mathbf{c}_0, \mathbf{c}_1, \pi)$, if $\mathsf{Verify}(crs, (\mathbf{A}_0, \mathbf{A}_1, \mathbf{c}_0, \mathbf{c}_1, \beta), \pi, \mathsf{label}) = 0$, return $\perp$. Otherwise, compute

$$\mathbf{t} = \begin{pmatrix} \mathbf{s}_0 \\ 1 \\ \mathbf{w} \end{pmatrix} \leftarrow \mathsf{Solve}(\mathbf{A}_0, \mathbf{R}_0, \mathbf{c}_0),$$

and return $\mathbf{w} \in \mathbb{Z}_q^{n_2}$ (note that a valid π ensures that $\mathbf{t}$ has the right form).

Correctness. By Lemma 2.3, we have that $\|\mathbf{e}_0\|, \|\mathbf{e}_1\| \leq \alpha q \sqrt{m}$ hold with overwhelming probability. Thus, it is enough to set $\beta \geq \alpha q \sqrt{m}$ for the NIZK proof to work. By Lemma 2.24, we have that $s_1(\mathbf{R}_0) \leq \sqrt{m} \cdot \omega(\sqrt{\log n})$, and the Solve algorithm can recover $\mathbf{t}$ from any $\mathbf{y} = \mathbf{A}_0^t \mathbf{t} + \mathbf{e}_0$ as long as $\|\mathbf{e}_0\| \cdot \sqrt{m} \cdot \omega(\sqrt{\log n}) \leq q$. Thus, we can set the parameters appropriately to satisfy the correctness. Besides, for the hardness of the LWE assumption, we need $\alpha q \geq 2\sqrt{n_1}$. In order to obtain an ϵ-approximate SPH function, we require $\beta \leq \sqrt{q}/4$, $\sqrt{mq}/4 \cdot \omega(\sqrt{\log n}) \leq q$ and $\alpha \gamma m < \epsilon/8$, where $\gamma \geq 4\sqrt{mq}$ is the parameter for ASPH in Sect. 6.4.6. In all, fix $\epsilon \in (0, 1/2)$, we can set the parameters $m, \alpha, \beta, q, \gamma$ as follows (where $c \geq 0$ is a real such that q is a prime) for both correctness and security:

$$\begin{array}{ll} m = O(n \log n), & \beta > 16m\sqrt{mn}/\epsilon \\ q = 16\beta^2 + c & \alpha = 2\sqrt{n}/q \\ \gamma = 4\sqrt{mq} & \end{array} \tag{6.1}$$

.

In practice, given a target length of session keys, one can first choose an appropriate ECC scheme, and then set other parameters to satisfy Eq. (6.1). For example, the Reed-Muller code with $\ell = 1024$ can be used to encode a 176-bit session key with $\epsilon = 1/32$, and thus is far enough to establish a 128-bit session key. In the setting of $\mathcal{P} = \{-\alpha q + 1, \ldots, \alpha q - 1\}^7$ (i.e., $n_2 = 7$), one can set $n_1 \approx 2^{11}, m \approx 2^{19}, \alpha \approx 2^{-83.5}, \beta \approx 2^{43}$ and $q \approx 2^{90}$, which provides about 105-bit security by the lwe-estimator [5]. We note that there are many tradeoffs between the parameters, and it is possible to give a more tight parameter for any targeted security level. One can also reduce the parameters by using a careful proof of Lemma 2.13 with smaller γ.
Security. For any $C = (\mathbf{c}_0, \mathbf{c}_1, \pi) \leftarrow \mathsf{Enc}(\mathsf{pk}, \mathsf{label}, \mathbf{w})$, let r be the corresponding random coins which includes $(\mathbf{s}_0, \mathbf{s}_1, \mathbf{e}_0, \mathbf{e}_1)$ for generating $(\mathbf{c}_0, \mathbf{c}_1)$, and the randomness used for generating π. We define functions (f, g) as follows:

- The function f takes $(\mathsf{pk}, \mathbf{w}, r)$ as inputs, computes $(\mathbf{c}_0, \mathbf{c}_1)$ with random coins r, and returns $(\mathbf{c}_0, \mathbf{c}_1)$, i.e., $(\mathbf{c}_0, \mathbf{c}_1) = f(\mathsf{pk}, \mathbf{w}, r)$;
- The function g takes $(\mathsf{pk}, \mathsf{label}, \mathbf{w}, r)$ as inputs, computes the Prove algorithm with random coins r and returns the result π, i.e., $\pi = g(\mathsf{pk}, \mathsf{label}, \mathbf{w}, r)$.

We fix the two functions (f, g) in the rest of Sect. 6.4.4 and have the following theorem for security:

Theorem 6.4 *Let $n = n_1 + n_2 + 1, m \in \mathbb{Z}, \alpha, \beta, \gamma \in \mathbb{R}$ and prime q be as in Eq. (6.1). If $\mathrm{LWE}_{n_1,q,\alpha}$ is hard, $(\mathsf{CRSGen}, \mathsf{Prove}, \mathsf{Verify})$ is a simulation-sound NIZK proof, then the scheme Π_{pke} is a splittable CCA-secure PKE scheme.*

Since Π_{pke} is essentially an instantiation of the Naor-Yung paradigm [90, 96] using a special LWE-based CPA scheme (similar to the ones in [71, 89]), and a SS-NIZK for a special relation R_{pke}, this theorem can be shown by adapting the proof techniques in [90, 96].

6.4.6 An Associated Approximate SPH

Fix a public key $\mathsf{pk} = (\mathbf{A}_0, \mathbf{A}_1, crs)$ of the PKE scheme Π_{pke}. Given any string $\mathsf{label} \in \{0,1\}^*$ and $C = (\mathbf{c}_0, \mathbf{c}_1, \pi)$, we say that (label, C) is a valid label-ciphertext pair with respect to pk if $\mathsf{Verify}(crs, (\mathbf{A}_0, \mathbf{A}_1, \mathbf{c}_0, \mathbf{c}_1, \beta), \pi, \mathsf{label}) = 1$. Let sets X, L and $\bar{L}$ be defined as in Sect. 6.4.2.1. Define the associated ASPH function $(K, \ell, \{\mathsf{H}_{\mathsf{hk}} : X \to \{0,1\}^{\ell}\}_{\mathsf{hk} \in K}, S, \mathsf{Proj} : K \to S)$ for Π_{pke} as follows:

- The hash key is an ℓ-tuple of vectors $\mathsf{hk} = (\mathbf{x}_1, \ldots, \mathbf{x}_\ell)$, where $\mathbf{x}_i \sim D_{\mathbb{Z}^m,\gamma}$. Write $\mathbf{A}_0^t = (\mathbf{B} \| \mathbf{U}) \in \mathbb{Z}_q^{m \times n}$ such that $\mathbf{B} \in \mathbb{Z}_q^{m \times n_1}$ and $\mathbf{U} \in \mathbb{Z}_q^{m \times (n_2+1)}$. Define the projection key $\mathsf{hp} = \mathsf{Proj}(\mathsf{hk}) = (\mathbf{u}_1, \ldots, \mathbf{u}_\ell)$, where $\mathbf{u}_i = \mathbf{B}^t \mathbf{x}_i$.
- $\mathsf{H}_{\mathsf{hk}}(x) = \mathsf{H}_{\mathsf{hk}}((\mathbf{c}_0, \mathbf{c}_1), \mathbf{w})$: Given $\mathsf{hk} = (\mathbf{x}_1, \ldots, \mathbf{x}_\ell)$ and $x = (\mathsf{label}, C, \mathbf{w}) \in X$ for some $C = (\mathbf{c}_0, \mathbf{c}_1, \pi)$, compute $z_i = \mathbf{x}_i^t \left(\mathbf{c}_0 - \mathbf{U} \begin{pmatrix} 1 \\ \mathbf{w} \end{pmatrix} \right)$ for $i \in \{1, \ldots, \ell\}$.

Then, treat each z_i as a number in $\{-(q-1)/2, \ldots, (q-1)/2\}$. If $z_i = 0$, then set $b_i \xleftarrow{\$} \{0, 1\}$. Else, set

$$b_i = \begin{cases} 0 \text{ if } z_i < 0 \\ 1 \text{ if } z_i > 0 \end{cases}.$$

Finally, return $\mathsf{H}_{\mathsf{hk}}((\mathbf{c}_0, \mathbf{c}_1), \mathbf{w}) = (b_1, \ldots, b_\ell)$.

- $\mathsf{Hash}(\mathsf{hp}, x, \mathbf{s}_0) = \mathsf{Hash}(\mathsf{hp}, ((\mathbf{c}_0, \mathbf{c}_1), \mathbf{w}), \mathbf{s}_0)$: Given $\mathsf{hp} = (\mathbf{u}_1, \ldots, \mathbf{u}_\ell)$, $x = (\mathsf{label}, (\mathbf{c}_0, \mathbf{c}_1, \pi), \mathbf{w}) \in L$ and $\mathbf{s}_0 \in \mathbb{Z}_q^{n_1}$ such that $\mathbf{c}_0 = \mathbf{B}\mathbf{s}_0 + \mathbf{U}\begin{pmatrix} 1 \\ \mathbf{w} \end{pmatrix} + \mathbf{e}_0$ for some $\mathbf{e}_0 \xleftarrow{\$} D_{\mathbb{Z}^m, \alpha q}$, compute $z'_i = \mathbf{u}_i^t \mathbf{s}_0$. Then, treat each z'_i as a number in $\{-(q-1)/2, \ldots, (q-1)/2\}$. If $z'_i = 0$, then set $b'_i \xleftarrow{\$} \{0, 1\}$. Else, set

$$b'_i = \begin{cases} 0 \text{ if } z_i < 0 \\ 1 \text{ if } z'_i > 0 \end{cases}.$$

Finally, return $\mathsf{Hash}(\mathsf{hp}, ((\mathbf{c}_0, \mathbf{c}_1), \mathbf{w}), \mathbf{s}_0) = (b'_1, \ldots, b'_\ell)$.

Theorem 6.5 *Let $\epsilon \in (0, 1/2)$, and let $n, m, q, \alpha, \beta, \gamma$ be as in Theorem 6.4. Let ℓ be polynomial in the security parameter κ. Then, $(K, \ell, \{\mathsf{H}_{\mathsf{hk}} : X \to \{0, 1\}^\ell\}_{\mathsf{hk} \in K}, S, \mathsf{Proj} : K \times C_{pk} \to S)$ is an ϵ-approximate SPH as in Definition 6.6.*

Proof Clearly, there are efficient algorithms for (1) sampling a hash key $\mathsf{hk} \xleftarrow{\$} K$, (2) computing $\mathsf{H}_{\mathsf{hk}}((\mathbf{c}_0, \mathbf{c}_1), \mathbf{w})$ for all $\mathsf{hk} \in K$ and all $x = (\mathsf{label}, C, \mathbf{w}) \in X$ with $C = (\mathbf{c}_0, \mathbf{c}_1, \pi)$ and (3) computing $\mathsf{hp} = \mathsf{Proj}(\mathsf{hk})$ for all $\mathsf{hk} \in K$. In addition, for any $x = (\mathsf{label}, C, \mathbf{w}) \in L$, the values $\mathsf{Hash}(\mathsf{hp}, ((\mathbf{c}_0, \mathbf{c}_1), \mathbf{w}), \mathbf{s}_0)$ can be efficiently computed, where $C = (\mathbf{c}_0, \mathbf{c}_1, \pi)$ and $\mathbf{c}_0$ is generated using the randomness $\mathbf{s}_0$. In the following, we show that the above construction also satisfies the approximate correctness and the smoothness given in Definition 6.6.

First, let $C = (\mathbf{c}_0, \mathbf{c}_1, \pi)$ be a ciphertext such that $\mathbf{c}_0 = \mathbf{B}\mathbf{s}_0 + \mathbf{U}\begin{pmatrix} 1 \\ \mathbf{w} \end{pmatrix} + \mathbf{e}_0$ for some $\mathbf{s}_0 \xleftarrow{\$} \mathbb{Z}_q^{n_1}$ and $\mathbf{e}_0 \xleftarrow{\$} D_{\mathbb{Z}^m, \alpha q}$. For any $i \in \{1, \ldots, \ell\}$, we have that $z_i = \mathbf{x}_i^t \left(\mathbf{c}_0 - \mathbf{U}\begin{pmatrix} 1 \\ \mathbf{w} \end{pmatrix}\right) = \mathbf{x}_i^t(\mathbf{B}\mathbf{s}_0 + \mathbf{e}_0) = \mathbf{u}_i^t\mathbf{s}_0 + \mathbf{x}_i^t\mathbf{e}_0$. This means that $|z_i - z'_i| \leq |\mathbf{x}_i^t\mathbf{e}_0| \leq \gamma\sqrt{m} \cdot \alpha q\sqrt{m} < \epsilon/2 \cdot q/4$ with overwhelming probability. Using the fact that $\mathbf{B} \in \mathbb{Z}_q^{m \times n_1}$ is statistically close to uniform, we have that $\mathbf{u}_i = \mathbf{B}^t\mathbf{x}_i$ is statistically close to uniform over $\mathbb{Z}_q^{n_1}$ for all $i \in \{1, \ldots, \ell\}$ by Lemma 2.11. Moreover, for any non-zero $\mathbf{s}_0 \in \mathbb{Z}_q^n$ (note that the probability that $\mathbf{s}_0 = 0$ is at most q^{-n_1}, which is negligible in κ), we have that $z'_i = \mathbf{u}_i^t\mathbf{s}_0$ is uniformly random. By a simple calculation, we have the probability that $b_i \neq b'_i$ is at most $\frac{\epsilon}{2}$. By a Chernoff bound, the Hamming distance between $\mathsf{H}_{\mathsf{hk}}((\mathbf{c}_0, \mathbf{c}_1), \mathbf{w}) = (b_1, \ldots, b_\ell)$ and $\mathsf{Hash}(\mathsf{hp}, ((\mathbf{c}_0, \mathbf{c}_1), \mathbf{w}), \mathbf{s}_0) = (b'_1, \ldots, b'_\ell)$ is at most $\epsilon\ell$ with overwhelming probability. This shows the approximate correctness.

Second, for any $C = (\mathbf{c}_0, \mathbf{c}_1, \pi)$ and $((\mathsf{label}, C), \mathbf{w}) \in X \backslash \bar{L}$, let $\mathbf{w}'$ be the decryption result of (label, C) using the secret key sk corresponding to pk (note that the validity of π ensures that the existence of $\mathbf{w}' \neq \perp$). By assumption, we know that

$\mathbf{w}' \neq \mathbf{w}$. Let $\mathbf{y} = \mathbf{c}_0 - \mathbf{U}\begin{pmatrix} 1 \\ \mathbf{w} \end{pmatrix} \in \mathbb{Z}_q^m$ and $\mathbf{y}' = \mathbf{c}_0 - \mathbf{U}\begin{pmatrix} 1 \\ \mathbf{w}' \end{pmatrix}$, we have $\mathsf{dist}(\mathbf{y}', \Lambda_q(\mathbf{B}^t)) \leq \beta \leq \frac{\sqrt{q}}{4}$ by the soundness of the NIZK proof π. Note that the matrix $\mathbf{U}$ is statistically close to uniform by Lemma 2.24. Hence, with overwhelming probability we always have that

$$\mathbf{y} = \mathbf{c}_0 - \mathbf{U}\begin{pmatrix} 1 \\ \mathbf{w} \end{pmatrix} \in Y = \left\{\tilde{\mathbf{y}} \in \mathbb{Z}_q^m : \forall a \in \mathbb{Z}_q \backslash \{0\}, \mathsf{dist}(a\tilde{\mathbf{y}}, \Lambda_q(\mathbf{B}^t)) \geq \sqrt{q}/4\right\}$$

for any $C = (\mathbf{c}_0, \mathbf{c}_1, \pi)$ and $((\mathsf{label}, C), \mathbf{w}) \in X \backslash \bar{L}$ by Lemma 2.1. In addition, if $z_i = \mathbf{x}_i^t \mathbf{y}$ is uniformly random over $\mathbb{Z}_q$, then by the definition the i-th bit b_i of $\mathsf{H}_{\mathsf{hk}}((\mathbf{c}_0, \mathbf{c}_1), \mathbf{w})$ is uniformly random over $\{0, 1\}$. Thus, for smoothness, it suffices to show that for any (even unbounded) function $h : \mathbb{Z}_q^{n_1 \times \ell} \rightarrow Y$, $\mathsf{hk} = (\mathbf{x}_1, \ldots, \mathbf{x}_\ell) \xleftarrow{\$} (D_{\mathbb{Z}^m, \gamma})^\ell$, $\mathsf{hp} = (\mathbf{B}^t\mathbf{x}_1, \ldots, \mathbf{B}^t\mathbf{x}_\ell) = \mathsf{Proj}(\mathsf{hk})$, $\mathbf{y} = h(\mathsf{hp})$, $\mathbf{z} = (\mathbf{x}_1^t\mathbf{y}, \ldots, \mathbf{x}_\ell^t\mathbf{y})$ and $\mathbf{z}' \xleftarrow{\$} \mathbb{Z}_q^\ell$, the statistical distance between $(\mathsf{hp}, \mathbf{z})$ and $(\mathsf{hp}, \mathbf{z}')$ is negligible in κ. Since $\gamma \geq 4\sqrt{mq}$ and $\mathbf{B} \in \mathbb{Z}_q^{m \times n_1}$ is statistically close to uniform, by Lemma 2.13 we have that for any function $h' : \mathbb{Z}_q^{n_1} \rightarrow Y$, the distribution of $(\mathbf{B}^t\mathbf{x}, \mathbf{x}^t\mathbf{y}')$ is statistically close to uniform over $\mathbb{Z}_q^{n_1} \times \mathbb{Z}_q$, where $\mathbf{x} \sim D_{\mathbb{Z}^m, \gamma}$ and $\mathbf{y}' = h'(\mathbf{B}^t\mathbf{x})$. Using the facts that Lemma 2.13 holds for arbitrary choice of $h' : \mathbb{Z}_q^{n_1} \rightarrow Y$ and that each $\mathbf{x}_i$ is independently chosen from $D_{\mathbb{Z}^m, \gamma}$, we have that $(\mathsf{hp}, \mathbf{z}) = ((\mathbf{B}^t\mathbf{x}_1, \ldots, \mathbf{B}^t\mathbf{x}_\ell), (\mathbf{x}_1^t\mathbf{y}, \ldots, \mathbf{x}_\ell^t\mathbf{y}))$ is statistically close to uniform by a standard hybrid argument. This completes the proof Theorem 6.5. □

6.4.7 *Achieving Simulation-Sound NIZK for R_{pke} on Lattices*

In this section, we will show how to construct an simulation-sound NIZK for R_{pke} from lattices in the random oracle model. Formally, let $n = n_1 + n_2 + 1, m, q \in \mathbb{Z}$ be defined as in Sect. 6.4.5. We begin by defining a variant relation R'_{pke} of R_{pke} (in the l_∞ form):

$$R'_{pke} := \left\{ \begin{array}{l} ((\mathbf{A}_0, \mathbf{A}_1, \mathbf{c}_0, \mathbf{c}_1, \zeta), (\mathbf{s}_0, \mathbf{s}_1, \mathbf{w})) : \|\mathbf{w}\|_\infty \leq \zeta \, \wedge \\ \qquad \left\| \mathbf{c}_0 - \mathbf{A}_0^t \begin{pmatrix} \mathbf{s}_0 \\ 1 \\ \mathbf{w} \end{pmatrix} \right\|_\infty \leq \zeta \wedge \left\| \mathbf{c}_1 - \mathbf{A}_1^t \begin{pmatrix} \mathbf{s}_1 \\ 1 \\ \mathbf{w} \end{pmatrix} \right\|_\infty \leq \zeta \end{array} \right\},$$

where $\mathbf{A}_0, \mathbf{A}_1 \in \mathbb{Z}_q^{n \times m}$, $\mathbf{c}_0, \mathbf{c}_1 \in \mathbb{Z}_q^m$, $\zeta \in \mathbb{R}$, $\mathbf{s}_0, \mathbf{s}_1 \in \mathbb{Z}_q^{n_1}$ and $\mathbf{w} \in \mathbb{Z}_q^{n_2}$. Write $\mathbf{A}_0^t = (\mathbf{B}_0 \| \mathbf{U}_0) \in \mathbb{Z}_q^{m \times n_1} \times \mathbb{Z}_q^{m \times (n_2+1)}$. Note that for large enough $m = O(n_1 \log q)$, the rows of a uniformly random $\mathbf{B}_0 \in \mathbb{Z}_q^{m \times n_1}$ generate $\mathbb{Z}_q^{n_1}$ with overwhelming probability. By the duality [88], one can compute a parity check matrix $\mathbf{G}_0 \in \mathbb{Z}_q^{(m-n_1) \times m}$ such that (1) the columns of $\mathbf{G}_0$ generate $\mathbb{Z}_q^{m-n_1}$, and (2) $\mathbf{G}_0\mathbf{B}_0 = \mathbf{0}$. Now, let vector $\mathbf{e}_0 \in \mathbb{Z}^m$ satisfy

$$\mathbf{c}_0 = \mathbf{A}_0^t \begin{pmatrix} \mathbf{s}_0 \\ 1 \\ \mathbf{w} \end{pmatrix} + \mathbf{e}_0 = \mathbf{B}_0\mathbf{s}_0 + \mathbf{U}_0 \begin{pmatrix} 1 \\ \mathbf{w} \end{pmatrix} + \mathbf{e}_0. \tag{6.2}$$

By multiplying Eq. (6.2) with matrix $\mathbf{G}_0$ and rearranging the terms, we have the equation $\mathbf{D}_0\mathbf{w} + \mathbf{G}_0\mathbf{e}_0 = \mathbf{b}_0$, where $(\mathbf{a}_0\|\mathbf{D}_0) = \mathbf{G}_0\mathbf{U}_0 \in \mathbb{Z}_q^{(m-n_1)\times(1+n_2)}$, and $\mathbf{b}_0 = \mathbf{G}_0\mathbf{c}_0 - \mathbf{a}_0 \in \mathbb{Z}_q^{m-n_1}$. Similarly, by letting $\mathbf{A}_1^t = (\mathbf{B}_1\|\mathbf{U}_1)$ and $\mathbf{c}_1 = \mathbf{B}_1\mathbf{s}_1 + \mathbf{U}_1 \begin{pmatrix} 1 \\ \mathbf{w} \end{pmatrix} + \mathbf{e}_1$, we can compute an equation $\mathbf{D}_1\mathbf{w} + \mathbf{G}_1\mathbf{e}_1 = \mathbf{b}_1$, where $\mathbf{G}_1 \in \mathbb{Z}_q^{(m-n_1)\times m}$ is a parity check matrix for $\mathbf{B}_1$, $(\mathbf{a}_1\|\mathbf{D}_1) = \mathbf{G}_1\mathbf{U}_1 \in \mathbb{Z}_q^{(m-n_1)\times(1+n_2)}$, and $\mathbf{b}_1 = \mathbf{G}_1\mathbf{c}_1 - \mathbf{a}_1 \in \mathbb{Z}_q^{m-n_1}$. As in [77, 82], in order to show $((\mathbf{A}_0, \mathbf{A}_1, \mathbf{c}_0, \mathbf{c}_1, \zeta), (\mathbf{s}_0, \mathbf{s}_1, \mathbf{w})) \in \mathbf{R}'_{pke}$, it is enough to prove that there exists $(\mathbf{w}, \mathbf{e}_0, \mathbf{e}_1)$ such that $((\mathbf{D}_0, \mathbf{G}_0, \mathbf{D}_1, \mathbf{G}_1, \mathbf{b}_0, \mathbf{b}_1, \zeta), (\mathbf{w}, \mathbf{e}_0, \mathbf{e}_1)) \in \tilde{R}'_{pke}$:

$$\tilde{R}'_{pke} := \left\{ \begin{array}{l} ((\mathbf{D}_0, \mathbf{G}_0, \mathbf{D}_1, \mathbf{G}_1, \mathbf{b}_0, \mathbf{b}_1, \zeta), (\mathbf{w}, \mathbf{e}_0, \mathbf{e}_1)) : \\ \qquad \begin{pmatrix} \mathbf{D}_0 & \mathbf{G}_0 & \mathbf{0} \\ \mathbf{D}_1 & \mathbf{0} & \mathbf{G}_1 \end{pmatrix} \begin{pmatrix} \mathbf{w} \\ \mathbf{e}_0 \\ \mathbf{e}_1 \end{pmatrix} = \begin{pmatrix} \mathbf{b}_0 \\ \mathbf{b}_1 \end{pmatrix} \wedge \\ \qquad \|\mathbf{w}\|_\infty \le \zeta \wedge \|\mathbf{e}_0\|_\infty \le \zeta \wedge \|\mathbf{e}_1\|_\infty \le \zeta \end{array} \right\},$$

which is essentially a special case of the ISIS relation R_{ISIS} (in the $_\infty$ norm):

$$R_{ISIS} := \{((\mathbf{M}, \mathbf{b}, \zeta), \mathbf{x}) : \mathbf{M}\mathbf{x} = \mathbf{b} \wedge \|\mathbf{x}\|_\infty \le \zeta\}.$$

Notice that if there is a three-round public-coin honest-verifier zero-knowledge (HVZK) proof for the relation R_{ISIS}, one can obtain an NIZK proof for R_{ISIS} by applying the Fiat-Shamir transform [48] in the random oracle model [22]. Moreover, if the basic protocol additionally has the *quasi unique responses* property [19, 45, 49], the literature [19, 20, 45] shows that the resulting NIZK proof derived from the Fiat-Shamir transform meets the simulation-soundness needed for constructing CCA-secure PKE via the Naor-Yung paradigm [90, 96]. Fortunately, we do have an efficient three-round public-coin HVZK proof with quasi unique responses in [80],[8] which is extended from the Stern protocol [99] and has the same structure as the latter. Specifically, the protocol [80] has three messages (a, e, z), where a consists of several commitments sent by the prover, e is the challenge sent by the verifier, and the third message z (i.e., the response) consists of the openings to the commitments specified by the challenge e.

Note that the quasi unique responses property [45, 49] essentially requires that it is computationally infeasible for an adversary to output (a, e, z) and (a, e, z') such that both (a, e, z) and (a, e, z') are valid. Thus, if, as is usually the case, the

[8] More precisely, the authors [80] showed a zero-knowledge proof of knowledge for R_{ISIS}, which has a constant soundness error about 2/3. By repeating the basic protocol $t = \omega(\log n)$ times in parallel, we can obtain a desired public-coin HVZK proof for R_{ISIS} with negligible soundness error [79, 81].

parameters of the commitment scheme are priorly fixed for all users, the protocol in [80] naturally has the quasi unique responses property by the binding property of the commitment scheme. In other words, the NIZK proof for R_{ISIS} [79, 81] (and thus for $\tilde{R}'_{pke}$) obtained by applying the Fiat-Shamir transform to the protocol in [80] suffices for the above PKE scheme (where labels can be incorporated into the input of the hash function used for the transformation).

Finally, we clarify that the protocol [80] is designed for R_{ISIS} in the l_∞ norm, while the l_2 norm is used in Sect. 6.4.5. This problem can be easily fixed by setting $\zeta = \alpha q \cdot \omega(\sqrt{\log n})$ in the NIZK proof, and setting the parameter β in Eq. (6.1) such that $\beta \geq 2\zeta\sqrt{n}$ holds, since (1) for $\mathbf{e}_0, \mathbf{e}_1 \xleftarrow{\$} D_{\mathbb{Z}^m,\alpha q}$, both $\Pr[\|\mathbf{e}_0\|_\infty \geq \zeta]$, $\Pr[\|\mathbf{e}_1\|_\infty \geq \zeta]$ are negligible in n by [55, Lemma. 4.2]; and (2) $\mathcal{P} = \{-\alpha q + 1, \ldots, \alpha q - 1\}^{n_2}$ in the PKE scheme Π_{pke}. By [80], the resulting NIZK can be achieved with total communication cost $\log_2 \beta \cdot \tilde{O}(m \log q)$.

6.5 Background and Further Reading

One approach for achieving authentication in KE protocols is to explicitly authenticate the exchanged messages between the involved parties by using some cryptographic primitives (e.g., signatures, or MACs), which usually incurs additional computation and communication overheads with respect to the basic KE protocol, and complicates the understanding of the KE protocol. This includes several well-known protocols such as IKE [65, 73], SIGMA [74], SSL [50], TLS [26, 40, 56, 76, 86], as well as the standard in German electronic identity cards, namely, EAC [37], and the standardized protocols OPACITY [38] and PLAID [39]. Another line of designing AKEs follows the idea of MTI [85] and MQV [87],[9] which aims at providing implicit authentication by directly utilizing the algebraic structure of DH problems (e.g., HMQV [75] and OAKE [104]). All the above AKEs are based on classic hard problems, such as factoring, the RSA problem or the computational/decisional DH problem. Since these hard problems are vulnerable to quantum computers [97] and as we are moving into the era of quantum computing, it is very appealing to find other counterparts based on problems believed to be resistant to quantum attacks. For instance, post-quantum AKE is considered of high priority by NIST [33].

Since the work of Katz et al. [71], there are mainly four papers focusing on designing AKEs from lattices [23, 51, 52, 92]. At a high level, all of them are following generic transformations from key encapsulation mechanisms (KEM) to AKEs. Concretely, Fujioka et al. [51] proposed a generic construction of AKE from KEMs, which can be proven secure in the CK model. Informally, they showed that if there is a CCA-secure KEM with high min-entropy keys and a family of pseudorandom functions (PRF), then there is a secure AKE protocol in the standard model. Thus, by using existing lattice-based CCA-secure KEMs such as [91, 94], it is possible to

[9]Note that MQV has been widely standardized by ANS [102, 103], ISO/IEC [68] and IEEE [67], and recommended by NIST and NSA Suite B [10].

construct lattice-based AKE protocols in the standard model. However, as the authors commented, their construction was just of theoretic interest due to huge public keys and the lack of an efficient and direct construction of PRFs from (Ring-)LWE. Later, the paper [52] tried to get a practical AKE protocol by improving the efficiency of the generic framework in [51], and showed that one-way CCA-secure KEMs were enough to get AKEs in the random oracle model. The two protocols in [51, 52] share some similarities such as having two-pass messages, and involving three encryptions (i.e., two encryptions under each party's static public key and one encryption under an ephemeral public key). However, the use of the random oracle heuristic makes the protocol in [52] more efficient than that in [51]. Specifically, the protocol in [52] requires exchanging seven ring elements when instantiated with the CPA-secure encryption from Ring-LWE [83] by first transforming it into a CCA-secure one with the Fujisaki-Okamoto transformation.

Peikert [92] presented an efficient KEM based on Ring-LWE, which was then transformed into an AKE protocol by using the same structure as SIGMA [74]. Similar to the SIGMA protocol, the resulting protocol had three-pass messages and was proven SK-secure [32] in the random oracle model. For the computation overheads, Peikert's protocol involved one KEM, two signatures and two MACs. By treating the KEM in [92] as a DH-like KE protocol, Bos et al. [23] integrated it into the transport layer security (TLS) protocol by directly using signatures to provide explicit authentication. Actually, the authors used traditional digital signatures such as RSA and ECDSA, and thus their protocol was not a pure post-quantum AKE. As for the security, the protocol in [23] was proven secure in the authenticated and confidential channel establishment (ACCE) security model [69] (which is based on the BR model, but has many differences to capture entity authentication and channel security).

Gong et al. [60] first considered the problem of resisting off-line attacks in the "PKI model" where the server also has a public key in addition to a password. A formal treatment on this model was provided by Halevi and Krawczyk [64]. At CRYPTO 1993, Bellovin and Merritt [17] considered the setting where only a password is shared between users, and proposed a PAKE with heuristic security arguments. Formal security models for PAKE were provided in [13, 24]. Goldreich and Lindell [58] showed a PAKE solution in the plain model, which does not support concurrent executions of the protocol by the same user. As a special case of secure multiparty computations, PAKEs supporting concurrent executions in the plain model were studied in [8, 28, 61]. All the protocols in [8, 28, 58, 61] are inefficient in terms of both computation and communication. In the setting where all users share a common reference string, Katz et al. [70] provided a practical three-round PAKE based on the DDH assumption, which was later generalized and abstracted out by Gennaro and Lindell [53] to obtain a PAKE framework from PKE with associated SPH [35]. Canetti et al. [30] considered the security of PAKE within the framework of universal composability (UC) [27], and showed that an extension of the KOY/GL protocol was secure in the UC model.

References

1. Abdalla, M., Benhamouda, F., MacKenzie, P.: Security of the J-PAKE password-authenticated key exchange protocol. In: IEEE S&P 2015, pp. 571–587 (2015). https://doi.org/10.1109/SP.2015.41
2. Abdalla, M., Benhamouda, F., Pointcheval, D.: Disjunctions for hash proof systems: new constructions and applications. In: Oswald, E., Fischlin, M. (eds.) EUROCRYPT 2015, LNCS, vol. 9057, pp. 69–100. Springer, Heidelberg (2015)
3. Abdalla, M., Chevalier, C., Pointcheval, D.: Smooth projective hashing for conditionally extractable commitments. In: Halevi, S. (ed.) CRYPTO 2009, LNCS, vol. 5677, pp. 671–689. Springer, Heidelberg (2009). https://doi.org/10.1007/978-3-642-03356-8_39
4. Abe, M., Cui, Y., Imai, H., Kiltz, E.: Efficient hybrid encryption from ID-based encryption. Des. Codes Crypt. **54**(3), 205–240 (2010)
5. Albrecht, M.R., Player, R., Scott, S.: On the concrete hardness of learning with errors. J. Math. Cryptol. **9**, 169–203 (2015)
6. Alkim, E., Ducas, L., Pöppelmann, T., Schwabe, P.: Post-quantum key exchange-a new hope. In: USENIX Security Symposium, vol. 2016 (2016)
7. Bai, S., Galbraith, S.D.: An improved compression technique for signatures based on learning with errors. In: CT-RSA, pp. 28–47 (2014)
8. Barak, B., Canetti, R., Lindell, Y., Pass, R., Rabin, T.: Secure computation without authentication. In: Shoup, V. (ed.) CRYPTO 2005, LNCS, vol. 3621, pp. 361–377. Springer, Heidelberg (2005)
9. Barak, B., Impagliazzo, R., Wigderson, A.: Extracting randomness using few independent sources. SIAM J. Comput. **36**(4), 1095–1118 (2006)
10. Barker, E., Chen, L., Roginsky, A., Smid, M.: Recommendation for pair-wise key establishment schemes using discrete logarithm cryptography. NIST Spec. Publ. **800**, 56A (2013)
11. Barker, E., Roginsky, A.: Recommendation for the entropy sources used for random bit generation. Draft NIST Special Publication 800-90B, August 2012
12. Bellare, M., Neven, G.: Multi-signatures in the plain public-key model and a general forking lemma. In: Proceedings of the 13th ACM conference on Computer and Communications Security, CCS '06, pp. 390–399. ACM (2006)
13. Bellare, M., Pointcheval, D., Rogaway, P.: Authenticated key exchange secure against dictionary attacks. In: Preneel, B. (ed.) EUROCRYPT 2000. LNCS, vol. 1807, pp. 139–155. Springer, Heidelberg (2000). https://doi.org/10.1007/3-540-45539-6_11
14. Bellare, M., Rogaway, P.: Random oracles are practical: a paradigm for designing efficient protocols. In: Proceedings of the 1st ACM Conference on Computer and Communications Security, CCS '93, pp. 62–73. ACM (1993). https://doi.org/10.1145/168588.168596. http://doi.acm.org/10.1145/168588.168596
15. Bellare, M., Rogaway, P.: Entity authentication and key distribution. In: Stinson, D. (ed.) Advances in Cryptology - CRYPTO '93. Lecture Notes in Computer Science, vol. 773, pp. 232–249. Springer, Heidelberg (1994)
16. Bellare, M., Yung, M.: Certifying cryptographic tools: the case of trapdoor permutations. In: Brickell, E.F. (ed.) CRYPTO '92, LNCS, vol. 740, pp. 442–460. Springer, Heidelberg (1992)
17. Bellovin, S.M., Merritt, M.: Encrypted key exchange: password-based protocols secure against dictionary attacks. In: Proceedings 1992 IEEE Computer Society Symposium on Research in Security and Privacy, pp. 72–84 (1992). https://doi.org/10.1109/RISP.1992.213269
18. Benhamouda, F., Blazy, O., Chevalier, C., Pointcheval, D., Vergnaud, D.: New techniques for SPHFs and efficient one-round PAKE protocols. In: Canetti, R., Garay, J.A. (eds.) CRYPTO 2013, LNCS, vol. 8042, pp. 449–475. Springer, Berlin, Heidelberg (2013)
19. Bernhard, D., Fischlin, M., Warinschi, B.: Adaptive proofs of knowledge in the random oracle model. In: Katz, J. (ed.) PKC 2015, LNCS, vol. 9020, pp. 629–649. Springer, Heidelberg (2015)

20. Bernhard, D., Pereira, O., Warinschi, B.: How not to prove yourself: pitfalls of the Fiat-Shamir heuristic and applications to helios. In: Wang, X., Sako, K. (eds.) ASIACRYPT 2012, LNCS, vol. 7658, pp. 626–643. Springer, Heidelberg (2012)
21. Blake-Wilson, S., Johnson, D., Menezes, A.: Key agreement protocols and their security analysis. In: Proceedings of the 6th IMA International Conference on Cryptography and Coding, pp. 30–45. Springer, London, UK (1997). http://dl.acm.org/citation.cfm?id=647993.742138
22. Blum, M., Feldman, P., Micali, S.: Non-interactive zero-knowledge and its applications. In: STOC '88, pp. 103–112. ACM (1988)
23. Bos, J.W., Costello, C., Naehrig, M., Stebila, D.: Post-quantum key exchange for the TLS protocol from the ring learning with errors problem. Cryptology ePrint Archive, Report 2014/599 (2014)
24. Boyko, V., MacKenzie, P., Patel, S.: Provably secure password-authenticated key exchange using Diffie-Hellman. In: Preneel, B. (ed.) EUROCRYPT 2000. LNCS, vol. 1807, pp. 156–171. Springer, Heidelberg (2000). https://doi.org/10.1007/3-540-45539-6_12
25. Bresson, E., Chevassut, O., Pointcheval, D.: Security proofs for an efficient password-based key exchange. In: CCS 2003, pp. 241–250. ACM (2003). https://doi.org/10.1145/948109.948142
26. Brzuska, C., Fischlin, M., Smart, N.P., Warinschi, B., Williams, S.C.: Less is more: relaxed yet composable security notions for key exchange. Int. J. Inf. Sec. **12**(4), 267–297 (2013)
27. Canetti, R.: Universally composable security: a new paradigm for cryptographic protocols. In: FOCS '01, pp. 136–145. IEEE (2001)
28. Canetti, R., Goyal, V., Jain, A.: Concurrent secure computation with optimal query complexity. In: Gennaro, R., Robshaw, M. (eds.) CRYPTO 2015. LNCS, vol. 9216, pp. 43–62. Springer, Heidelberg (2015). https://doi.org/10.1007/978-3-662-48000-7_3
29. Canetti, R., Halevi, S., Katz, J.: Chosen-ciphertext security from identity-based encryption. In: Cachin, C., Camenisch, J. (eds.) Advances in Cryptology - EUROCRYPT 2004. Lecture Notes in Computer Science, vol. 3027, pp. 207–222. Springer, Berlin/Heidelberg (2004)
30. Canetti, R., Halevi, S., Katz, J., Lindell, Y., MacKenzie, P.: Universally composable password-based key exchange. In: Cramer, R. (ed.) EUROCRYPT 2005. LNCS, vol. 3494, pp. 404–421. Springer, Heidelberg (2005). https://doi.org/10.1007/11426639_24
31. Canetti, R., Krawczyk, H.: Analysis of key-exchange protocols and their use for building secure channels. In: Pfitzmann, B. (ed.) Advances in Cryptology - EUROCRYPT 2001. Lecture Notes in Computer Science, vol. 2045, pp. 453–474. Springer, Heidelberg (2001)
32. Canetti, R., Krawczyk, H.: Security analysis of IKE's signature-based key-exchange protocol. In: Yung, M. (ed.) Advances in Cryptology - CRYPTO 2002. Lecture Notes in Computer Science, vol. 2442, pp. 143–161. Springer, Heidelberg (2002)
33. Chen, L.: Practical impacts on qutumn computing. Quantum-Safe-Crypto Workshop at the European Telecommunications Standards Institute (2013). http://docbox.etsi.org/Workshop/2013/201309_CRYPTO/S05_DEPLOYMENT/NIST_CHEN.pdf
34. Chor, B., Goldreich, O.: Unbiased bits from sources of weak randomness and probabilistic communication complexity. In: FOCS, pp. 429–442 (1985)
35. Cramer, R., Shoup, V.: Universal hash proofs and a paradigm for adaptive chosen ciphertext secure public-key encryption. In: Knudsen, L.R. (ed.) EUROCRYPT 2002, LNCS, vol. 2332, pp. 45–64. Springer, Heidelberg (2002)
36. Cremers, C., Feltz, M.: Beyond eCK: perfect forward secrecy under actor compromise and ephemeral-key reveal. In: Foresti, S., Yung, M., Martinelli, F. (eds.) Computer Security - ESORICS 2012. Lecture Notes in Computer Science, vol. 7459, pp. 734–751. Springer, Heidelberg (2012)
37. Dagdelen, Ö., Fischlin, M.: Security analysis of the extended access control protocol for machine readable travel documents. In: ISC, pp. 54–68 (2010)
38. Dagdelen, Ö., Fischlin, M., Gagliardoni, T., Marson, G.A., Mittelbach, A., Onete, C.: A cryptographic analysis of OPACITY - (extended abstract). In: ESORICS, pp. 345–362 (2013)

39. Degabriele, J.P., Fehr, V., Fischlin, M., Gagliardoni, T., Günther, F., Marson, G.A., Mittelbach, A., Paterson, K.G.: Unpicking PLAID. In: Chen, L., Mitchell, C. (eds.) Security Standardisation Research. Lecture Notes in Computer Science, vol. 8893, pp. 1–25. Springer International Publishing (2014)
40. Dierks, T.: The Transport Layer Security (TLS) Protocol Version 1.2 (2008)
41. Diffie, W., Hellman, M.: New directions in cryptography. IEEE Trans. Inf. Theory **22**(6), 644–654 (1976)
42. Ding, J., Xie, X., Lin, X.: A simple provably secure key exchange scheme based on the learning with errors problem. Cryptology ePrint Archive, Report 2012/688 (2012)
43. Dodis, Y., Gennaro, R., Hastad, J., Krawczyk, H., Rabin, T.: Randomness extraction and key derivation using the CBC, Cascade and HMAC modes. In: Franklin, M. (ed.) Advances in Cryptology - CRYPTO 2004. Lecture Notes in Computer Science, vol. 3152, pp. 494–510. Springer, Heidelberg (2004)
44. Ducas, L., Durmus, A., Lepoint, T., Lyubashevsky, V.: Lattice signatures and bimodal gaussians. In: Canetti, R., Garay, J. (eds.) Advances in Cryptology - CRYPTO 2013. Lecture Notes in Computer Science, vol. 8042, pp. 40–56. Springer, Heidelberg (2013)
45. Faust, S., Kohlweiss, M., Marson, G.A., Venturi, D.: On the non-malleability of the Fiat-Shamir transform. In: Galbraith, S., Nandi, M. (eds.) INDOCRYPT 2012, LNCS, vol. 7668, pp. 60–79. Springer, Heidelberg (2012)
46. Faust, S., Mukherjee, P., Nielsen, J.B., Venturi, D.: Continuous non-malleable codes. In: Lindell, Y. (ed.) TCC 2014, LNCS, vol. 8349, pp. 465–488. Springer, Heidelberg (2014)
47. Feige, U., Lapidot, D., Shamir, A.: Multiple non-interactive zero knowledge proofs based on a single random string. In: FOCS '90, pp. 308–317. IEEE (1990)
48. Fiat, A., Shamir, A.: How to prove yourself: practical solutions to identification and signature problems. In: Odlyzko, A. (ed.) Advances in Cryptology - CRYPTO '86. Lecture Notes in Computer Science, vol. 263, pp. 186–194. Springer, Heidelberg (1987)
49. Fischlin, M.: Communication-efficient non-interactive proofs of knowledge with online extractors. In: Shoup, V. (ed.) Advances in Cryptology - CRYPTO 2005. Lecture Notes in Computer Science, vol. 3621, pp. 152–168. Springer, Berlin/Heidelberg (2005)
50. Freier, A.: The SSL Protocol Version 3.0 (1996). http://wp.netscape.com/eng/ssl3/draft302.txt
51. Fujioka, A., Suzuki, K., Xagawa, K., Yoneyama, K.: Strongly secure authenticated key exchange from factoring, codes, and lattices. In: Fischlin, M., Buchmann, J., Manulis, M. (eds.) Public Key Cryptography - PKC 2012. Lecture Notes in Computer Science, vol. 7293, pp. 467–484. Springer, Heidelberg (2012)
52. Fujioka, A., Suzuki, K., Xagawa, K., Yoneyama, K.: Practical and post-quantum authenticated key exchange from one-way secure key encapsulation mechanism. In: Proceedings of the 8th ACM SIGSAC Symposium on Information, Computer and Communications Security, ASIA CCS '13, pp. 83–94. ACM (2013)
53. Gennaro, R., Lindell, Y.: A framework for password-based authenticated key exchange. In: Biham, E. (ed.) EUROCRYPT 2003, LNCS, vol. 2656, pp. 524–543. Springer, Heidelberg (2003)
54. Gennaro, R., Shoup, V.: A note on an encryption scheme of Kurosawa and Desmedt. Cryptology ePrint Archive, Report 2004/194 (2004)
55. Gentry, C., Peikert, C., Vaikuntanathan, V.: Trapdoors for hard lattices and new cryptographic constructions. In: Proceedings of the 40th annual ACM symposium on Theory of computing, STOC '08, pp. 197–206. ACM (2008)
56. Giesen, F., Kohlar, F., Stebila, D.: On the security of TLS renegotiation. In: ACM Conference on Computer and Communications Security – CCS '13, pp. 387–398 (2013)
57. Goldreich, O.: Basing non-interactive zero-knowledge on (enhanced) trapdoor permutations: the state of the art. In: Goldreich, O. (ed.) Studies in Complexity and Cryptography, LNCS, vol. 6550, pp. 406–421. Springer, Heidelberg (2011)
58. Goldreich, O., Lindell, Y.: Session-key generation using human passwords only. In: Kilian, J. (ed.) CRYPTO 2001. LNCS, vol. 2139, pp. 408–432. Springer, Heidelberg (2001). https://doi.org/10.1007/3-540-44647-8_24

59. Goldwasser, S., Kalai, Y.T., Peikert, C., Vaikuntanathan, V.: Robustness of the learning with errors assumption. In: Innovations in Computer Science, pp. 230–240 (2010)
60. Gong, L., Lomas, M.A., Needham, R.M., Saltzer, J.H.: Protecting poorly chosen secrets from guessing attacks. IEEE J. Sel. Areas Commun. **11**(5), 648–656 (1993). https://doi.org/10.1109/49.223865
61. Goyal, V., Jain, A., Ostrovsky, R.: Password-authenticated session-key generation on the internet in the plain model. In: Rabin, T. (ed.) CRYPTO 2010, LNCS, vol. 6223, pp. 277–294. Springer, Heidelberg (2010)
62. Groce, A., Katz, J.: A new framework for efficient password-based authenticated key exchange. In: CCS 2010, pp. 516–525. ACM (2010). https://doi.org/10.1145/1866307.1866365
63. Güneysu, T., Lyubashevsky, V., Pöppelmann, T.: Practical lattice-based cryptography: a signature scheme for embedded systems. In: CHES, pp. 530–547 (2012)
64. Halevi, S., Krawczyk, H.: Public-key cryptography and password protocols. ACM Trans. Inf. Syst. Secur. **2**(3), 230–268 (1999). https://doi.org/10.1145/322510.322514
65. Harkins, D., Carrel, D., et al.: The internet key exchange (IKE). Tech. rep., RFC 2409, November (1998)
66. Hoffstein, J., Pipher, J., Schanck, J.M., Silverman, J.H., Whyte, W.: Practical signatures from the partial Fourier recovery problem. In: ACNS, pp. 476–493 (2014)
67. IEEE 1363: IEEE std 1363-2000: standard specifications for public key cryptography. IEEE, August 2000
68. ISO/IEC: 11770-3:2008 information technology – security techniques – key management – part 3: Mechanisms using asymmetric techniques
69. Jager, T., Kohlar, F., Schäge, S., Schwenk, J.: On the security of TLS-DHE in the standard model. In: Safavi-Naini, R., Canetti, R. (eds.) Advances in Cryptology - CRYPTO 2012. Lecture Notes in Computer Science, vol. 7417, pp. 273–293. Springer, Heidelberg (2012)
70. Katz, J., Ostrovsky, R., Yung, M.: Efficient and secure authenticated key exchange using weak passwords. J. ACM **57**(1), 3:1–3:39 (2009). https://doi.org/10.1145/1613676.1613679
71. Katz, J., Vaikuntanathan, V.: Smooth projective hashing and password-based authenticated key exchange from lattices. In: Matsui, M. (ed.) Advances in Cryptology - ASIACRYPT 2009. Lecture Notes in Computer Science, vol. 5912, pp. 636–652. Springer, Heidelberg (2009)
72. Katz, J., Vaikuntanathan, V.: Round-optimal password-based authenticated key exchange. In: Ishai, Y. (ed.) TCC 2011, LNCS, vol. 6597, pp. 293–310. Springer, Heidelberg (2011)
73. Kaufman, C., Hoffman, P., Nir, Y., Eronen, P.: Internet key exchange protocol version 2 (IKEv2). Tech. rep., RFC 5996, September (2010)
74. Krawczyk, H.: SIGMA: the 'SIGn-and-MAc' approach to authenticated Diffie-Hellman and its use in the IKE protocols. In: Boneh, D. (ed.) Advances in Cryptology - CRYPTO 2003. Lecture Notes in Computer Science, vol. 2729, pp. 400–425. Springer, Heidelberg (2003)
75. Krawczyk, H.: HMQV: a high-performance secure Diffie-Hellman protocol. In: Shoup, V. (ed.) Advances in Cryptology - CRYPTO 2005. Lecture Notes in Computer Science, vol. 3621, pp. 546–566. Springer, Heidelberg (2005)
76. Krawczyk, H., Paterson, K., Wee, H.: On the security of the TLS protocol: a systematic analysis. In: Canetti, R., Garay, J. (eds.) Advances in Cryptology - CRYPTO 2013. Lecture Notes in Computer Science, vol. 8042, pp. 429–448. Springer, Heidelberg (2013)
77. Laguillaumie, F., Langlois, A., Libert, B., Stehlé, D.: Lattice-based group signatures with logarithmic signature size. In: Sako, K., Sarkar, P. (eds.) Advances in Cryptology - ASIACRYPT 2013. Lecture Notes in Computer Science, vol. 8270, pp. 41–61. Springer, Heidelberg (2013)
78. LaMacchia, B.A., Lauter, K.E., Mityagin, A.: Stronger security of authenticated key exchange. In: ProvSec, pp. 1–16 (2007)
79. Libert, B., Mouhartem, F., Nguyen, K.: A lattice-based group signature scheme with message-dependent opening. In: Manulis, M., Sadeghi, A.R., Schneider, S. (eds.) ACNS 2016, pp. 137–155. Springer International Publishing (2016)
80. Ling, S., Nguyen, K., Stehlé, D., Wang, H.: Improved zero-knowledge proofs of knowledge for the ISIS problem, and applications. In: Kurosawa, K., Hanaoka, G. (eds.) Public-Key

Cryptography - PKC 2013. Lecture Notes in Computer Science, vol. 7778, pp. 107–124. Springer, Heidelberg (2013)
81. Ling, S., Nguyen, K., Wang, H.: Group signatures from lattices: simpler, tighter, shorter, ring-based. In: Public-Key Cryptography – PKC 2015, Lecture Notes in Computer Science. Springer, Heidelberg (2015)
82. Lyubashevsky, V.: Lattice signatures without trapdoors. In: Pointcheval, D., Johansson, T. (eds.) Advances in Cryptology - EUROCRYPT 2012. Lecture Notes in Computer Science, vol. 7237, pp. 738–755. Springer, Heidelberg (2012)
83. Lyubashevsky, V., Peikert, C., Regev, O.: On ideal lattices and learning with errors over rings. In: Gilbert, H. (ed.) Advances in Cryptology - EUROCRYPT 2010. Lecture Notes in Computer Science, vol. 6110, pp. 1–23. Springer, Berlin/Heidelberg (2010)
84. MacKenzie, P., Patel, S., Swaminathan, R.: Password-authenticated key exchange based on RSA. In: Okamoto, T. (ed.) ASIACRYPT 2000, LNCS, vol. 1976, pp. 599–613. Springer, Heidelberg (2000). https://doi.org/10.1007/3-540-44448-3_46
85. Matsumoto, T., Takashima, Y.: On seeking smart public-key-distribution systems. IEICE TRANSACTIONS (1976–1990) **69**(2), 99–106 (1986)
86. Mavrogiannopoulos, N., Vercauteren, F., Velichkov, V., Preneel, B.: A cross-protocol attack on the TLS protocol. In: ACM Conference on Computer and Communications Security – CCS '12, pp. 62–72 (2012)
87. Menezes, A., Qu, M., Vanstone, S.: Some new key agreement protocols providing mutual implicit authentication. In: Selected Areas in Cryptography, pp. 22–32 (1995)
88. Micciancio, D., Mol, P.: Pseudorandom knapsacks and the sample complexity of LWE search-to-decision reductions. In: Rogaway, P. (ed.) Advances in Cryptology - CRYPTO 2011. Lecture Notes in Computer Science, vol. 6841, pp. 465–484. Springer, Heidelberg (2011)
89. Micciancio, D., Peikert, C.: Trapdoors for lattices: simpler, tighter, faster, smaller. In: Pointcheval, D., Johansson, T. (eds.) Advances in Cryptology - EUROCRYPT 2012. Lecture Notes in Computer Science, vol. 7237, pp. 700–718. Springer, Heidelberg (2012)
90. Naor, M., Yung, M.: Public-key cryptosystems provably secure against chosen ciphertext attacks. In: STOC 1990, pp. 427–437. ACM (1990). https://doi.org/10.1145/100216.100273
91. Peikert, C.: Public-key cryptosystems from the worst-case shortest vector problem: extended abstract. In: Proceedings of the 41st Annual ACM Symposium on Theory of Computing, STOC '09, pp. 333–342. ACM (2009)
92. Peikert, C.: Lattice cryptography for the internet. In: Mosca, M. (ed.) Post-Quantum Cryptography. Lecture Notes in Computer Science, vol. 8772, pp. 197–219. Springer International Publishing (2014)
93. Peikert, C., Vaikuntanathan, V., Waters, B.: A framework for efficient and composable oblivious transfer. In: Wagner, D. (ed.) CRYPTO 2008, LNCS, vol. 5157, pp. 554–571. Springer, Heidelberg (2008)
94. Peikert, C., Waters, B.: Lossy trapdoor functions and their applications. In: Proceedings of the 40th Annual ACM Symposium on Theory of Computing, STOC '08, pp. 187–196. ACM (2008)
95. Rosen, A., Segev, G.: Chosen-ciphertext security via correlated products. In: Reingold, O. (ed.) Theory of Cryptography. Lecture Notes in Computer Science, vol. 5444, pp. 419–436. Springer, Berlin/Heidelberg (2009)
96. Sahai, A.: Non-malleable non-interactive zero knowledge and adaptive chosen-ciphertext security. In: FOCS '99, pp. 543–553. IEEE Computer Society (1999)
97. Shor, P.: Polynomial-time algorithms for prime factorization and discrete logarithms on a quantum computer. SIAM J. Comput. **26**(5), 1484–1509 (1997)
98. Stehlé, D., Steinfeld, R.: Making NTRU as secure as worst-case problems over ideal lattices. In: Paterson, K. (ed.) Advances in Cryptology - EUROCRYPT 2011. Lecture Notes in Computer Science, vol. 6632, pp. 27–47. Springer, Berlin/Heidelberg (2011)
99. Stern, J.: A new paradigm for public key identification. IEEE Trans. Inf. Theory **42**(6), 1757–1768 (1996)
100. Trevisan, L.: Extractors and pseudorandom generators. J. ACM **48**(4), 860–879 (2001)

101. Trevisan, L., Vadhan, S.: Extracting randomness from samplable distributions. In: Proceedings of the 41st Annual Symposium on Foundations of Computer Science, FOCS '00, pp. 32–42. IEEE Computer Society, Washington, DC, USA (2000). http://dl.acm.org/citation.cfm?id=795666.796582
102. X9.42-2001, A.: Public key cryptography for the financial services industry: agreement of symmetric keys using discrete logarithm cryptography
103. X9.63-2001, A.: Public key cryptography for the financial services industry: key agreement and key transport using elliptic curve cryptography
104. Yao, A.C.C., Zhao, Y.: OAKE: a new family of implicitly authenticated Diffie-Hellman protocols. In: Proceedings of the 2013 ACM SIGSAC Conference on Computer and Communications Security, CCS '13, pp. 1113–1128. ACM (2013)
105. Zhang, J., Chen, Y., Zhang, Z.: Programmable hash functions from lattices: short signatures and IBEs with small key sizes. In: Robshaw, M., Katz, J. (eds.) Advances in Cryptology - CRYPTO 2016, pp. 303–332. Springer, Heidelberg (2016)
106. Zhang, J., Yu, Y.: Two-round PAKE from approximate SPH and instantiations from lattices. In: Takagi, T., Peyrin, T. (eds.) Advances in Cryptology – ASIACRYPT 2017, pp. 37–67. Springer (2017)
107. Zhang, J., Zhang, Z., Ding, J., Snook, M., Dagdelen, Ö.: Authenticated key exchange from ideal lattices. In: Oswald, E., Fischlin, M. (eds.) Advances in Cryptology - EUROCRYPT 2015, pp. 719–751. Springer, Heidelberg (2015)
108. Zhang, R.: Tweaking TBE/IBE to PKE transforms with chameleon hash functions. In: Katz, J., Yung, M. (eds.) Applied Cryptography and Network Security. Lecture Notes in Computer Science, vol. 4521, pp. 323–339. Springer, Heidelberg (2007)

Chapter 7
Digital Signatures

Abstract Digital signatures are a cryptographic analogue of handwritten signatures, but provide more features and stronger security guarantees than the latter. In a signature scheme, a user with a private signing key can authenticate messages or documents by generating a signature such that anyone with the public verification key can be convinced that the message or documents indeed come from the signer and have not been altered during the transmission process. Digital signatures have been widely used in many applications, and more importantly they are the foundation of public key infrastructure. In this chapter, we focus on constructing digital signatures from lattices. After given the necessary background in Sect. 7.1, we present the first digital signature scheme from lattices due to Gentry et al. in Sect. 7.2. Then, we show how to construct short signature from lattices in the standard model in Sect. 7.3. Finally, we give a group signature from lattices in Sect. 7.4.

7.1 Definition

The basic idea of digital signatures was first introduced in the seminal work of Diffie and Hellman [29]. There are typically two kinds of users, namely, signers and verifiers in a digital signature scheme. The signer can run a key generation algorithm to get a pair of a public verification key and a private signing key and can use the private signing key to sign any message of his choice in a way that any users holding the public verification key can be convinced that the message is originated from the signer. The intuitive security requirement for digital signature scheme is that no probabilistic polynomial time (PPT) adversary which is given only the public verification key can sign a message on behalf of the signer. The first concrete construction of digital signature scheme was suggested by Rivest, Shamir and Adleman [60], and the first formal security notion of digital signature was given by Goldwasser, Micali and Rivest [35]. Since then, many effects have been devoted to the construction and analysis of various digital signature schemes from different hardness assumptions.

J. Zhang and Z. Zhang, *Lattice-Based Cryptosystems*,
https://doi.org/10.1007/978-981-15-8427-5_7

Definition 7.1 (*Digital Signatures*) A digital signature scheme $\Pi_{\text{sig}} = (\mathsf{KeyGen}, \mathsf{Sign}, \mathsf{Verify})$ consists of three PPT algorithms:

- $\mathsf{KeyGen}(\kappa)$. Taking the security parameter κ as input, the algorithm outputs a verification key vk and a secret signing key sk, i.e., $(vk, sk) \leftarrow \mathsf{KeyGen}(1^\kappa)$.
- $\mathsf{Sign}(sk, M)$. The signing algorithm takes vk, sk and a message $M \in \{0, 1\}^*$ as inputs, outputs a signature σ on M, briefly denoted as $\sigma \leftarrow \mathsf{Sign}(sk, M)$.
- $\mathsf{Verify}(vk, M, \sigma)$. The verification algorithm takes vk, message $M \in \{0, 1\}^*$ and a string $\sigma \in \{0, 1\}^*$ as inputs, outputs 1 if σ is a valid signature on M, else outputs 0, denoted as $1/0 \leftarrow \mathsf{Verify}(vk, M, \sigma)$.

For correctness, we require that for any $(vk, sk) \leftarrow \mathsf{KeyGen}(1^\kappa)$, any message $M \in \{0, 1\}^*$ and any $\sigma \leftarrow \mathsf{Sign}(sk, M)$, the equation $\mathsf{Verify}(vk, M, \sigma) = 1$ holds with overwhelming probability, where the probability is taken over the choices of the random coins used in KeyGen, Sign and Verify.

The standard security notion for digital signature scheme is the existential unforgeability against chosen message attacks (EUF-CMA), which (informally) says that any PPT forger, after seeing valid signatures on a polynomial number of adaptively chosen messages, cannot create a valid signature on a new message. Formally, consider the following game between a challenger C and a forger $\mathcal{F}$:

KeyGen. The challenger C first runs $(vk, sk) \leftarrow \mathsf{KeyGen}(1^\kappa)$ with the security parameter κ. Then, it gives the verification key vk to the forger $\mathcal{F}$ and keeps the signing secret key sk to itself.

Signing. The forger $\mathcal{F}$ is allowed to ask the signature on any fresh message M. The challenger C computes and sends $\sigma \leftarrow \mathsf{Sign}(sk, M)$ to the forger $\mathcal{F}$. The forger can repeat this any polynomial times.

Forge. $\mathcal{F}$ outputs a message-signature pair (M^*, σ^*). Let Q be the set of all messages queried by $\mathcal{F}$ in the signing phase. If $M^* \notin Q$ and $\mathsf{Verify}(vk, M^*, \sigma^*) = 1$, the challenger C outputs 1, else outputs 0.

We say that $\mathcal{F}$ wins the game if the challenger C outputs 1. The advantage of $\mathcal{F}$ in the above security game is defined as $\text{Adv}^{\text{euf-cma}}_{\Pi_{\text{sig}}, \mathcal{F}}(1^\kappa) = \Pr[C \text{ outputs } 1]$.

Definition 7.2 (*EUF-CMA Security*) A signature scheme Π_{sig} is said to be existentially unforgeable against chosen message attacks (EUF-CMA) if the advantage $\text{Adv}^{\text{euf-cma}}_{\Pi_{\text{sig}}, \mathcal{F}}(1^\kappa)$ is negligible in the security parameter κ for any PPT forger $\mathcal{F}$.

The above security game can be modified to define the strongly EUF-CMA (sEUF-CMA) security by relaxing the success condition of the adversary in the sense that the pair (M^*, σ^*) is not produced by the signing queries. Specifically, the adversary is allowed to output a new signature on a message that has been used in previous signing queries. In the security game of existential unforgeability against non-adaptive chosen message attacks, $\mathcal{F}$ is asked to output all the messages $\{M_1, \ldots, M_Q\}$ for signing queries before seeing the verification key vk and is given vk and the signatures $\{\sigma_1, \ldots, \sigma_Q\}$ on all the queried messages at the same time (i.e., there is no adaptive signing query phase). The resulting security notion defined using the modified game as in Definition 7.2 is denoted as EUF-naCMA. One can transform an EUF-naCMA secure signature scheme into an EUF-CMA secure one [11, 31] by using chameleon hash functions [45].

The traditional digital signatures are very useful in many applications since it can provide very strong authenticity to the signer, and to the documents or messages that have been signed by the signer. However, this also means that the identity of the signer is always known to the verifier which is not desirable in some applications with privacy requirement on protecting the identity of the signer. In 1991, Chaum and van Heyst [27] first introduced a special kind of digital signatures, i.e., group signatures. Specifically, in a group signature, each group member has a private key that is certified with its identity by the group manager. By using its private key, each group member is able to sign messages on behalf of the group without compromising its identity to the signature verifier. Group signatures provide users a nice tradeoff between authenticity and anonymity (i.e., given a signature, the verifier is assured that someone in the group signed a message, but cannot determine which member of the group signed). However, such a functionality allows malicious group members to damage the whole group without being detected, e.g., signing some unauthorized/illegal messages. To avoid this, the group manager usually has a secret key which can be used to break anonymity.

Several real-life applications require properties of group signatures. For example, in trusted computing, a trusted platform module (TPM) usually has to attest certain statements w.r.t. the current configurations of the host device to a remote party (i.e., the verifier) via a signature on corresponding messages. After the attestation, the verifier is assured that some remote device that contains a TPM authorized the messages. For user privacy, the signature is often required not to reveal the identity of the TPM. In fact, a variant of group signatures (namely, direct anonymous attestation (DAA) [22, 28]) has been implemented in TPM 1.2 [40] and TPM 2.0 [41] by the Trusted Computing Group. Another promising application is vehicle safety communications [42], where group signatures can protect the privacy of users so that a broadcast message does not reveal the current location/speed of the vehicle. Besides, other applications of group signatures are found in anonymous communications, e-commerce systems, etc.

Definition 7.3 (*Group Signatures*) A (static) group signature scheme $\Pi_{\mathrm{gs}} = (\mathsf{KeyGen}, \mathsf{Sign}, \mathsf{Verify}, \mathsf{Open})$ consists of a tuple of four probabilistic polynomial time (PPT) algorithms:

- $\mathsf{KeyGen}(1^n, 1^N)$: Take the security parameter n and the maximum number of group members N as inputs, output the group public key gpk, the group manager secret key $gmsk$ and a vector of users' keys $\mathbf{gsk} = (gsk_1, \ldots, gsk_N)$, where gsk_j is the j-th user's secret key for $j \in \{1, \ldots, N\}$.
- $\mathsf{Sign}(gpk, gsk_j, M)$: Take the group public key gpk, the j-th user's secret key gsk_j, and a message $M \in \{0, 1\}^*$ as inputs, output a signature σ of M.
- $\mathsf{Verify}(gpk, M, \sigma)$: Take the group public key gpk, a message $M \in \{0, 1\}^*$ and a string σ as inputs, return 1 if σ is a valid signature of M, else return 0.

Experiment $\mathbf{Exp}_{\mathcal{GS},\mathcal{A}}^{\mathrm{anon}}(n, N)$
$(gpk, gmsk, \mathbf{gsk}) \leftarrow \mathsf{KeyGen}(1^n, 1^N)$
$(st, i_0, i_1, M^*) \leftarrow \mathcal{A}^{\mathsf{Open}(\cdot,\cdot)}(gpk, \mathbf{gsk})$
$b \xleftarrow{\$} \{0, 1\}$
$\sigma^* \leftarrow \mathsf{Sign}(gpk, gsk_{i_b}, M^*)$
$b' \leftarrow \mathcal{A}^{\mathsf{Open}(\cdot,\cdot)}(st, \sigma^*)$
If $b = b'$ return 1, else return 0

Experiment $\mathbf{Exp}_{\mathcal{GS},\mathcal{A}}^{\mathrm{trace}}(n, N)$
$(gpk, gmsk, \mathbf{gsk}) \leftarrow \mathsf{KeyGen}(1^n, 1^N)$
$(M^*, \sigma^*) \leftarrow \mathcal{A}^{\mathsf{Sign}(\cdot,\cdot),\mathbf{Corrupt}(\cdot)}(gpk, gmsk)$
If $\mathsf{Verify}(gpk, M^*, \sigma^*) = 0$ then return 0
If $\mathsf{Open}(gmsk, M^*, \sigma^*) = \perp$ then return 1
If $\exists j^* \in \{1, \ldots, N\}$ such that
$\mathsf{Open}(gpk, gmsk, M^*, \sigma^*) = j^*$ and $j^* \notin C$,
and (j^*, M^*) was not queried to $\mathsf{Sign}(\cdot, \cdot)$ by $\mathcal{A}$,
then return 1, else return 0

Fig. 7.1 Security games for group signatures

- $\mathsf{Open}(gpk, gmsk, M, \sigma)$: Take the group public key gpk, the group manager secret key $gmsk$, a message $M \in \{0, 1\}^*$, and a valid signature σ of M as inputs, output an index $j \in \{1, \ldots, N\}$ or a special symbol $\perp$ in case of opening failure.

For correctness, we require that for any $(gpk, gmsk, \mathbf{gsk}) \leftarrow \mathsf{KeyGen}(1^n, 1^N)$, any $j \in \{1, \ldots, N\}$, any message $M \in \{0, 1\}^*$, and any $\sigma \leftarrow \mathsf{Sign}(gpk, gsk_j, M)$, the following conditions hold with overwhelming probability:

$$\mathsf{Verify}(gpk, M, \sigma) = 1 \textit{ and } \mathsf{Open}(gpk, gmsk, M, \sigma) = j$$

.

There are two security notions for group signatures: anonymity and traceability [7]. The first notion, informally, says that anyone without the group manager secret key cannot determine the owner of a valid signature. The second notion says a set C of group members cannot collude to create a valid signature such that the Open algorithm fails to trace back to one of them. In particular, this notion implies that any non-group member cannot create a valid signature.

Definition 7.4 (*Full anonymity*) For any (static) group signature scheme $\mathcal{GS}$, we associate to an adversary $\mathcal{A}$ against the full anonymity of $\mathcal{GS}$ experiment $\mathbf{Exp}_{\mathcal{GS},\mathcal{A}}^{\mathrm{anon}}(n, N)$ in the left-side of Fig. 7.1, where the $\mathsf{Open}(\cdot, \cdot)$ oracle takes a valid message-signature pair (M, σ) as inputs, outputs the index of the user whose secret key is used to create σ. In the guess phase, the adversary $\mathcal{A}$ is not allowed to make an Open query with inputs (M^*, σ^*). We define the advantage of $\mathcal{A}$ in the experiment as

$$\mathrm{Adv}_{\mathcal{GS},\mathcal{A}}^{\mathrm{anon}}(n, N) = \left| \Pr[\mathbf{Exp}_{\mathcal{GS},\mathcal{A}}^{\mathrm{anon}}(n, N) = 1] - \frac{1}{2} \right| .$$

A group signature $\mathcal{GS}$ is said to be fully anonymous if the advantage $\mathrm{Adv}_{\mathcal{GS},\mathcal{A}}^{\mathrm{anon}}(n, N)$ is negligible in n, N for any PPT adversary $\mathcal{A}$.

In a weak definition of anonymity (i.e., CPA-anonymity), the adversary is not given access to an open oracle. In this paper, we first present a CPA-anonymous scheme, then we extend it to satisfy full/CCA anonymity.

Definition 7.5 (*Full traceability*) For any (static) group signature scheme $\mathcal{GS}$, we associate to an adversary $\mathcal{A}$ against the full traceability of $\mathcal{GS}$ experiment $\mathbf{Exp}^{\text{trace}}_{\mathcal{GS},\mathcal{A}}(n, N)$ in the right-side of Fig. 7.1, where the $\mathsf{Sign}(\cdot,\cdot)$ oracle takes a user index i and a message M as inputs, returns a signature of M by using gsk_i. The $\mathbf{Corrupt}(\cdot)$ oracle takes a user index i as input, returns gsk_i, and C is a set of user indexes that $\mathcal{A}$ submitted to the $\mathbf{Corrupt}(\cdot)$ oracle. The advantage of $\mathcal{A}$ in the experiment is defined as

$$\mathrm{Adv}^{\text{trace}}_{\mathcal{GS},\mathcal{A}}(n, N) = \Pr[\mathbf{Exp}^{\text{trace}}_{\mathcal{GS},\mathcal{A}}(n, N) = 1].$$

A group signature $\mathcal{GS}$ is said to be fully traceable if the advantage $\mathrm{Adv}^{\text{trace}}_{\mathcal{GS},\mathcal{A}}(n, N)$ is negligible in n, N for any PPT adversary $\mathcal{A}$.

7.2 Signatures in the Random Oracle Model

The full-domain hash (FDH) signature scheme is a generic construction of digital signature in the random oracle model [9], which can be instantiated by using any trapdoor permutation. Due to the special algebraic structures of lattices, we currently still do not know how to construct trapdoor permutations from lattices. In 2008, Gentry, Peikert and Vaikuntanathan [34] proposed a family of trapdoor preimage sampleable one-way functions, and adapted the idea of FDH to obtain a digital signature from lattices.

7.2.1 The Gentry-Peikert-Vaikuntanathan Signature Scheme

let κ be the security parameter. Let $n, m, q > 0$ be integers, and $\alpha, s > 0$ be reals. Let $\mathsf{TrapGen}$ and $\mathsf{SampleD}$ be the trapdoor generation and discrete Gaussian sampler algorithms given in Lemma 2.24. Let $H : \{0,1\}^* \to \mathbb{Z}_q^n$ be a hash function, which is modeled as a random oracle. The description of the GPV signature scheme $\Pi_{\text{sig}} = (\mathsf{KeyGen}, \mathsf{Sign}, \mathsf{Verify})$ with parameters (n, m, q, s, α) is given as follows:

- $\mathsf{KeyGen}(1^\kappa)$: given the security parameter κ as input, compute

$$(\mathbf{A}, \mathbf{R}) \leftarrow \mathsf{TrapGen}(1^n, 1^m, q, \mathbf{I}_n),$$

  where $\mathbf{I}_n$ is the identity matrix. Return the verification key $\mathsf{vk} = \mathbf{A}$ and the signing key $\mathsf{sk} = (\mathbf{A}, \mathbf{R})$;
- $\mathsf{Sign}(\mathsf{sk}, m)$: given the signing key $\mathsf{sk} = (\mathbf{A}, \mathbf{R})$ and a message $m \in \{0,1\}^*$, compute $\mathbf{u} = \mathrm{H}(m) \in \mathbb{Z}_q^n$ and $\mathbf{x} \leftarrow \mathsf{SampleD}(\mathbf{R}, \mathbf{A}, \mathbf{I}_n, \mathbf{u}, s)$ such that the distribution of $\mathbf{x} \in \mathbb{Z}^m$ is statistically close to $D_{\mathbb{Z}^m, s}$ conditioned on $\mathbf{A}\mathbf{x} = \mathbf{u}$, return the signature $\sigma = \mathbf{x}$.

- $\mathsf{Verify}(\mathsf{vk}, m, \sigma)$: given the verification key $\mathsf{vk} = \mathbf{A}$, a message $m \in \{0, 1\}^*$ and a signature $\sigma = \mathbf{x}$, compute $\mathbf{u} = \mathrm{H}(m) \in \mathbb{Z}_q^n$. Return 1 if $\mathbf{A}\mathbf{x} = \mathbf{u}$ and $\|\mathbf{x}\| \leq s\sqrt{m}$, otherwise return 0.

Note that if $\sigma = \mathbf{x} \in \mathbb{Z}^m$ is a signature on $m \in \{0, 1\}^*$ output by the signing algorithm, then we have that $\mathbf{A}\mathbf{x} = H(m) = \mathbf{u}$ and that the distribution of $\mathbf{x}$ is statistically close to $D_{\mathbb{Z}^m,s}$ under appropriate choice of parameters by Lemma 2.24. This means that $|\mathbf{x}| \leq s\sqrt{m}$ except with negligible probability by Lemma 2.3. In other words, the above signature scheme is correct.

7.2.2 Security

We have the following theorem for the security of the GPV signature scheme [34].

Theorem 7.1 *If the* $\mathrm{SIS}_{m,n,q,2s\sqrt{m}}$ *problem is hard, then the GPV signature scheme is sEUF-CMA-secure.*

Proof Let $\mathcal{A}$ be any adversary that can break the sEUF-CMA security of the GPV signature scheme with advantage ϵ. It suffices to construct an algorithm $\mathcal{B}$ that can solve the $\mathrm{SIS}_{m,n,q,2s\sqrt{m}}$ problem almost with the same advantage ϵ. Formally, given a $\mathrm{SIS}_{m,n,q,2s\sqrt{m}}$ challenge instance $\mathbf{A} \in \mathbb{Z}_q^{n\times m}$, $\mathcal{B}$ maintains a list L for the random oracle queries to H, sends $\mathsf{vk} = \mathbf{A}$ to $\mathcal{A}$ and answers the queries from $\mathcal{A}$ as follows:

Queries to H. When receiving a query $m \in \{0, 1\}^*$ to H, the challenger C first checks if there is a tuple $(m, \mathbf{x}, \mathbf{u}) \in \{0, 1\}^* \times \mathbb{Z}_q^n$ in the list L, and adds a tuple $(m, \mathbf{x}, \mathbf{u}) \in \{0, 1\}^* \times \mathbb{Z}_q^n$ into the list L by randomly choosing $\mathbf{x} \xleftarrow{\$} D_{\mathbb{Z}^m,s}$ and computing $\mathbf{u} = \mathbf{A}\mathbf{x}$ if there is no such tuple in L. Then, return $\mathbf{u} \in \mathbb{Z}_q^n$.

Signing Queries. The adversary is allowed to query the signature for any message $m \in \{0, 1\}^*$. If there is a tuple $(m, \mathbf{x}, \mathbf{u}) \in \{0, 1\}^* \times \mathbb{Z}_q^n$ in the list L, return $\mathbf{x} \in \mathbb{Z}^m$ to the adversary. Otherwise, randomly choose $\mathbf{x} \xleftarrow{\$} D_{\mathbb{Z}^m,s}$ and compute $\mathbf{u} = \mathbf{A}\mathbf{x}$. Then, add the tuple $(m, \mathbf{x}, \mathbf{u}) \in \{0, 1\}^* \times \mathbb{Z}_q^n$ into the list L, and return $\mathbf{x} \in \mathbb{Z}^m$ to the adversary.

After receiving a forgery $(m^*, \sigma^* = \mathbf{x}^*)$ from the adversary $\mathcal{A}$, if $\mathcal{A}$ has not made a query m^* to H, the algorithm $\mathcal{B}$ makes a query with m^* to H on behalf of $\mathcal{A}$. Then, if there is a tuple $(m^*, \mathbf{x}' \neq \mathbf{x}^*, \mathbf{u}^*)$ in the list L, $\mathcal{B}$ returns $\mathbf{x}^* - \mathbf{x}'$ as its own solution, otherwise aborts.

Now, it suffices to show that the probability that there exists a tuple $(m^*, \mathbf{x}' \neq \mathbf{x}^*, \mathbf{u}^*)$ in the list L is at least $\epsilon - \mathsf{negl}(\kappa)$. First, by Lemma 2.11 we have that the distribution of $\mathbf{u} = \mathbf{A}\mathbf{x}$ for $\mathbf{x} \xleftarrow{\$} D_{\mathbb{Z}^m,s}$ is statistically close to uniform over $\mathbb{Z}_q^n$, which means that the simulation of the H queries is almost identical to that in the real security game. Second, by Lemma 2.24 we have that the distribution of the verification key and the signatures is also statistically to that in the real security game. This means that $\mathcal{A}$ will output a valid forgery $(m^*, \sigma^* = \mathbf{x}^*)$ with advantage at least $\epsilon - \mathsf{negl}(\kappa)$. We distinguish the following two cases:

- If $\mathcal{A}$ has made a signing query on m^* to obtain a signature σ', there must exist a tuple $(m^*, \mathbf{x}', \mathbf{u})$ in the list L such that $\mathbf{A}\mathbf{x}' = \mathbf{u}$ and $\|\mathbf{x}'\| \leq s\sqrt{m}$ with overwhelming probability. By the assumption that $(m^*, \sigma^* = \mathbf{x}^*)$ is a valid forgery, we must have that $\mathbf{A}\mathbf{x}^* = \mathbf{u}$ and $\mathbf{x}^* \neq \mathbf{x}'$;
- If $\mathcal{A}$ has not made a signing query on m^*, by the assumption that $(m^*, \sigma^* = \mathbf{x}^*)$ is a valid forgery, there must exist a tuple $(m^*, \mathbf{x}', \mathbf{u})$ in the list L such that $\mathbf{A}\mathbf{x}' = \mathbf{u} = \mathbf{A}\mathbf{x}^*$ and $\|\mathbf{x}'\|, \|\mathbf{x}*\| \leq s\sqrt{m}$ with overwhelming probability. Since $\mathbf{x}'$ is chosen from $D_{\mathbb{Z}^m,s}$, we have that the probability that $\mathbf{x}' = \mathbf{x}^*$ is negligible by Lemmas 2.9 and 2.10.

In all, we have that $\mathcal{B}$ can find a solution $\mathbf{x}^* - \mathbf{x}' \in \mathbb{Z}^m$ such that $\mathbf{A}(\mathbf{x}^* - \mathbf{x}') = \mathbf{0}$ and $\|\mathbf{x}^* - \mathbf{x}'\| \leq 2s\sqrt{m}$ with probability negligible close to ϵ. This completes the proof. □

7.3 Short Signatures in the Standard Model

In this section, we show how to construct signatures in the standard model. First, we will give a generic construction of signatures from lattice-based PHFs. Then, we will show how to combine two different PHFs to obtain a lattice-based short signature with tighter security.

7.3.1 Design Rational

In this section, we outline the idea on how to construct a generic signature scheme Π_{sig} from lattice-based PHFs in the standard model. Let $n, \bar{m}, m', \ell, q$ be some positive integers, and let $m = \bar{m} + m'$. Given a lattice-based PHF $\mathcal{H} = \{\mathrm{H}_K\}$ from $\{0, 1\}^\ell$ to $\mathbb{Z}_q^{n\times m'}$, let $\mathbf{B} \in \mathbb{Z}_q^{n\times m'}$ be a trapdoor matrix that is compatible with $\mathcal{H}$. Then, the verification key of the generic signature scheme Π_{sig} consists of a uniformly distributed (trapdoor) matrix $\mathbf{A} \in \mathbb{Z}_q^{n\times \bar{m}}$, a uniformly random vector $\mathbf{u} \in \mathbb{Z}_q^n$ and a random key K for $\mathcal{H}$, i.e., $vk = (\mathbf{A}, \mathbf{u}, K)$. The signing key is a trapdoor $\mathbf{R}$ of $\mathbf{A}$ that allows to sample short vector $\mathbf{e}$ satisfying $\mathbf{A}\mathbf{e} = \mathbf{v}$ for any vector $\mathbf{v} \in \mathbb{Z}_q^n$. Given a message $M \in \{0, 1\}^\ell$, the signing algorithm first computes $\mathbf{A}_M = (\mathbf{A}\|\mathrm{H}_K(M)) \in \mathbb{Z}_q^{n\times m}$, and then uses the trapdoor $\mathbf{R}$ to sample a short vector $\mathbf{e} \in \mathbb{Z}^m$ satisfying $\mathbf{A}_M\mathbf{e} = \mathbf{u}$ by employing the sampling algorithms in [26, 34, 55]. Finally, it returns $\sigma = \mathbf{e}$ as the signature on the message M. The verifier accepts $\sigma = \mathbf{e}$ as a valid signature on M if and only if $\mathbf{e}$ is short and $\mathbf{A}_M\mathbf{e} = \mathbf{u}$. The correctness of the generic scheme Π_{sig} is guaranteed by the nice properties of the sampling algorithms in [34, 55].

In addition, if $\mathcal{H} = \{\mathrm{H}_K\}$ is a $(1, v, \beta)$-PHF for some integer v and real β, we can show that under the ISIS assumption, Π_{sig} is existentially unforgeable against adaptive chosen message attacks (EUF-CMA) in the standard model as long as the forger $\mathcal{F}$ makes at most $Q \leq v$ signing queries. Intuitively, given an ISIS challenge instance $(\hat{\mathbf{A}}, \hat{\mathbf{u}})$ in the security reduction, the challenger first generates a trapdoor mode key K' for $\mathcal{H}$ by using $(\hat{\mathbf{A}}, \mathbf{B})$. Then, it defines $vk = (\hat{\mathbf{A}}, \hat{\mathbf{u}}, K')$ and keeps the trapdoor td of K' private. For message M_i in the i-th signing query, we have $\mathbf{A}_{M_i} = (\hat{\mathbf{A}} \| \mathrm{H}_{K'}(M_i)) = (\hat{\mathbf{A}} \| \hat{\mathbf{A}}\mathbf{R}_{M_i} + \mathbf{S}_{M_i}\mathbf{B}) \in \mathbb{Z}_q^{n \times m}$. By the programmability of $\mathcal{H}$, with a certain probability we have that $\mathbf{S}_{M_i}$ is invertible for all the Q signing messages $\{M_i\}_{i \in \{1,\ldots,Q\}}$, but $\mathbf{S}_{M^*} = \mathbf{0}$ for the forged message M^*. In this case, the challenger can use $\mathbf{R}_{M_i}$ to perfectly answer the signing queries and use the forged message-signature pair (M^*, σ^*) to solve the ISIS problem by the equation $\mathbf{u} = \mathbf{A}_{M^*}\sigma^* = \hat{\mathbf{A}}(\mathbf{I}_{\bar{m}} \| \mathbf{R}_{M^*})\sigma^*$.

7.3.2 A Short Signature Scheme with Short Verification Key

Let integers $\ell, n, m', v, q \in \mathbb{Z}$, $\beta \in \mathbb{R}$ be some polynomials in the security parameter κ, and let $k = \lceil \log_2 q \rceil$. Let $\mathcal{H} = (\mathcal{H}.\mathrm{Gen}, \mathcal{H}.\mathrm{Eval})$ be any $(1, v, \beta)$-PHF from $\{0,1\}^\ell$ to $\mathbb{Z}_q^{n \times m'}$. Let $\bar{m} = O(n \log q)$, $m = \bar{m} + m'$, and large enough $s > \max(\beta, \sqrt{m}) \cdot \omega(\sqrt{\log n}) \in \mathbb{R}$ be the system parameters. Our generic signature scheme $\Pi_{\mathrm{sig}} = (\mathsf{KeyGen}, \mathsf{Sign}, \mathsf{Verify})$ is defined as follows.

$\mathsf{KeyGen}(1^\kappa)$: Given a security parameter κ, compute $(\mathbf{A}, \mathbf{R}) \leftarrow \mathsf{TrapGen}(1^n, 1^{\bar{m}}, q, \mathbf{I}_n)$ such that $\mathbf{A} \in \mathbb{Z}_q^{n \times \bar{m}}$, $\mathbf{R} = \mathbb{Z}_q^{(\bar{m}-nk) \times nk}$, and randomly choose $\mathbf{u} \xleftarrow{\$} \mathbb{Z}_q^n$. Then, compute $K \leftarrow \mathcal{H}.\mathrm{Gen}(1^\kappa)$, and return a pair of verification key and secret signing key $(vk, sk) = ((\mathbf{A}, \mathbf{u}, K), \mathbf{R})$.

$\mathsf{Sign}(sk, M \in \{0,1\}^\ell)$: Given $sk = \mathbf{R}$ and any message M, compute $\mathbf{A}_M = (\mathbf{A} \| \mathrm{H}_K(M)) \in \mathbb{Z}_q^{n \times m}$, where $\mathrm{H}_K(M) = \mathcal{H}.\mathrm{Eval}(K, M) \in \mathbb{Z}_q^{n \times m'}$. Then, compute $\mathbf{e} \leftarrow \mathsf{SampleD}(\mathbf{R}, \mathbf{A}_M, \mathbf{I}_n, \mathbf{u}, s)$, and return the signature $\sigma = \mathbf{e}$.

$\mathsf{Verify}(vk, M, \sigma)$: Given vk, a message M and a vector $\sigma = \mathbf{e}$, compute $\mathbf{A}_M = (\mathbf{A} \| \mathrm{H}_K(M)) \in \mathbb{Z}_q^{n \times m}$, where $\mathrm{H}_K(M) = \mathcal{H}.\mathrm{Eval}(K, M) \in \mathbb{Z}_q^{n \times m'}$. Return 1 if $\|\mathbf{e}\| \leq s\sqrt{m}$ and $\mathbf{A}_M\mathbf{e} = \mathbf{u}$, else return 0.

The correctness of the scheme Π_{sig} can be easily checked. Besides, the schemes with linear verification keys in [16, 55] can be seen as instantiations of Π_{sig} with the Type-I PHF construction in Theorem 4.2.[1] Since the size of the verification key is mainly determined by the key size of $\mathcal{H}$, one can instantiate $\mathcal{H}$ with the Type-II PHF construction in Definition 4.5 to obtain a signature scheme with verification keys consisting of a logarithmic number of matrices. As for the security, we have the following theorem:

Theorem 7.2 *Let $\ell, n, \bar{m}, m', q \in \mathbb{Z}$ and $\bar{\beta}, \beta, s \in \mathbb{R}$ be some polynomials in the security parameter κ, and let $m = \bar{m} + m'$. Let $\mathcal{H} = (\mathcal{H}.\mathrm{Gen}, \mathcal{H}.\mathrm{Eval})$ be a $(1, v, \beta, \gamma, \delta)$-PHF from $\{0,1\}^\ell$ to $\mathbb{Z}_q^{n \times m'}$ with $\gamma = \mathsf{negl}(\kappa)$ and noticeable $\delta > 0$.*

[1] Note that the scheme in [16] used a syndrome $\mathbf{u} = \mathbf{0}$, we prefer to use a random chosen syndrome $\mathbf{u} \xleftarrow{\$} \mathbb{Z}_q^n$ as that in [55] for simplifying the security analysis.

Then, for large enough $\bar{m} = O(n \log q)$ *and* $s > \max(\beta, \sqrt{m}) \cdot \omega(\sqrt{\log n}) \in \mathbb{R}$*, if there exists a PPT forger* $\mathcal{F}$ *breaking the* EUF-CMA *security of* Π_{sig} *with non-negligible probability* $\epsilon > 0$ *and making at most* $Q \leq v$ *signing queries, there exists an algorithm* $\mathcal{B}$ *solving the* $\text{ISIS}_{q,\bar{m},\bar{\beta}}$ *problem for* $\bar{\beta} = \beta s \sqrt{m} \cdot \omega(\sqrt{\log n})$ *with probability at least* $\epsilon' \geq \epsilon\delta - \mathsf{negl}(\kappa)$.

Let Π^*_{sig} denote the signature scheme obtained by instantiating Π_{sig} with the Type-II PHF construction in Definition 4.5. Then, the verification key of Π^*_{sig} has $O(\log n)$ matrices and each signature of Π^*_{sig} consists of a single lattice vector.

Corollary 7.1 *Let* $n, q \in \mathbb{Z}$ *be polynomials in the security parameter* κ*. Let* $\bar{m} = O(n \log q)$, $v = \mathsf{poly}(n)$ *and* $\ell = n$*. If there exists a PPT forger* $\mathcal{F}$ *breaking the* EUF-CMA *security of* Π^*_{sig} *with non-negligible probability* ϵ *and making at most* $Q \leq v$ *signing queries, then there exists an algorithm* $\mathcal{B}$ *solving the* $\text{ISIS}_{q,\bar{m},\bar{\beta}}$ *problem for* $\bar{\beta} = v^2 \cdot \tilde{O}(n^{5.5})$ *with probability at least* $\epsilon' \geq \frac{\epsilon}{16nv^2} - \mathsf{negl}(\kappa)$.

7.3.3 An Improved Short Signature Scheme from Weaker Assumption

Compared to prior constructions in [4, 11, 31], the Π^*_{sig} only has a reduction loss about $16nQ^2$, which does not depend on the forger's success probability ϵ. However, because of $v \geq Q$, this improvement requires the $\text{ISIS}_{q,\bar{m},\bar{\beta}}$ problem to be hard for $\bar{\beta} = Q^2 \cdot \tilde{O}(n^{5.5})$, which means that the modulus q should be bigger than $Q^2 \cdot \tilde{O}(n^{5.5})$. Even though q is still a polynomial of n in an asymptotic sense, it might be very large in practice. In this section, we further remove the direct dependency on Q from $\bar{\beta}$ by introducing a short tag about $O(\log Q)$ bits to each signature. For example, this only increases about 30 bits to each signature for a number $Q = 2^{30}$ of the forger's signing queries.

At a high level, the basic idea is to relax the requirement on a $(1, v, \beta)$-PHF $\mathcal{H} = \{\text{H}_K\}$ so that a much smaller $v = \omega(\log n)$ can be used by employing a simple weak PHF $\mathcal{H}' = \{\text{H}'_{K'}\}$ (recall that $v \geq Q$ is required in the scheme Π_{sig}). Concretely, for each message M to be signed, instead of using $\text{H}_K(M)$ in the signing algorithm of Π_{sig}, we choose a short random tag $\mathbf{t}$ and compute $\text{H}'_{K'}(\mathbf{t}) + \text{H}_K(M)$ to generate the signature on M. Thus, if the trapdoor keys of both PHFs are generated by using the same "generators" $\mathbf{A}$ and $\mathbf{G}$, we have that $\text{H}'_{K'}(\mathbf{t}) + \text{H}_K(M) = \mathbf{A}(\mathbf{R}'_{\mathbf{t}} + \mathbf{R}_M) + (\mathbf{S}'_{\mathbf{t}} + \mathbf{S}_M)\mathbf{G}$, where $\text{H}'_{K'}(\mathbf{t}) = \mathbf{A}\mathbf{R}'_{\mathbf{t}} + \mathbf{S}'_{\mathbf{t}}\mathbf{G}$ and $\text{H}_K(M) = \mathbf{A}\mathbf{R}_M + \mathbf{S}_M\mathbf{G}$. Moreover, if we can ensure that $\mathbf{S}'_{\mathbf{t}} + \mathbf{S}_M \in \mathcal{I}_n$ when $\mathbf{S}'_{\mathbf{t}} \in \mathcal{I}_n$ or $\mathbf{S}_M \in \mathcal{I}_n$, then $\mathbf{S}_M$ is not required to be invertible for all the Q signing messages. In particular, $v = \omega(\log n)$ can be used as long as the probability that $\mathbf{S}'_{\mathbf{t}} + \mathbf{S}_M \in \mathcal{I}_n$ is invertible for all the Q signing messages, but $\mathbf{S}'_{\mathbf{t}^*} + \mathbf{S}_{M^*} = \mathbf{0}$ for the forged signature on the pair $(\mathbf{t}^*, M^*)$, is noticeable.

Actually, the weak PHF $\mathcal{H}'$ and the $(1, v, \beta)$-PHF $\mathcal{H} = (\mathcal{H}.\text{Gen}, \mathcal{H}.\text{Eval})$ are, respectively, the first instantiated Type-I PHF $\mathcal{H}'$ in Theorem 4.3 and the Type-II PHF $\mathcal{H} = (\mathcal{H}.\text{Gen}, \mathcal{H}.\text{Eval})$ given in Definition 4.5. Since $\mathcal{H}'$ is very simple, one directly plug its construction into the signature scheme Π'_{sig}. Specifically, let $n, q \in \mathbb{Z}$ be some polynomials in the security parameter κ, and let $k = \lceil \log_2 q \rceil$, $\bar{m} = O(n \log q)$, $m = \bar{m} + nk$ and $s = \tilde{O}(n^{2.5}) \in \mathbb{R}$. Let $H : \mathbb{Z}_q^n \to \mathbb{Z}_q^{n \times n}$ be the FRD encoding in [2] such that for any vector $\mathbf{v} = (v, 0 \ldots, 0)^t, \mathbf{v}_1, \mathbf{v}_2 \in \mathbb{Z}_q^n$, we have that $H(\mathbf{v}) = v\mathbf{I}_n$ and $H(\mathbf{v}_1) + H(\mathbf{v}_2) = H(\mathbf{v}_1 + \mathbf{v}_2)$ hold. For any $\mathbf{t} \in \{0, 1\}^\ell$ with $\ell < n$, we naturally treat it as a vector in $\mathbb{Z}_q^n$ by appending it $(n - \ell)$ zero coordinates. The weak PHF $\mathcal{H}'$ from $\{0, 1\}^\ell$ to $\mathbb{Z}_q^{n \times nk}$ has a form of $\mathrm{H}'_{K'}(\mathbf{t}) = \mathbf{A}_0 + H(\mathbf{t})\mathbf{G}$, where $K' = \mathbf{A}_0$. We restrict the domain of $\mathcal{H}'$ to be $\{0\} \times \{0, 1\}^\ell$ for $\ell \le n - 1$ such that $\mathbf{S}'_{\mathbf{t}} + \mathbf{S}_M$ is invertible when $(\mathbf{S}'_{\mathbf{t}}, \mathbf{S}_M) \neq (\mathbf{0}, \mathbf{0})$. Our signature scheme $\Pi'_{\text{sig}} =$ (KeyGen, Sign, Verify) is defined as follows:

KeyGen(1^κ): Given a security parameter κ, compute $(\mathbf{A}, \mathbf{R}) \leftarrow$ TrapGen$(1^n, 1^{\bar{m}}, q, \mathbf{I}_n)$ such that $\mathbf{A} \in \mathbb{Z}_q^{n \times \bar{m}}$, $\mathbf{R} = \mathbb{Z}_q^{(\bar{m} - nk) \times nk}$. Randomly choose $\mathbf{A}_0 \xleftarrow{\$} \mathbb{Z}_q^{n \times nk}$ and $\mathbf{u} \xleftarrow{\$} \mathbb{Z}_q^n$. Finally, compute $K \leftarrow \mathcal{H}.\text{Gen}(1^\kappa)$, and return $(vk, sk) = ((\mathbf{A}, \mathbf{A}_0, \mathbf{u}, K), \mathbf{R})$.

Sign$(sk, M \in \{0, 1\}^n)$: Given the secret key sk and a message M, randomly choose $\mathbf{t} \xleftarrow{\$} \{0, 1\}^\ell$, and compute $\mathbf{A}_{M,\mathbf{t}} = (\mathbf{A} \| (\mathbf{A}_0 + H(0\|\mathbf{t})\mathbf{G}) + \mathrm{H}_K(M)) \in \mathbb{Z}_q^{n \times m}$, where $\mathrm{H}_K(M) = \mathcal{H}.\text{Eval}(K, M) \in \mathbb{Z}_q^{n \times nk}$. Then, compute $\mathbf{e} \leftarrow$ SampleD$(\mathbf{R}, \mathbf{A}_{M,\mathbf{t}}, \mathbf{I}_n, \mathbf{u}, s)$, and return the signature $\sigma = (\mathbf{e}, \mathbf{t})$.

Verify(vk, M, σ): Given vk, message M and $\sigma = (\mathbf{e}, \mathbf{t})$, compute $\mathbf{A}_{M,\mathbf{t}} = (\mathbf{A} \| (\mathbf{A}_0 + H(0\|\mathbf{t})\mathbf{G}) + \mathrm{H}_K(M)) \in \mathbb{Z}_q^{n \times m}$, where $\mathrm{H}_K(M) = \mathcal{H}.\text{Eval}(K, M) \in \mathbb{Z}_q^{n \times nk}$. Return 1 if $\|\mathbf{e}\| \le s\sqrt{m}$ and $\mathbf{A}_{M,\mathbf{t}}\mathbf{e} = \mathbf{u}$. Otherwise, return 0.

Since $\mathbf{R}$ is a $\mathbf{G}$-trapdoor of $\mathbf{A}$, by padding with zero rows it can be extended to a $\mathbf{G}$-trapdoor for $\mathbf{A}_{M,\mathbf{t}}$ with the same quality $s_1(\mathbf{R}) \le \sqrt{m} \cdot \omega(\sqrt{\log n})$. Since $s = \tilde{O}(n^{2.5}) > s_1(\mathbf{R}) \cdot \omega(\sqrt{\log n})$, the vector $\mathbf{e}$ output by SampleD follows the distribution $D_{\mathbb{Z}^m, s}$ satisfying $\mathbf{A}_{M,\mathbf{t}}\mathbf{e} = \mathbf{u}$. In other words, $\|\mathbf{e}\| \le s\sqrt{m}$ holds with overwhelming probability by Lemma 2.3. This shows that Π'_{sig} is correct.

Note that if we set $v = \omega(\log n)$, the key K only has $\mu = O(\log n)$ number of matrices and each signature consists of a vector plus a short ℓ-bit tag. We have the following theorem for security.

Theorem 7.3 *Let $\ell, \bar{m}, n, q, v \in \mathbb{Z}$ be polynomials in the security parameter κ. For appropriate choices of $\ell = O(\log n)$ and $v = \omega(\log n)$, if there exists a PPT forger $\mathcal{F}$ breaking the* EUF-CMA *security of Π'_{sig} with non-negligible probability ϵ and making at most $Q =$ poly(n) signing queries, there exists an algorithm $\mathcal{B}$ solving the* $\text{ISIS}_{q,\bar{m},\bar{\beta}}$ *problem for $\bar{\beta} = \tilde{O}(n^{5.5})$ with probability at least $\epsilon' \ge \frac{\epsilon}{16 \cdot 2^\ell n v^2} - \text{negl}(\kappa) = \frac{\epsilon}{Q \cdot \tilde{O}(n)}$.*

Proof We now give the construction of algorithm $\mathcal{B}$, which simulates the attack environment for $\mathcal{F}$, and solves the $\text{ISIS}_{q,\bar{m},\bar{\beta}}$ problem with probability at least $\frac{\epsilon}{\tilde{O}(n^2)}$. Formally, $\mathcal{B}$ first randomly chooses a vector $\mathbf{t}' \xleftarrow{\$} \{0, 1\}^\ell$ and hopes that $\mathcal{F}$ will output a forged signature with tag $\mathbf{t}^* = \mathbf{t}'$. Then, $\mathcal{B}$ simulates the EUF-CMA game as follows:

KeyGen. Given an $\mathrm{ISIS}_{q,\bar{m},\bar{\beta}}$ challenge instance $(\mathbf{A}, \mathbf{u}) \in \mathbb{Z}_q^{n\times\bar{m}} \times \mathbb{Z}_q^n$, the algorithm $\mathcal{B}$ first randomly chooses $\mathbf{R}_0 \xleftarrow{\$} (D_{\mathbb{Z}^{\bar{m}},\omega(\sqrt{\log n})})^{nk}$, and computes $\mathbf{A}_0 = \mathbf{A}\mathbf{R}_0 - H(0\|\mathbf{t}')\mathbf{G}$. Then, compute $(K', td) \leftarrow \mathcal{H}.\mathrm{TrapGen}(1^\kappa, \mathbf{A}, \mathbf{G})$ as in Theorem 4.4. Finally, set $vk = (\mathbf{A}, \mathbf{A}_0, \mathbf{u}, K')$ and keep $(\mathbf{R}_0, td)$ private.

Signing. Given a message M, the algorithm $\mathcal{B}$ first randomly chooses a tag $\mathbf{t} \xleftarrow{\$} \{0, 1\}^\ell$. If $\mathbf{t}$ has been used in answering the signatures for more than v messages, $\mathcal{B}$ aborts. Otherwise, $\mathcal{B}$ computes $(\mathbf{R}_M, \mathbf{S}_M) = \mathcal{H}.\mathrm{TrapEval}(td, K', M)$ as in Theorem 4.4. Then, we have $\mathbf{A}_{M,\mathbf{t}} = (\mathbf{A}\|(\mathbf{A}_0 + H(0\|\mathbf{t})\mathbf{G}) + \mathrm{H}_{K'}(M)) = (\mathbf{A}\|\mathbf{A}(\mathbf{R}_0 + \mathbf{R}_M) + (H(0\|\mathbf{t}) - H(0\|\mathbf{t}') + \mathbf{S}_M)\mathbf{G})$. Since $\mathbf{S}_M = b\mathbf{I}_n = H(b\|0)$ for some $b \in \{-1, 0, 1\}$, we have that $\hat{\mathbf{S}} = H(0\|\mathbf{t}) - H(0\|\mathbf{t}') + \mathbf{S}_M = H(b\|(\mathbf{t} - \mathbf{t}'))$ holds by the homomorphic property of the FRD encoding H in [2]. $\mathcal{B}$ distinguishes the following two cases:

- $\mathbf{t} \neq \mathbf{t}'$ or $(\mathbf{t} = \mathbf{t}' \wedge b \neq 0)$: In both cases, we have that $\hat{\mathbf{S}}$ is invertible. In other words, $\hat{\mathbf{R}} = \mathbf{R}_0 + \mathbf{R}_M$ is a $\mathbf{G}$-trapdoor for $\mathbf{A}_{M,\mathbf{t}}$. Since $s_1(\mathbf{R}_0) \leq \sqrt{m} \cdot \omega(\sqrt{\log n})$ by Lemma 2.6 and $s_1(\mathbf{R}_M) \leq \tilde{O}(n^{2.5})$, we have $s_1(\hat{\mathbf{R}}) \leq \tilde{O}(n^{2.5})$. Then, compute $\mathbf{e} \leftarrow \mathsf{SampleD}(\hat{\mathbf{R}}, \mathbf{A}_{M,\mathbf{t}}, \hat{\mathbf{S}}, \mathbf{u}, s)$, and return the signature $\sigma = (\mathbf{e}, \mathbf{t})$. If we set an appropriate $s = \tilde{O}(n^{2.5}) \geq s_1(\hat{\mathbf{R}}) \cdot \omega(\sqrt{\log n})$, then $\mathcal{B}$ can generate a valid signature on M with overwhelming probability by Lemma 2.24.
- $\mathbf{t} = \mathbf{t}' \wedge b = 0$: $\mathcal{B}$ aborts.

Forge. After making at most Q signing queries, $\mathcal{F}$ outputs a forged signature $\sigma^* = (\mathbf{e}^*, \mathbf{t}^*)$ on message $M^* \in \{0, 1\}^n$ such that $\|\mathbf{e}^*\| \leq s\sqrt{m}$ and $\mathbf{A}_{M^*,\mathbf{t}^*}\mathbf{e}^* = \mathbf{u}$, where $\mathbf{A}_{M^*,\mathbf{t}^*} = (\mathbf{A}\|(\mathbf{A}_0 + H(0\|\mathbf{t}^*)\mathbf{G}) + \mathrm{H}_K(M^*)) \in \mathbb{Z}_q^{n\times m}$. The algorithm $\mathcal{B}$ computes $(\mathbf{R}_{M^*}, \mathbf{S}_{M^*}) = \mathcal{H}.\mathrm{TrapEval}(td, K', M^*)$ and aborts the simulation if $\mathbf{t}^* \neq \mathbf{t}'$ or $\mathbf{S}_{M^*} \neq \mathbf{0}$. Else, we have $\mathbf{A}_{M^*,\mathbf{t}^*} = (\mathbf{A}\|\mathbf{A}(\mathbf{R}_0 + \mathbf{R}_{M^*})) = (\mathbf{A}\|\mathbf{A}\hat{\mathbf{R}})$, where $\hat{\mathbf{R}} = \mathbf{R}_0 + \mathbf{R}_{M^*}$. Finally, $\mathcal{B}$ outputs $\hat{\mathbf{e}} = (\mathbf{I}_{\bar{m}}\|\hat{\mathbf{R}})\mathbf{e}^*$ as its own solution.

By the definition of the $\mathrm{ISIS}_{q,\bar{m},\bar{\beta}}$ problem, $(\mathbf{A}, \mathbf{u})$ is uniformly distributed over $\mathbb{Z}_q^{n\times\bar{m}} \times \mathbb{Z}_q^n$. Since $\mathbf{R}_0 \xleftarrow{\$} (D_{\mathbb{Z}^{\bar{m}},\omega(\sqrt{\log n})})^{nk}$, we have that $\mathbf{A}_0 \in \mathbb{Z}_q^{n\times nk}$ is statistically close to uniform over $\mathbb{Z}_q^{n\times nk}$ by Lemma 2.11. In addition, by Theorem 4.4 the simulated key K' is statistically close to the real key K. Thus, the distribution of the simulated verification key vk is statistically close to that of the real one.

Let $M_1, \ldots, M_u$ be all the messages in answering the signing queries that $\mathcal{B}$ happens to use the same tag $\mathbf{t} = \mathbf{t}'$, and let $(\mathbf{R}_{M_i}, \mathbf{S}_{M_i}) = \mathcal{H}.\mathrm{TrapEval}(td, K', M_i)$ for $i \in \{1, \ldots, u\}$. Then, the algorithm $\mathcal{B}$ will abort in the simulation if and only if either of the following two conditions hold:

- Some tag $\mathbf{t}$ is used in answering the signatures for more than v messages,
- $\mathbf{S}_{M_i}$ is not invertible for some $i \in \{1, \ldots, u\}$, or $\mathbf{S}_{M^*} \neq \mathbf{0}$, or $\mathbf{t}^* \neq \mathbf{t}'$.

Since the forger $\mathcal{F}$ will make at most $Q = \mathsf{poly}(n)$ signing queries, we can choose $\ell = O(\log n)$ such that $\frac{Q}{2^\ell} \leq \frac{1}{2}$. Note that $\mathcal{B}$ always randomly chooses a tag $\mathbf{t} \xleftarrow{\$} \{0, 1\}^\ell$ for each signing message, the probability that $\mathcal{B}$ uses any tag $\mathbf{t}$ in answering the signatures for more than v messages is less than $Q^2 \cdot (\frac{Q}{2^\ell})^v$ by a similar analysis in [43], which is negligible by the setting of $v = \omega(\log n)$. In particular, the probability that $\mathcal{B}$ will use the same tag $\mathbf{t} = \mathbf{t}'$ in answering the signatures for $u \geq v$ messages is also negligible. Conditioned on $u \leq v$, the probability that $\mathbf{S}_{M_i}$ is invertible for all $i \in \{1, \ldots, u\}$ and $\mathbf{S}_{M^*} = \mathbf{0}$ (using the fact that $M^* \notin \{M_1, \ldots, M_u\}$) is at least $\delta = \frac{1}{16nv^2} - \mathsf{negl}(\kappa)$ by Theorem 4.4. Note that $\mathbf{t}'$ is randomly chosen and is statistically hidden from $\mathcal{F}$, the probability $\Pr[\mathbf{t}^* = \mathbf{t}']$ is at least $\frac{1}{2^\ell} - \mathsf{negl}(\kappa)$. Thus, if the forger $\mathcal{F}$ can attack the EUF-CMA security of Π'_{sig} with probability ϵ in the real game, then it will also output a valid forgery $(\mathbf{M}^*, \mathbf{e}^*)$ in the simulated game with probability at least $(\epsilon - Q^2(\frac{Q}{2^\ell})^v) \cdot \delta \cdot (\frac{1}{2^\ell} - \mathsf{negl}(\kappa)) = \frac{\epsilon}{2^\ell \cdot 16nv^2} - \mathsf{negl}(\kappa) = \frac{\epsilon}{Q \cdot \tilde{O}(n)}$ (note that $\mathcal{F}$'s success probability ϵ might be correlated with the first abort condition).

Now, we show that $\hat{\mathbf{e}} = (\mathbf{I}_{\bar{m}} \| \hat{\mathbf{R}})\mathbf{e}^*$ is a valid solution to the $\text{ISIS}_{q,\bar{m},\bar{\beta}}$ instance $(\mathbf{A}, \mathbf{u})$. By the conditions in the verification algorithm, we have that $\mathbf{A}_{M^*,\mathbf{t}^*}\mathbf{e}^* = \mathbf{u}$ and $\|\mathbf{e}^*\| \leq s\sqrt{m}$. Since $s_1(\mathbf{R}_0) \leq \sqrt{m} \cdot \omega(\sqrt{\log n})$ by Lemma 2.6 and $s_1(\mathbf{R}_{M^*}) \leq \beta = \tilde{O}(n^{2.5})$ by Theorem 4.4, we have that $\|\hat{\mathbf{e}}\| \leq \tilde{O}(n^{2.5}) \cdot s\sqrt{m} = \tilde{O}(n^{5.5}) = \bar{\beta}$. This finally completes the proof. □

7.4 Group Signatures

In this chapter, we focus on constructing group signatures from lattices.

7.4.1 Design Rational

We begin with a brief introduction to the BMW paradigm [7]. Roughly speaking, the group manager first generates the group public key gpk and group manager secret key $gmsk$. For a user with identity $i \in \{1, \ldots, N\}$ (recall that N is the maximum number of group members and is fixed at the system setup), the group manager computes the user's secret key gsk_i corresponding to an encoded "public key" $H(gpk, i)$, where H is an encoding function that (uniquely) encodes the group user's identity i in gsk_i. When signing a message m, the group user proves to the verifier that he has a secret key gsk_i for some $i \in \{1, \ldots, N\}$ (i.e., to prove that he is a legal member of the group). The hardness of this general paradigm usually lies in the choices of an appropriate encoding function $H(gpk, i)$ and a compatible non-interactive zero-knowledge (NIZK) for the membership relations determined by $H(gpk, i)$.

Gordon et al. [37] used a simple projective encoding function $H(gpk, i)$ and a NIZK extended from [56] to construct the first lattice-based group signature. Informally, the group public key gpk consists of N independent public keys of the GPV signature [34], i.e., $gpk = (pk_1, \ldots, pk_N)$ where pk_j is an integer matrix over $\mathbb{Z}_q$ for some positive $q \in \mathbb{Z}$, and all $j \in \{1, \ldots, N\}$. The encoding function simply outputs the i-th element of gpk, i.e., $H(gpk, i) := pk_i$. Due to the particular choice of $H(gpk, i)$, both the group public key and the signature of [37] have a size linear in N.

At Asiacrypt'13, by using an efficient encoding function inspired by Boyen's lattice-based signature [16] and a NIZK derived from [52], Laguillaumie et al. [46] proposed a more efficient lattice-based group signature. Roughly speaking, the group public key gpk consists of $\ell = \lfloor \log N \rfloor + 1$ independent matrices over $\mathbb{Z}_q$, i.e., $gpk = (\mathbf{A}_1, \ldots, \mathbf{A}_\ell)$. The encoding function is defined as $H(gpk, i) := \sum_{j=1}^{\ell} i_j \mathbf{A}_j$, where $(i_1, \ldots, i_l) \in \mathbb{Z}_2^\ell$ is the binary decomposition of i. We also note that Langlois et al. [47] constructed a lattice-based group signature with verifier-local revocation by using the same identity encoding function but a different NIZK from [50]. Both schemes [46, 47] decreased the sizes of the group public key and the signature to proportional to $\log N$.

In order to obtain better efficiency, we can use a more efficient and compact way to encode the group member's identity, by building upon the encoding technique introduced by Agrawal et al. [2] for identity-based encryption (IBE). Let the group public key gpk consist of three matrices over $\mathbb{Z}_q^{n \times m}$ for some positive integers n, m, q, i.e., $gpk = (\mathbf{A}_1, \mathbf{A}_{2,1}, \mathbf{A}_{2,2})$. We define $H(gpk, i) = \hat{\mathbf{A}}_i := (\mathbf{A}_1 \| \mathbf{A}_{2,1} + G(i)\mathbf{A}_{2,2})$ where $G(\cdot)$ is a function from $\mathbb{Z}_N$ to $\mathbb{Z}_q^{n \times n}$. Then, the secret key of user i is a short basis of the classical q-ary lattice Λ_i determined by $H(gpk, i) = \hat{\mathbf{A}}_i$, where $\Lambda_i := \{\mathbf{e} \in \mathbb{Z}^m \ \ s.t. \ \hat{\mathbf{A}}_i \mathbf{e} = \mathbf{0} \bmod q\}$. When signing a message, user i samples a short vector $\mathbf{e}_i$ from Λ_i (by using the short basis of Λ_i) and encrypts it using Regev's encryption [59]. Then, he proves to the verifier that $\mathbf{e}_i$ is a short vector in a lattice determined by $H(gpk, i)$ for some $i \in \{1, \ldots, N\}$. However, there is no known lattice-based NIZK suitable for the membership relation determined by $H(gpk, i)$.

Fortunately, since the maximum number of group members N is always bounded by a polynomial in the security parameter n, we actually do not need an encoding function as powerful as for IBE [2], where there are possibly exponentially many users. We simplify the encoding function by defining $H(gpk, i) := (\mathbf{A}_1 \| \mathbf{A}_{2,1} + i\mathbf{A}_{2,2})$. (A similar combination of matrices has been used in a different way in [3, 14, 33] to construct functional encryption.) Namely, the identity function $G(i) := i$ is used instead of a function $G : \mathbb{Z}_N \to \mathbb{Z}_q^{n \times n}$. For collision resistance, we require that $N < q$. Since N is usually fixed at the system setup in group signatures for static groups such as [37, 46, 47], one can simply set q big enough (but still a polynomial in n) to satisfy the requirement. Hereafter, we assume that $N < q$ always holds.

Basically, this encoding function provides two main benefits: (1) Only three matrices are needed for the encoding function, which provides a short group public key. By comparison, there are respectively at least $O(N)$ and $O(\log N)$ matrices needed in [37] and [46]; (2) It gives a simple membership relation, which allows to construct an efficient NIZK proof for the relation. Recall that the secret key of user i is a short basis $\mathbf{T}_i$ of the q-ary lattice Λ_i determined by $\hat{\mathbf{A}}_i = (\mathbf{A}_1 \| \mathbf{A}_{2,1} + i\mathbf{A}_{2,2})$. To sign a message, user i first samples a short vector $(\mathbf{x}_1, \mathbf{x}_2)$ by using Gentry et al.'s Gaussian sampling algorithm [34] such that $\mathbf{A}_1\mathbf{x}_1 + (\mathbf{A}_{2,1} + i\mathbf{A}_{2,2})\mathbf{x}_2 = \mathbf{0} \bmod q$. Then, he generates an LWE-based encryption $\mathbf{c}$ of $\mathbf{x}_1$. The final signature σ consists of $\mathbf{c}$, $\mathbf{x}_2$, a proof π_1 that $\mathbf{c}$ encrypts $\mathbf{x}_1$ correctly, and a proof π_2 that there exists a tuple $(\mathbf{x}_1, i)$ satisfying $\mathbf{A}_1\mathbf{x}_1 + i\mathbf{A}_{2,2}\mathbf{x}_2 = -\mathbf{A}_{2,1}\mathbf{x}_2 \bmod q$, namely, $\sigma = (\mathbf{c}, \mathbf{x}_2, \pi_1, \pi_2)$.

The nice properties of the sampling algorithm in [26, 34] guarantee that the public $\mathbf{x}_2$ is statistically indistinguishable for all user $i \in \{1, \ldots, N\}$, namely, the verifier cannot determine the signer's identity i solely from $\mathbf{x}_2$, however, he can efficiently determine it from $(\mathbf{x}_1, \mathbf{x}_2)$, that's why we choose to encrypt $\mathbf{x}_1$. The proof of π_1 can be generated by using the duality of LWE and Small Integer Solutions (SIS) [54], and the NIZK proof for SIS [52] in a standard way. Thanks to the new identity encoding function $H(gpk, i)$ and the public $\mathbf{x}_2$, we manage to design a NIZK proof (i.e., π_2) for the statement $\mathbf{A}_1\mathbf{x}_1 + i\mathbf{A}_{2,2}\mathbf{x}_2 = -\mathbf{A}_{2,1}\mathbf{x}_2 \bmod q$ based on the hardness of SIS.

Formally, we first introduce a problem called split-SIS, which is a variant of SIS. Given a split-SIS instance $\mathbf{A}_1, \mathbf{A}_{2,2} \in \mathbb{Z}_q^{n\times m}$, the algorithm is asked to output a triple $(\mathbf{x}_1, \mathbf{x}_2, h)$ such that $\mathbf{x}_1, \mathbf{x}_2 \in \mathbb{Z}^m$ have small norms, and $h < q = poly(n)$ is a positive integer satisfying $\mathbf{A}_1\mathbf{x}_1 + h\mathbf{A}_{2,2}\mathbf{x}_2 = \mathbf{0} \bmod q$. We first show that the split-SIS problem (associated with an appropriate solution space) is polynomially equivalent to the standard SIS problem. Then, we derive a family of hash functions

$$\mathcal{H} = \left\{ \begin{array}{r} f_{\mathbf{A}_1,\mathbf{A}_{2,2}}(\mathbf{x}_1, \mathbf{x}_2, h) = (\mathbf{A}_1\mathbf{x}_1 + h\mathbf{A}_{2,2}\mathbf{x}_2 \bmod q, \mathbf{x}_2) : \\ (\mathbf{x}_1, \mathbf{x}_2, h) \in \mathbb{Z}^m \times \mathbb{Z}^m \times \mathbb{Z} \end{array} \right\}_{\mathbf{A}_1,\mathbf{A}_{2,2}\in\mathbb{Z}_q^{n\times m}}$$

from the split-SIS problem and prove that the hash function family $\mathcal{H}$ with appropriate domain is *one-way*, *collision-resistant*, and *statistically hiding* with respect to the third input (i.e., h). Combining those useful properties with the observation that $\mathbf{A}_1\mathbf{x}_1 + h\mathbf{A}_{2,2}\mathbf{x}_2 = (\mathbf{A}_1 \| \mathbf{A}_{2,2}\mathbf{x}_2)(\mathbf{x}_1; h) \bmod q$, we manage to adapt a Σ-protocol for $\mathcal{H}$ from existing protocols for standard ISIS problems [46, 51, 52], which can in turn be transformed into a NIZK using the Fiat-Shamir transformation in the random oracle model. This finally helps us obtain a lattice-based group signature scheme with $O(tm \log q)$-bit signature, where the repetition parameter $t = \omega(\log n)$ is due to the above NIZK as in [46, 47].

In order to open a signature $\sigma = (\mathbf{c}, \mathbf{x}_2, \pi_1, \pi_2)$, the group manager only has to decrypt $\mathbf{c}$ to obtain $\mathbf{x}_1$ and computes an integer $h < q$ satisfying $\mathbf{A}_1\mathbf{x}_1 + h\mathbf{A}_{2,2}\mathbf{x}_2 = -\mathbf{A}_{2,1}\mathbf{x}_2 \bmod q$. Note that such an integer is unique if $\mathbf{A}_{2,2}\mathbf{x}_2 \neq \mathbf{0} \bmod q$ for prime q. Replacing the CPA-encryption of $\mathbf{x}_1$ with a CCA one (i.e., by applying the CHK transformation [25] to the IBE [2]), we obtain a CCA-anonymous group signature at a minimal price of doubling the sizes of the group public key and the signature.

7.4.2 Non-interactive Zero-Knowledge Proofs of Knowledge

In 2013, Laguillaumie et al. [46] adapted the protocol of [51, 52] to obtain a zero-knowledge proof of knowledge for the ISIS problem in the random oracle model. Concretely, there is a non-interactive zero-knowledge proof of knowledge (NIZKPoK) for the ISIS relations

$$R_{\text{ISIS}} = \{(\mathbf{A}, \mathbf{y}, \beta; \mathbf{x}) \in \mathbb{Z}_q^{n\times m} \times \mathbb{Z}_q^n \times \mathbb{R} \times \mathbb{Z}^m : \mathbf{A}\mathbf{x} = \mathbf{y} \text{ and } \|\mathbf{x}\| \leq \beta\}.$$

In particular, there is a knowledge extractor which, given two valid proofs with the same commitment message but two different challenges, outputs a witness $\mathbf{x}'$ satisfying $\|\mathbf{x}'\| \leq O(\beta m^2)$ and $\mathbf{A}\mathbf{x}' = \mathbf{y}$. By using the duality between LWE and ISIS, there exists an NIZKPoK for the LWE relation:

$$R_{\mathrm{LWE}} = \{(\mathbf{A}, \mathbf{b}, \alpha; \mathbf{s}) \in \mathbb{Z}_q^{n\times m} \times \mathbb{Z}_q^m \times \mathbb{R} \times \mathbb{Z}_q^n : \|\mathbf{b} - \mathbf{A}^T\mathbf{s}\| \leq \alpha q\sqrt{m}\}.$$

Actually, as noted in [54], given a random matrix $\mathbf{A} \in \mathbb{Z}_q^{n\times m}$ such that the columns of $\mathbf{A}$ generate $\mathbb{Z}_q^n$ (this holds with overwhelming probability for a uniformly random $\mathbf{A} \in \mathbb{Z}_q^{n\times m}$), one can compute a matrix $\mathbf{G} \in \mathbb{Z}_q^{(m-n)\times m}$ such that 1) the columns of $\mathbf{G}$ generate $\mathbb{Z}_q^{m-n}$; 2) $\mathbf{G}\mathbf{A}^T = \mathbf{0}$. Thus, to prove $(\mathbf{A}, \mathbf{b}, \alpha; \mathbf{s}) \in R_{\mathrm{LWE}}$, one can instead prove the existence of $\mathbf{e}$ such that $\|\mathbf{e}\| \leq \alpha q\sqrt{m}$ and $\mathbf{G}\mathbf{e} = \mathbf{G}\mathbf{b}$. In particular, in the construction of our group signature we need to prove that for given $(\mathbf{A}, \mathbf{b}) \in \mathbb{Z}_q^{n\times m} \times \mathbb{Z}_q^m$, there exist short vectors $(\mathbf{e}, \mathbf{x})$ such that $\|\mathbf{e}\| \leq \alpha q\sqrt{m}$, $\|\mathbf{x}\| \leq \beta$ and $\mathbf{b} = \mathbf{A}^T\mathbf{s} + p\mathbf{e} + \mathbf{x}$ for some $\mathbf{s} \in \mathbb{Z}_q^n$, where $p \geq (\alpha q\sqrt{m} + \beta)m^2$. Similarly, this can also be achieved by proving the existence of short vectors $\mathbf{e}$ and $\mathbf{x}$ such that $p\mathbf{G}\mathbf{e} + \mathbf{G}\mathbf{x} = \mathbf{G}\mathbf{b}$ using the NIZKPoK for ISIS relations. Formally, denoting $\gamma = \max(\alpha q\sqrt{m}, \beta)$, there exists an NIZKPoK for the extended-LWE (eLWE) relations

$$\begin{aligned} R_{\mathrm{eLWE}} = \{(\mathbf{A}, \mathbf{b}, \gamma; \mathbf{s}, \mathbf{e}, \mathbf{x}) \in \mathbb{Z}_q^{n\times m} \times \mathbb{Z}_q^m \times \mathbb{R} \times \mathbb{Z}_q^n \times \mathbb{Z}^{2m} :\\ \mathbf{b} = \mathbf{A}^T\mathbf{s} + p\mathbf{e} + \mathbf{x} \text{ and } \|\mathbf{e}\| \leq \gamma \text{ and } \|\mathbf{x}\| \leq \gamma\}. \end{aligned}$$

7.4.3 Split-SIS Problems

Given uniformly random matrices $(\mathbf{A}_1, \mathbf{A}_2) \in \mathbb{Z}_q^{n\times m} \times \mathbb{Z}_q^{n\times m}$, integer $N = N(n)$ and $\beta = \beta(n)$, an algorithm solving the split-$\mathrm{SIS}_{q,m,\beta,N}$ problem is asked to output a tuple $(\mathbf{x} = (\mathbf{x}_1; \mathbf{x}_2), h) \in \mathbb{Z}^{2m} \times \mathbb{Z}$ such that

- $\mathbf{x}_1 \neq 0$ or $h\mathbf{x}_2 \neq 0$
- $\|\mathbf{x}\| \leq \beta$, $h \in [N]$, and $\mathbf{A}_1\mathbf{x}_1 + h\mathbf{A}_2\mathbf{x}_2 = 0 \bmod q$.

Recall that the standard $\mathrm{SIS}_{q,m',\beta}$ problem asks an algorithm to find a root of the hash function $f_{\mathbf{A}}(\mathbf{x}) = \mathbf{A}\mathbf{x} = \mathbf{0} \bmod q$ for a uniformly chosen matrix $\mathbf{A}$ and a "narrow" domain $\hat{D}_{m',\beta} := \{\mathbf{x} \in \mathbb{Z}^{m'} : \|\mathbf{x}\| \leq \beta\}$. While for the split-$\mathrm{SIS}_{q,m,\beta,N}$ problem, the algorithm is allowed to "modify" the function by defining $f_{\mathbf{A}'}(\mathbf{x}') = \mathbf{A}'\mathbf{x}' \bmod q$ for $\mathbf{A}' = (\mathbf{A}_1\|\mathbf{A}_2\mathbf{x}_2)$ with arbitrarily $\mathbf{x}_2 \in \hat{D}_{m,\beta}$, and outputs a root $\mathbf{x}' = (\mathbf{x}_1, h) \in \hat{D}_{m,\beta} \times [N]$. Intuitively, the split-$\mathrm{SIS}_{q,m,\beta,N}$ problem is not harder than $\mathrm{SIS}_{q,2m,\beta}$ problem. Since if $\mathbf{x} = (\mathbf{x}_1, \mathbf{x}_2)$ is a solution of the $\mathrm{SIS}_{q,2m,\beta}$ instance $\mathbf{A} = (\mathbf{A}_1\|\mathbf{A}_2)$, $(\mathbf{x}, 1)$ is a solution of the split-$\mathrm{SIS}_{q,m,\beta,N}$ instance $(\mathbf{A}_1, \mathbf{A}_2)$ with $N \geq 1$.

However, for prime $q = q(n)$, and $N = N(n) < q$ of a polynomial in n, we show in the following theorem that the split-$\mathrm{SIS}_{q,m,\beta,N}$ problem is at least as hard as $\mathrm{SIS}_{q,2m,\beta}$. Thus, the average-case hardness of the split-SIS problem is based on the worst-case hardness of SIVP by Lemma 2.14.

Theorem 7.4 (Hardness of Split-SIS Problems) *For any polynomial $m = m(n)$, $\beta = \beta(n)$, $N = N(n)$, and any prime $q \geq \beta \cdot \omega(\sqrt{n \log n}) > N$, the split-SIS$_{q,m,\beta,N}$ problem is polynomially equivalent to SIS$_{q,2m,\beta}$ problem. In particular, the average-case split-SIS$_{q,m,\beta,N}$ is as hard as approximating the SIVP problem in the worst-case to within certain $\gamma = \beta \cdot \widetilde{O}(\sqrt{n})$ factors.*

Proof The direction from split-SIS$_{q,m,\beta,N}$ to SIS$_{q,2m,\beta}$ is obvious. We now prove the other direction. Assume that there is an algorithm $\mathcal{A}$ that solves split-SIS$_{q,m,\beta,N}$ with probability ϵ, we now construct an algorithm $\mathcal{B}$ that solves SIS$_{q,2m,\beta}$ with probability at least ϵ/N (recall that N is a polynomial in n). Formally, given a SIS$_{q,2m,\beta}$ instance $\hat{\mathbf{A}} = (\hat{\mathbf{A}}_1 \| \hat{\mathbf{A}}_2) \in \mathbb{Z}_q^{n \times 2m}$, $\mathcal{B}$ randomly chooses an integer $h^* \xleftarrow{\$} [N]$. If $h^* = 0$, $\mathcal{B}$ sets $\mathbf{A} = \hat{\mathbf{A}}$. Otherwise, $\mathcal{B}$ sets $\mathbf{A} = (h^* \hat{\mathbf{A}}_1 \| \hat{\mathbf{A}}_2)$. Since q is a prime and $N < q$ (i.e., $h^* \neq 0$ is invertible in $\mathbb{Z}_q$), we have that $\mathbf{A}$ is uniformly distributed over $\mathbb{Z}_q^{n \times 2m}$. Then, $\mathcal{B}$ gives $\mathbf{A} = (\mathbf{A}_1 \| \mathbf{A}_2)$ to $\mathcal{A}$, and obtains a solution $(\mathbf{x} = (\mathbf{x}_1; \mathbf{x}_2), h) \in \mathbb{Z}^{2m} \times [N]$ satisfying $\mathbf{A}_1 \mathbf{x}_1 + h \mathbf{A}_2 \mathbf{x}_2 = \mathbf{0}$. If $h^* \neq h$, $\mathcal{B}$ aborts. (Since h^* is randomly chosen from $[N]$, the probability $\Pr[h^* = h]$ is at least $1/N$.) Otherwise, $\mathcal{B}$ returns $\mathbf{y} = (\mathbf{x}_1; \mathbf{0})$ if $h^* = 0$, else returns $\mathbf{y} = \mathbf{x}$. The first claim follows from the fact that $\mathbf{y} \neq 0$, $\|\mathbf{y}\| \leq \beta$ and $\hat{\mathbf{A}}\mathbf{y} = 0$. Combining this with Lemma 2.14, the second claim follows. □

7.4.3.1 A Family of Hash Functions from Split-SIS Problems

We define a new family of hash functions based on the split-SIS problem, which plays a key role in reducing the sizes of the group public key and the signature in the construction. Formally, for integers n, m, prime q and polynomial $\beta = \beta(n) \geq \omega(\sqrt{\log m})$, $N = N(n) < q$, we define $D_{m,\beta} = \{\mathbf{x} \xleftarrow{\$} D_{\mathbb{Z}^m,\beta} : \|\mathbf{x}\| \leq \beta\sqrt{m}\}$, and a hash function family $\mathcal{H}_{n,m,q,\beta,N} = \{f_{\mathbf{A}} : D_{m,\beta,N} \to \mathbb{Z}_q^n \times D_{m,\beta}\}_{\mathbf{A} \in \mathbb{Z}_q^{n \times 2m}}$, where $D_{m,\beta,N} := D_{m,\beta} \times D_{m,\beta} \times [N]$. For index $\mathbf{A} = (\mathbf{A}_1 \| \mathbf{A}_2) \in \mathbb{Z}_q^{n \times 2m}$, and input $(\mathbf{x}_1, \mathbf{x}_2, h) \in D_{m,\beta,N}$, the hash value $f_{\mathbf{A}}(\mathbf{x}_1, \mathbf{x}_2, h) := (\mathbf{A}_1 \mathbf{x}_1 + h\mathbf{A}_2 \mathbf{x}_2, \mathbf{x}_2) \in \mathbb{Z}_q^n \times D_{m,\beta}$. In the following, we show three properties of $\mathcal{H}_{n,m,q,\beta,N}$, which are useful to construct zero-knowledge proofs for the function in $\mathcal{H}_{n,m,q,\beta,N}$.

Theorem 7.5 (One-Wayness) *For parameters $m > 2n \log q$, $\beta = \beta(n) > 2 \cdot \omega(\sqrt{\log m})$, prime $q = q(n)$, and polynomial $N = N(n) < q$, if the split-SIS$_{q,m,\sqrt{5m}\beta,N}$ problem is hard, then the family of hash functions $\mathcal{H}_{n,m,q,\beta,N}$ is one-way.*

Proof Assume that there is an algorithm $\mathcal{A}$ that breaks the one-wayness of $\mathcal{H}_{n,m,q,\beta,N}$, we construct an algorithm $\mathcal{B}$ that solves the split-SIS$_{q,m,\sqrt{5m}\beta,N}$ problem. Actually, given a split-SIS$_{q,m,\sqrt{5m}\beta,N}$ instance $\mathbf{A}=(\mathbf{A}_1 \| \mathbf{A}_2) \in \mathbb{Z}_q^{n \times 2m}$, $\mathcal{B}$ randomly chooses $(\mathbf{x}_1, \mathbf{x}_2) \in D_{\mathbb{Z}^m,\beta} \times D_{\mathbb{Z}^m,\beta}$ and $h \xleftarrow{\$} [N]$ and computes $\mathbf{y} = f_{\mathbf{A}}(\mathbf{x}_1, \mathbf{x}_2, h) = (\mathbf{A}_1 \mathbf{x}_1 + h\mathbf{A}_2 \mathbf{x}_2, \mathbf{x}_2)$. Then, it gives $(\mathbf{A}, \mathbf{y})$ to $\mathcal{A}$ and obtains $(\mathbf{x}'_1, \mathbf{x}'_2, h')$ satisfying $(\mathbf{A}_1 \mathbf{x}'_1 + h'\mathbf{A}_2 \mathbf{x}'_2, \mathbf{x}'_2) = \mathbf{y}$. Finally, if $h \geq h'$, $\mathcal{B}$ outputs $(\hat{\mathbf{x}}_1, \hat{\mathbf{x}}_2, \hat{h}) = (\mathbf{x}_1 - \mathbf{x}'_1, \mathbf{x}_2, h - h')$. Else, $\mathcal{B}$ outputs $(\hat{\mathbf{x}}_1, \hat{\mathbf{x}}_2, \hat{h}) = (\mathbf{x}'_1 - \mathbf{x}_1, \mathbf{x}_2, h' - h)$.

It is easy to check that $\mathbf{A}_1\hat{\mathbf{x}}_1 + \hat{h}\mathbf{A}_2\hat{\mathbf{x}}_2 = \mathbf{0} \bmod q$, $\hat{h} \in [N]$, and $\|(\hat{\mathbf{x}}_1; \hat{\mathbf{x}}_2)\| \leq \sqrt{5m}\beta$ with overwhelming probability by the standard tail inequality of the Gaussian distribution $D_{\mathbb{Z}^m,\beta}$. We finish this proof by showing that $\Pr[\hat{\mathbf{x}}_1 = \mathbf{0}]$ is negligible in n. Note that $\mathcal{A}$ can only obtain the information about $\mathbf{x}_1$ from $\mathbf{A}_1\mathbf{x}_1$. By [34, Lem. 5.2], this only leaks the distribution $\mathbf{t} + D_{\Lambda_q^{\perp}(\mathbf{A}_1),\beta,-\mathbf{t}}$ for any $\mathbf{t}$ satisfying $\mathbf{A}_1\mathbf{t} = \mathbf{A}_1\mathbf{x}_1$. Namely, $\mathbf{x}_1$ should be uniformly distributed over $\mathbf{t} + D_{\Lambda_q^{\perp}(\mathbf{A}_1),\beta,-\mathbf{t}}$ from the view of $\mathcal{A}$. Combining this with [58, Lem. 2.16], we have $\Pr[\mathbf{x}_1 = \mathbf{x}_1']$ is negligible in n. In other words, we have $\Pr[\hat{\mathbf{x}} \neq \mathbf{0}] = 1 - negl(n)$, which completes the proof. □

Since $f_{\mathbf{A}}(\mathbf{x}_1, \mathbf{0}, h) = f_{\mathbf{A}}(\mathbf{x}_1, \mathbf{0}, 0)$ holds for all $h \in [N]$, the function $\mathcal{H}_{n,m,q,\beta,N}$ with domain $D_{m,\beta,N} := D_{m,\beta} \times D_{m,\beta} \times [N]$ are not collision-resistant. However, if we slightly restrict the domain of $\mathcal{H}_{n,m,q,\beta,N}$ to exclude the above trivial case, we can prove that the family of $\mathcal{H}_{n,m,q,\beta,N}$ is collision-resistant. Formally, we slightly restrict the domain of $\mathcal{H}_{n,m,q,\beta,N}$ to be $D'_{m,\beta,N} = \{(\mathbf{x}_1, \mathbf{x}_2, h) \in D_{m,\beta,N} : \mathbf{x}_2 \neq \mathbf{0}\}$.

Theorem 7.6 (Collision-Resistance) *For parameter $m = m(n)$, $\beta = \beta(n)$, prime $q = q(n)$ and polynomial $N = N(n) < q$, if the split-SIS$_{q,m,\sqrt{5m}\beta,N}$ problem is hard, then the family of hash functions $\mathcal{H}_{n,m,q,\beta,N}$ with domain $D'_{m,\beta,N}$ is collision-resistant.*

Proof Assume there is a PPT algorithm $\mathcal{A}$ that can find collisions of $\mathcal{H}_{n,m,q,\beta,N}$ with non-negligible probability ϵ, we construct an algorithm $\mathcal{B}$ solving split-SIS$_{q,m,\sqrt{5m}\beta,N}$ with the same probability. Concretely, after obtaining a split-SIS$_{q,m,\sqrt{5m}\beta,N}$ instance $\mathbf{A} = (\mathbf{A}_1\|\mathbf{A}_2)$, $\mathcal{B}$ directly gives $\mathbf{A}$ to $\mathcal{A}$ and obtains a pair of collisions $(\mathbf{x}_1, \mathbf{x}_2, h) \in D'_{m,\beta,N}$ and $(\mathbf{x}_1', \mathbf{x}_2', h') \in D'_{m,\beta,N}$ satisfying $(\mathbf{x}_1, \mathbf{x}_2, h) \neq (\mathbf{x}_1', \mathbf{x}_2', h')$ and $f_{\mathbf{A}}(\mathbf{x}_1, \mathbf{x}_2, h) = f_{\mathbf{A}}(\mathbf{x}_1', \mathbf{x}_2', h')$. Note that in this case, we must have $\mathbf{x}_2 = \mathbf{x}_2' \neq \mathbf{0}$. If $h \geq h'$, $\mathcal{B}$ returns $(\hat{\mathbf{x}}_1, \hat{\mathbf{x}}_2, \hat{h}) = (\mathbf{x}_1 - \mathbf{x}_1', \mathbf{x}_2, h - h')$, else it returns $(\hat{\mathbf{x}}_1, \hat{\mathbf{x}}_2, \hat{h}) = (\mathbf{x}_1' - \mathbf{x}_1, \mathbf{x}_2, h' - h)$. By the assumption that $(\mathbf{x}_1, \mathbf{x}_2, h) \neq (\mathbf{x}_1', \mathbf{x}_2', h')$, the inequality $(\hat{\mathbf{x}}_1, \hat{h}) \neq \mathbf{0}$ holds in both cases, i.e., we always have $\hat{\mathbf{x}}_1 \neq \mathbf{0}$ or $\hat{h}\hat{\mathbf{x}}_2 \neq \mathbf{0}$. The claim follows from the fact that $\|(\hat{\mathbf{x}}_1; \hat{\mathbf{x}}_2)\| \leq \sqrt{5m}\beta$ and $\hat{h} \in [N]$. □

Finally, we show that the family of hash functions $\mathcal{H}_{n,m,q,\beta,N}$ statistically hides its third input.

Theorem 7.7 *Let parameter $m > 2n\log q$, $\beta = \beta(n) > \omega(\sqrt{\log m})$, prime $q = q(n)$, and polynomial $N = N(n)$. Then, for a randomly chosen $\mathbf{A} = (\mathbf{A}_1\|\mathbf{A}_2) \in \mathbb{Z}_q^{n\times 2m}$, and arbitrarily $\mathbf{x}_2$ with norm $\|\mathbf{x}_2\| \leq \beta\sqrt{m}$, the statistical distance between the following two distributions:*

$$\{(\mathbf{A}, f_{\mathbf{A}}(\mathbf{x}_1, \mathbf{x}_2, h), h) : \mathbf{x}_1 \xleftarrow{\$} D_{m,\beta}, h \xleftarrow{\$} [N]\}$$

and

$$\{(\mathbf{A}, (\mathbf{u}, \mathbf{x}_2), h) : \mathbf{u} \xleftarrow{\$} \mathbb{Z}_q^n, h \xleftarrow{\$} [N]\}$$

is negligible in n.

Proof Since the second output of $f_{\mathbf{A}}(\mathbf{x}_1, \mathbf{x}_2, h)$ (i.e., $\mathbf{x}_2$) is independent from the choices of h, we only have to show that, for arbitrary $\mathbf{x}_2$ and h, the distribution

$\{\mathbf{A}_1\mathbf{x}_1 + h\mathbf{A}_2\mathbf{x}_2 : \mathbf{x}_1 \xleftarrow{\$} D_{m,\beta}\}$ is statistically close to uniform over $\mathbb{Z}_q^n$. Actually, using the fact that $\beta \geq \omega(\sqrt{\log m})$ together with Lemma 2.11, we have that the distribution of $\mathbf{A}_1\mathbf{x}_1$ is statistically close to uniform over $\mathbb{Z}_q^n$ when $\mathbf{x}_1 \xleftarrow{\$} D_{\mathbb{Z}^m,\beta}$. The claim of this theorem follows from the fact that the statistical distance between $D_{\mathbb{Z}^m,\beta}$ and $D_{m,\beta}$ is negligible, and that the distribution $\{\mathbf{u} + h\mathbf{A}_2\mathbf{x}_2 : \mathbf{u} \xleftarrow{\$} \mathbb{Z}_q^n\}$ is exactly the uniform distribution over $\mathbb{Z}_q^n$ for arbitrary $\mathbf{x}_2 \in D_{m,\beta}$, $h \in [N]$. □

7.4.3.2 Zero-Knowledge Proof of Knowledge for the Hash Functions

In this subsection, we present a proof of knowledge protocol for the family of hash functions $\mathcal{H}_{n,m,q,\beta,N}$. Concretely, given a matrix $\mathbf{A} = (\mathbf{A}_1 \| \mathbf{A}_2)$, a vector $\mathbf{y} = (\mathbf{y}_1, \mathbf{y}_2) \in \mathbb{Z}_q^n \times \mathbb{Z}^m$ with $0 < \|\mathbf{y}_2\| \leq \beta\sqrt{m}$, the prover can generate a proof of knowledge of $\mathbf{x} = (\mathbf{x}_1, \mathbf{x}_2, h) \in \mathbb{Z}^{2m+1}$ satisfying $\|\mathbf{x}_1\| \leq \beta\sqrt{m}$, $h \in [N]$ and $f_{\mathbf{A}}(\mathbf{x}_1, \mathbf{x}_2, h) = (\mathbf{A}\mathbf{x}_1 + h\mathbf{A}_2\mathbf{x}_2, \mathbf{x}_2) = \mathbf{y}$. Since $\mathbf{x}_2$ must be equal to $\mathbf{y}_2$, the protocol is actually a proof of knowledge for the relation

$$R_{\text{split-SIS}} = \{(\mathbf{A}, \mathbf{y}, \beta, N; \mathbf{x}_1, h) \in \mathbb{Z}_q^{n\times 2m} \times (\mathbb{Z}_q^n \times \mathbb{Z}^m) \times \mathbb{R} \times \mathbb{Z} \times \mathbb{Z}^m \times \mathbb{Z} : \\ \mathbf{A}_1\mathbf{x}_1 + h\mathbf{A}_2\mathbf{y}_2 = \mathbf{y}_1, \|\mathbf{x}_1\| \leq \beta\sqrt{m} \text{ and } h \in [N]\}.$$

Intuitively, we can adapt a variant of the protocols for ISIS relations in [46, 51, 52], since one can rewrite $\mathbf{y}_1 = \mathbf{A}_1\mathbf{x}_1 + h\mathbf{A}_2\mathbf{y}_2 = (\mathbf{A}_1\|\mathbf{A}_2\mathbf{y}_2)(\mathbf{x}_1; h)$. However, this may not work when $N \gg \beta$. Since the basic idea of [46, 51, 52] is to use randomness from a "large width" distribution (compared to the distribution of the witness) to hide the distribution of the witness, the width of the randomness distribution should be sufficiently larger than N, which might lead to a proof without soundness guarantee.

Fortunately, we can borrow the "bit-decomposition" technique from [19–21] to deal with large N. The idea is to decompose $h \in [N]$ into a vector of small elements, and then prove the existence of such a vector for h. Formally, for any $h \in [N]$, we compute the representation of h in base $\bar{\beta} = \lfloor \beta \rfloor$, namely, a ℓ-dimension vector $\mathbf{v}_h = (v_0, \ldots, v_{\ell-1}) \in \mathbb{Z}^\ell$ such that $0 \leq v_i \leq \beta - 1$ and $h = \sum_{i=0}^{\ell-1} v_i \bar{\beta}^i$, where $\ell = \lceil \log_{\bar{\beta}} N \rceil$. Denote $\mathbf{b} = \mathbf{A}_2\mathbf{y}_2$, compute $\mathbf{D} = (\mathbf{b}, \bar{\beta}\mathbf{b}, \ldots, \bar{\beta}^{\ell-1}\mathbf{b}) \in \mathbb{Z}_q^{n\times\ell}$. It is easy to check that for any vector $\mathbf{e} \in \mathbb{Z}^\ell$, there exists a $h' \in \mathbb{Z}_q$ such that $\mathbf{D}\mathbf{e} = h'\mathbf{b} \bmod q$. ($h' \in \mathbb{Z}_q$ is unique if $\mathbf{b} \neq \mathbf{0}$.) In particular, we have that $\mathbf{y}_1 = \hat{\mathbf{A}}\hat{\mathbf{x}}$, where $\hat{\mathbf{A}} = (\mathbf{A}_1\|\mathbf{D}) \in \mathbb{Z}_q^{n\times(m+\ell)}$, $\hat{\mathbf{x}} = (\mathbf{x}_1; \mathbf{v}_h) \in \mathbb{Z}^{m+\ell}$ and $\|\hat{\mathbf{x}}\| \leq \beta\sqrt{m+\ell}$. Since $\bar{\beta} > 2$ and N is a polynomial in n, we have $\ell \ll m$ and $\|\hat{\mathbf{x}}\| < \eta = \beta\sqrt{2m}$.

We first present a Σ-protocol for the function family $\mathcal{H}_{n,m,q,\beta,N}$, which repeats a basic protocol with single-bit challenge $t = \omega(\log n)$ times in parallel. As in [46, 52], the basic protocol makes use of the rejection sampling technique to achieve zero-knowledge. Formally, let $\gamma = \eta \cdot m^{1.5}$, denote $\zeta(\mathbf{z}, \mathbf{y}) = 1 - \min(\frac{D_{\mathbb{Z}^{m+\ell},\gamma}(\mathbf{z})}{M_l \cdot D_{\mathbb{Z}^{m+\ell},\mathbf{y},\gamma}(\mathbf{z})}, 1)$, where $\mathbf{y}, \mathbf{z} \in \mathbb{Z}^{m+\ell}$, and the constant $M_l \leq 1 + O(\frac{1}{m})$ is set according to Lemma 4.5 in [52], the protocol is depicted in Fig 7.2.

Prover		Verifier
	CRS: $\hat{\mathbf{A}} \in \mathbb{Z}_q^{n\times(m+\ell)}$, $\mathbf{y}_1 \in \mathbb{Z}_q^n$	
Private input: $\hat{\mathbf{x}} \in \mathbb{Z}^{m+\ell}$		
For $i \in \{0,\ldots,t-1\}$ $\mathbf{e}_i \xleftarrow{\$} D_{\mathbb{Z}^{m+\ell},\gamma}$ $\mathbf{u}_i = \hat{\mathbf{A}}\mathbf{e}_i$		
	$\mathbf{U} = (\mathbf{u}_0,\ldots,\mathbf{u}_{t-1})$ $\longrightarrow$	$\mathbf{c} \xleftarrow{\$} \{0,1\}^t$
$\mathbf{z}_i = \mathbf{e}_i + c_i\hat{\mathbf{x}}$ Set $\mathbf{z}_i = \perp$ with probability $\zeta(\mathbf{z}_i, c_i\hat{\mathbf{x}})$	$\longleftarrow$ $\mathbf{c} = (c_0,\ldots,c_{t-1})$	
	$\mathbf{Z} = (\mathbf{z}_0,\ldots,\mathbf{z}_{t-1})$ $\longrightarrow$	Set $d_i = 1$ if $\|\mathbf{z}_i\| \le 2\gamma\sqrt{m+\ell}$ and $\hat{\mathbf{A}}\mathbf{z}_i = \mathbf{u}_i + c_i\mathbf{y}_1$ Accept iff $\sum_i d_i \ge 0.65t$

Fig. 7.2 Σ-protocol for $R_{\text{split-SIS}}$

By [52, Th. 4.6] we have $\Pr[\mathbf{z}_i \neq \perp] \approx \frac{1}{M_l} = 1 - O(\frac{1}{m})$ for each $i \in \{0,\ldots,t-1\}$. In addition, $\Pr[\|\mathbf{z}_i\| \le 2\gamma\sqrt{m+\ell} \mid \mathbf{z}_i \neq \perp] = 1 - negl(m)$ by [52, Lem. 4.4]. A simple calculation shows that the completeness error of the protocol is at most $2^{-\Omega(t)}$ (when m is sufficiently large, e.g., $m > 100$). Besides, the protocol has the property of special honest-verifier zero-knowledge (HVZK). Namely, given a challenge c_i, there exists a simulator $\mathcal{S}$ that outputs a distribution $(\mathbf{u}_i, c_i, \mathbf{z}_i)$ statistically close to the real transcript distribution. Concretely, $\mathcal{S}$ first chooses $\mathbf{z}_i \xleftarrow{\$} D_{\mathbb{Z}^{m+\ell},\gamma}$ and computes $\mathbf{u}_i = \hat{\mathbf{A}}\mathbf{z}_i - c_i\mathbf{y}_1 \bmod q$. Then, it sets $\mathbf{z}_i = \perp$ with probability $1 - \frac{1}{M_l}$, and outputs $(\mathbf{u}_i, c_i, \mathbf{z}_i)$. By Theorem 7.7, the term $\hat{\mathbf{A}}\mathbf{z}_i (\bmod q)$ is statistically close to uniform over $\mathbb{Z}_q^n$, thus the distribution of $\mathbf{u}_i$ is statistically close to that in the real proof. Moreover, by [52, Th. 4.6], the distribution of $\mathbf{z}_i$ is also statistically close to that in the real transcripts.

Finally, since the binary challenges (i.e., $\mathbf{c}$) are used, the above protocol has the property of special soundness. Actually, given two transcripts $(\mathbf{U}, \mathbf{c}, \mathbf{Z})$ and $(\mathbf{U}, \mathbf{c}', \mathbf{Z}')$ with distinct challenges $\mathbf{c} \neq \mathbf{c}'$, one can extract a "weak" witness $\mathbf{x}' = \mathbf{z}_i - \mathbf{z}_i'$ for some i satisfying $\hat{\mathbf{A}}\mathbf{x}' = \mathbf{y}_1$ and $\|\mathbf{x}'\| \le 4\gamma\sqrt{2m}$.

Applying the "Fiat-Shamir Heuristic" [32] in a standard way, one can obtain an NIZKPoK by computing $\mathbf{c} = H(\rho, \mathbf{U})$, where $H : \{0,1\}^* \to \{0,1\}^t$ is modeled as a random oracle, and ρ represents all the other auxiliary inputs, e.g., a specified message M to be signed. Finally, due to the nice property of Σ-protocol, one can easily combine the protocol to prove EQ-relation, OR-relation and AND-relation.

7.4.4 A Group Signature from Split-SIS Problems

Let $\Pi_{\text{sig}} = $ (KeyGen, Sign, Verify) be a one-time strongly existentially unforgeable (ot-sEUF) signature scheme, and FRD : $\{0,1\}^* \to \mathbb{Z}_q^{n\times n}$ be an encoding with full-rank differences. Assume that the security parameter is n, and δ is a real such that $n^{1+\delta} > \lceil (n+1)\log q + n \rceil$, all other parameters $m, s, \alpha, \beta, \eta, p, q$ are determined as follows:

$$\begin{aligned}
&m = 6n^{1+\delta}\\
&s = m \cdot \omega(\log m)\\
&\beta = s\sqrt{3m} \cdot \omega(\sqrt{\log 3m}) = m^{1.5} \cdot \omega(\log^{1.5} m)\\
&p = m^{2.5}\beta = m^4 \cdot \omega(\log^{1.5} m)\\
&q = m^{2.5} \cdot \max(pm^{2.5} \cdot \omega(\log m), 4N) = m^{2.5}\max(m^{6.5} \cdot \omega(\log^{2.5} m), 4N)\\
&\alpha = 2\sqrt{m}/q\\
&\eta = \max(\beta, \alpha q\sqrt{m})\sqrt{2m} = m^2 \cdot \omega(\log^{1.5} m)
\end{aligned} \tag{7.1}$$

Now, we define group signature $\Pi_{\text{gs}} = (\mathsf{KeyGen}, \mathsf{Sign}, \mathsf{Verify}, \mathsf{Open})$ as follows:

- $\mathsf{KeyGen}(1^n, 1^N)$: Take the security parameter n and the maximum number of group members N as inputs, set an integer $m \in \mathbb{Z}$, primes $p, q \in \mathbb{Z}$, and $s, \alpha, \beta, \eta \in \mathbb{R}$ as in (7.1), and choose a hash function $H : \{0,1\}^* \to \{0,1\}^t$ (modeled as random oracle) for the NIZKPoK proof, where $t = \omega(\log n)$. Then, the algorithm proceeds as follows:
  1. Compute $(\mathbf{A}_{1,1}, \mathbf{T}_{\mathbf{A}_{1,1}}) \leftarrow \mathsf{BasisGen}(n, m, q)$, and randomly choose $\mathbf{A}_{1,2}$, $\mathbf{A}_{2,1}, \mathbf{A}_{2,2} \xleftarrow{\$} \mathbb{Z}_q^{n\times m}$. Denote $\mathbf{A}_1 = (\mathbf{A}_{1,1}\|\mathbf{A}_{1,2}) \in \mathbb{Z}_q^{n\times 2m}$.
  2. Randomly choose $\mathbf{B}_{2,1} \xleftarrow{\$} \mathbb{Z}_q^{n\times m}$, compute $(\mathbf{B}_1, \mathbf{T}_{\mathbf{B}_1}) \leftarrow \mathsf{SuperSamp}(\mathbf{A}_{1,1}, -\mathbf{A}_{1,2}\mathbf{B}_{2,1}^T, q)$ and $(\mathbf{B}_{2,2}, \mathbf{T}_{\mathbf{B}_{2,2}}) \leftarrow \mathsf{SuperSamp}(\mathbf{A}_{1,2}, \mathbf{0}, q)$. It is easy to check that for any matrix $\mathbf{C}$, we have $\mathbf{A}_1(\mathbf{B}_1\|\mathbf{B}_{2,1} + \mathbf{C}\mathbf{B}_{2,2})^T = \mathbf{0}$.
  3. For $j = 1, \ldots, N$, define $\bar{\mathbf{A}}_j = (\mathbf{A}_1\|\mathbf{A}_{2,1} + j\mathbf{A}_{2,2}) \in \mathbb{Z}_q^{n\times 3m}$, extract a random basis $\mathbf{T}_{\bar{\mathbf{A}}_j} \leftarrow \mathsf{ExtRndBasis}(\bar{\mathbf{A}}_j, \mathbf{T}_{\mathbf{A}_{1,1}}, s)$.
  4. Define the group public key $\mathsf{gpk} = \{\mathbf{A}_{1,1}, \mathbf{A}_{1,2}, \mathbf{A}_{2,1}, \mathbf{A}_{2,2}, \mathbf{B}_1, \mathbf{B}_{2,1}, \mathbf{B}_{2,2}\}$, the group manager secret key $\mathsf{gmsk} = \mathbf{T}_{\mathbf{B}_1}$, and the group member's secret key $\mathbf{gsk} = \{\mathsf{gsk}_j = \mathbf{T}_{\bar{\mathbf{A}}_j}\}_{j\in\{1,\ldots,N\}}$.
- $\mathsf{Sign}(\mathsf{gpk}, \mathsf{gsk}_j, M)$: Take the group public key $\mathsf{gpk} = \{\mathbf{A}_{1,1}, \mathbf{A}_{1,2}, \mathbf{A}_{2,1}, \mathbf{A}_{2,2}, \mathbf{B}_1, \mathbf{B}_{2,1}, \mathbf{B}_{2,2}\}$, the j-th user's secret key $\mathsf{gsk}_j = \mathbf{T}_{\bar{\mathbf{A}}_j}$, and a message $M \in \{0,1\}^*$, proceed as follows:
  1. Compute $(\mathbf{x}_1, \mathbf{x}_2) \leftarrow \mathsf{SamplePre}(\bar{\mathbf{A}}_j, \mathbf{T}_{\bar{\mathbf{A}}_j}, \mathbf{0}, \beta)$, where $\mathbf{x}_1 \in D_{\mathbb{Z}^{2m},\beta}, \mathbf{x}_2 \in D_{\mathbb{Z}^m,\beta}$.
  2. Generate a one-time signature verification and signing keys $(\mathsf{svk}, \mathsf{ssk}) \leftarrow \Pi_{\text{sig}}.\mathsf{KeyGen}(1^\kappa)$, denote $\hat{\mathbf{B}}_{\mathsf{svk}} = (\mathbf{B}_1\|\mathbf{B}_{2,1} + \mathsf{FRD}(\mathsf{svk})\mathbf{B}_{2,2}) \in \mathbb{Z}_q^{n\times 2m}$. Randomly choose $\mathbf{s} \xleftarrow{\$} \mathbb{Z}_q^n$, $\mathbf{e} \xleftarrow{\$} \chi_\alpha^m$, and $\mathbf{R} \xleftarrow{\$} \{0,1\}^{m\times m}$, compute

$$\mathbf{c} = \hat{\mathbf{B}}_{\mathsf{svk}}^T\mathbf{s} + p\hat{\mathbf{e}} + \mathbf{x}_1, \text{ where } \hat{\mathbf{e}} = \begin{pmatrix}\mathbf{e}\\ \mathbf{R}^T\mathbf{e}\end{pmatrix}$$

  3. Generate a NIZKPoK proof π_1 of $(\hat{\mathbf{e}}, \mathbf{x}_1)$ so that $(\hat{\mathbf{B}}_{\mathsf{svk}}, \mathbf{c}, \eta; \mathbf{s}, \hat{\mathbf{e}}, \mathbf{x}_1) \in R_{\text{eLWE}}$.

4. Let $\bar{\beta} := \lfloor \beta \rfloor$ and $\ell = \lceil \log_{\bar{\beta}} N \rceil$, define $\mathbf{b} = \mathbf{A}_{2,2}\mathbf{x}_2$, and $\mathbf{D} = (\mathbf{b}, \bar{\beta}\mathbf{b}, \ldots, \bar{\beta}^{\ell-1}\mathbf{b}) \in \mathbb{Z}_q^{n\times\ell}$. Generate a NIZKPoK π_2 of $\mathbf{x}_1$, $\hat{\mathbf{e}} = (\mathbf{e}; \mathbf{R}^T\mathbf{e})$, and $\mathbf{v}_j = (v_0, \ldots, v_{\ell-1}) \in \mathbb{Z}_{\bar{\beta}}^{\ell}$ of $j \in [N]$ such that,

$$\begin{aligned} \mathbf{A}_1\mathbf{c} + \mathbf{A}_{2,1}\mathbf{x}_2 &= (p\mathbf{A}_1)\hat{\mathbf{e}} - \mathbf{D}\mathbf{v}_j, \text{ and} \\ \mathbf{A}_1\mathbf{c} &= (p\mathbf{A}_1)\hat{\mathbf{e}} + \mathbf{A}_1\mathbf{x}_1, \end{aligned} \tag{7.2}$$

where the challenge is computed by $H(\mathsf{svk}, \mathbf{c}, \mathbf{x}_2, \pi_1, M, \mathbf{Com})$, and $\mathbf{Com}$ is the commitment message for the NIZKPoK proof of π_2.
5. Denote $\sigma_1 = (\mathbf{c}, \mathbf{x}_2, \pi_1, \pi_2)$, compute $\sigma_2 = \Pi_{\text{sig}}.\mathsf{Sign}(\mathsf{ssk}, \sigma_1)$, return the group signature $\sigma = (\mathsf{svk}, \sigma_1, \sigma_2)$.

- $\mathsf{Verify}(gpk, M, \sigma)$: Parse $\sigma = (\mathsf{svk}, \sigma_1 = (\mathbf{c}, \mathbf{x}_2, \pi_1, \pi_2), \sigma_2)$, return 1 if $\Pi_{\text{sig}}.\mathsf{Verify}(\mathsf{svk}, \sigma_1, \sigma_2) = 1$, $\|\mathbf{x}_2\| \le \beta\sqrt{m}$, $\mathbf{A}_{2,2}\mathbf{x}_2 \ne \mathbf{0}$, and the proofs π_1, π_2 are valid, else return 0.
- $\mathsf{Open}(gpk, gmsk, M, \sigma)$: Parse $\mathsf{gpk} = \{\mathbf{A}_{1,1}, \mathbf{A}_{1,2}, \mathbf{A}_{2,1}, \mathbf{A}_{2,2}, \mathbf{B}_1, \mathbf{B}_{2,1}, \mathbf{B}_{2,2}\}$ and $\mathsf{gmsk} = \mathbf{T}_{\mathbf{B}_1}$. Denote $\hat{\mathbf{B}}_{\mathsf{svk}} = (\mathbf{B}_1 \| \mathbf{B}_{2,1} + \mathsf{FRD}(\mathsf{svk})\mathbf{B}_{2,2}) \in \mathbb{Z}_q^{n\times 2m}$, extract a basis $\mathbf{T}_{\hat{\mathbf{B}}_{\mathsf{svk}}} \leftarrow \mathsf{ExtRndBasis}(\hat{\mathbf{B}}_{\mathsf{svk}}, \mathbf{T}_{\mathbf{B}_1}, s\sqrt{m}\cdot\omega(\sqrt{\log m}))$. Compute $\mathbf{x}_1$ by decrypting $\mathbf{c}$ using $\mathbf{T}_{\hat{\mathbf{B}}_{\mathsf{svk}}}$, and let $\mathbf{y}_1 = \mathbf{A}_{2,2}\mathbf{x}_2$ and $\mathbf{y}_2 = -\mathbf{A}_1\mathbf{x}_1 - \mathbf{A}_{2,1}\mathbf{x}_2$. If $\mathbf{y}_1 \ne \mathbf{0}$ and there is a $j \in \mathbb{Z}_q^*$ such that $\mathbf{y}_2 = j \cdot \mathbf{y}_1$, return j, else $\perp$.

We summarize the correctness of $\mathcal{GS}$ in the following theorem:

Theorem 7.8 *Assume that n is the security parameter, and all other parameters $m, s, \alpha, \beta, \eta, p, q$ are functions of n defined as in Eq. 7.1, where p, q are primes. Then, the group signature Π_{gs} is correct, and the group public key and the signature have bit-length $O(nm\log q)$ and $O(tm\log q) + \ell_{\mathcal{SIG}}$, respectively, where $\ell_{\mathcal{SIG}}$ is the total bit-length of the verification key and signature (i.e., svk and σ_2) of the underlying one-time signature.*

Proof Since we set $m = 6n^{1+\delta} > \lceil 6n\log q + n\rceil$, the two algorithms $\mathsf{BasisGen}$ and $\mathsf{SuperSamp}$ can work correctly with overwhelming probability. In particular, we have $\|\tilde{\mathbf{T}}_{\mathbf{A}_1}\| \le O(\sqrt{m})$, and $\|\mathbf{T}_{\mathbf{B}}\| \le m^{1.5}\cdot O(\sqrt{\log m})$ by Lemma 2.19 and Lemma 2.22. By Lemma 2.20, we have $\|\widetilde{\mathbf{T}}_{\bar{\mathbf{A}}_i}\| \le s\sqrt{2m}$ for all $i \in \{1, \ldots, N\}$ with overwhelming probability. Since the group public key only contains seven matrices over $\mathbb{Z}_q^{n\times m}$, it has bit-size $7nm\log q = O(nm\log q)$.

For the Sign algorithm, since $\beta = s\sqrt{2m}\cdot\omega(\sqrt{\log 2m}) \ge \|\widetilde{\mathbf{T}}_{\bar{\mathbf{A}}_j}\|\cdot\omega(\sqrt{\log 2m})$, we have $\mathbf{x}_1, \mathbf{x}_2 \backsim D_{\mathbb{Z}^m,\beta}$ by the correctness of the $\mathsf{SamplePre}$ algorithm in [34], and $\|\mathbf{x}_i\| \le \beta\sqrt{m}$ with overwhelming probability. In addition, since $\mathbf{e}$ is chosen from χ_α, we have $\|\mathbf{e}\| \le \alpha q\sqrt{m}$ with overwhelming probability. By the choices of $\eta = \max(\beta, \alpha q)\sqrt{m}$, the algorithm can successfully generate the proofs π_1 and π_2. For the bit-length of the signature $\sigma = (\mathsf{svk}, \sigma_1, \sigma_2)$ where $\sigma_1 = (\mathbf{c}, \mathbf{x}_2, \pi_1, \pi_2)$, we know that both the bit-length of $\mathbf{c}$ and $\mathbf{x}_2$ are at most $2m\log q$ and $m\log q$, respectively. In addition, if we set the repetition parameter $t = \omega(\log n)$ for the proof π_1 and π_2, the bit-length of π_1 and π_2 are at most $(6m - n)t\log q$ and

$(4m + 2n + \ell)t \log q$, respectively. Thus, the total bit-length of the signature σ_1 is less than $3m \log q + (8m + n + \ell)t \log q = O(t(m + \log N) \log q) = O(tm \log q)$ since $\ell = \lceil \log_{\bar{\beta}} N \rceil \ll n$ and $m = O(n \log q)$.

Note that $\mathbf{x}_2 \hookleftarrow D_{\mathbb{Z}^m,\beta}$,[2] therefore $\Pr[\mathbf{A}_{2,2}\mathbf{x}_2 = \mathbf{0}] \leq O(q^{-n})$ by Lemma 2.11. Moreover, by the completeness of π_1 and π_2, the algorithm Verify will work correctly with overwhelming probability. As for the Open algorithm, we only have to show that we can correctly decrypt $\mathbf{x}_1$ from $\mathbf{c}$ by using $\mathbf{T}_{\mathbf{B}_{\mathsf{svk}}}$. Since $\mathbf{T}_{\mathbf{B}_{\mathsf{svk}}}^T \cdot \mathbf{c} = \mathbf{T}_{\mathbf{B}_{\mathsf{svk}}}^T(p\mathbf{e} + \mathbf{x}_1) \bmod q$ holds, one can expect that $\mathbf{T}_{\mathbf{B}_{\mathsf{svk}}}^T(p\mathbf{e} + \mathbf{x}_1) = (\mathbf{T}_{\mathbf{B}_{\mathsf{svk}}}^T \mathbf{c} \bmod q)$ holds over $\mathbb{Z}$ if $\|\mathbf{T}_{\mathbf{B}_{\mathsf{svk}}}^T(p\mathbf{e} + \mathbf{x}_1)\|_\infty < q/2$. Thus, one can solve $\hat{\mathbf{x}} = (p\mathbf{e} + \mathbf{x}_1)$ by Gaussian elimination over $\mathbb{Z}$ since $\mathbf{T}_{\mathbf{B}_{\mathsf{svk}}} \in \mathbb{Z}^{m \times m}$ is full-rank. Moreover, by the choices of p and β, we have $\|\mathbf{x}_1\|_\infty \leq \|\mathbf{x}_1\| < p$, therefore $\mathbf{x}_1$ can be recovered by computing $\hat{\mathbf{x}} \bmod p$ if we set the parameter such that $\|\mathbf{T}_{\mathbf{B}_{\mathsf{svk}}}^T(p\mathbf{e} + \mathbf{x}_1)\|_\infty < q/2$. □

7.4.5 Full Anonymity

Theorem 7.9 (Full Anonymity) *Under the LWE assumption, the group signature Π_{gs} is fully anonymous in the random oracle model.*

Proof We prove the theorem via a sequence of games. In game G_0, the challenger honestly creates the group public key $gpk = \{\mathbf{A}_{1,1}, \mathbf{A}_{1,2}, \mathbf{A}_{2,1}, \mathbf{A}_{2,2}, \mathbf{B}_1, \mathbf{B}_{2,1}, \mathbf{B}_{2,2}\}$, the group manager's secret key $gmsk = \mathbf{T}_{\mathbf{B}_1}$, and the group member's secret key $\mathbf{gsk} = \{\mathsf{gsk}_j = \mathbf{T}_{\bar{\mathbf{A}}_j}\}_{j \in \{1,\ldots,N\}}$ by running the KeyGen algorithm. Then, it gives $(\mathsf{gpk}, \mathbf{gsk})$ to the adversary, and obtains a message M, and two user indexes i_0, i_1. Finally, the challenger chooses a bit $b \xleftarrow{\$} \{0, 1\}$, computes $\sigma^* = (\mathsf{svk}^*, \sigma_1^* = (\mathbf{c}^*, \mathbf{x}_2^*, \pi_1^*, \pi_2^*), \sigma_2^*) \leftarrow \mathsf{Sign}(\mathsf{gpk}, \mathsf{gsk}_{i_b}, M)$ and returns σ^* to the adversary $\mathcal{A}$.

In Game G_1, the challenger behaves almost the same as in G_0, except that it uses the NIZKPoK simulators (by appropriately programming the random oracle) to generate π_1^*, π_2^*. By the property of the NIZKPoKs, G_1 is indistinguishable from G_0.

In Game G_2, the challenger behaves almost the same as in G_1, except that it generates a one-time signature key pair $(\mathsf{svk}^*, \mathsf{ssk}^*) \leftarrow \Pi_{\mathrm{sig}}.\mathsf{KeyGen}(1^\kappa)$ at the beginning of the experiment and aborts when $\mathcal{A}$ outputs a valid signature $\sigma = (\mathsf{svk}^*, \sigma_1, \sigma_2) \neq \sigma^*$, where σ^* is the challenge signature.

Lemma 7.1 *If the signature Π_{sig} is one-time strongly existentially unforgeable (ot-sEUF) for any PPT forger, then Game G_2 is computationally indistinguishable from Game G_1.*

---

[2]As noted by Cash et al. [26], the output distribution of the SamplePre algorithm in [34] is statistically close to the distribution $(\mathbf{x}_1, \mathbf{x}_2)$ that samples as follows: randomly choose $\mathbf{x}_2 \xleftarrow{\$} D_{\mathbb{Z}^m,s}$, and then compute $\mathbf{x}_1$ using $\mathbf{T}_{\mathbf{A}_1}$ to satisfy the condition $\mathbf{A}_1\mathbf{x}_1 + (\mathbf{A}_{2,1} + j\mathbf{A}_{2,2})\mathbf{x}_2 = \mathbf{0}$. We note that this is also the reason why directly publish $\mathbf{x}_2$ in the signature as in [46], since it leaks little information about j without $\mathbf{x}_1$.

Proof If there is an adversary $\mathcal{A}$ that outputs a valid signature $\sigma = (\mathsf{svk}^*, \sigma_1, \sigma_2) \neq \sigma^*$, we can construct a forger $\mathcal{F}$ that breaks the strong unforgeability of Π_{sig}. Actually, given a challenge verification key svk^* from the challenger of the ot-sEUF security game, $\mathcal{F}$ behaves almost the same as the challenger in G_2 to simulate the attack environment for $\mathcal{A}$, except that it uses svk^* in generating the challenge signature σ^* by making a signing query to its own signing oracle. Whenever $\mathcal{A}$ outputs a valid signature $\sigma = (\mathsf{svk}^*, \sigma_1, \sigma_2) \neq \sigma^*$, $\mathcal{F}$ returns (σ_1, σ_2) as its own forgery. Obviously, if $\mathcal{A}$ can output a valid signature with probability ϵ, $\mathcal{F}$ can successfully forge a signature for the underlying one-time signature almost with the same probability ϵ. □

In Game G_3, the challenger behaves almost the same as in G_2, except that it first chooses $\mathbf{x}_2$ from $D_{\mathbb{Z}^m,\beta}$, and then uses the trapdoor $\mathbf{T}_{\mathbf{A}_{1,1}}$ to extract $\mathbf{x}_1$ such that $\bar{\mathbf{A}}_{i_b}(\mathbf{x}_1; \mathbf{x}_2) = \mathbf{0}$. By the property of the SamplePre algorithm in [26, 34], we have that Game G_3 is statistically close to Game G_2.

In Game G_4, the challenger behaves almost the same as in G_3, except that it computes $\mathbf{c}^* = \mathbf{u} + \mathbf{x}_1$ for a randomly chosen $\mathbf{u} \xleftarrow{\$} \mathbb{Z}_q^{2m}$.

Lemma 7.2 *Under the LWE assumption, Game G_4 is computationally indistinguishable from Game G_3.*

Proof Assume that there is an algorithm $\mathcal{A}$ which distinguishes G_3 from G_4 with non-negligible probability. We will show that there is an algorithm $\mathcal{B}$ breaking the LWE assumption. Formally, given a LWE tuple $(\hat{\mathbf{B}}, \mathbf{u}) \in \mathbb{Z}_q^{n\times m} \times \mathbb{Z}_q^m$, $\mathcal{B}$ sets $\mathbf{B}_1 = p\hat{\mathbf{B}}$. Randomly choose $\mathbf{A}_{1,2}, \mathbf{A}_{2,1}, \mathbf{A}_{2,2} \xleftarrow{\$} \mathbb{Z}_q^{n\times m}$ and $\mathbf{R} \in \{0,1\}^{m\times m}$, generate a one-time signature pair $(\mathsf{svk}^*, \mathsf{ssk}^*) \leftarrow \Pi_{\text{sig}}.\mathsf{KeyGen}(1^\kappa)$. Then, it computes the public parameters $(\mathbf{B}_{2,2}, \mathbf{T}_{\mathbf{B}_{2,2}}) \leftarrow \mathsf{SuperSamp}(\mathbf{A}_{1,2}, \mathbf{0}, q)$, $\mathbf{B}_{2,1} = \mathbf{B}_1\mathbf{R} - \mathsf{FRD}(\mathsf{svk}^*)\mathbf{B}_{2,2}$, and $(\mathbf{A}_{1,1}, \mathbf{T}_{\mathbf{A}_{1,1}}) \leftarrow \mathsf{SuperSamp}(\mathbf{B}_1, -\mathbf{B}_{2,1}\mathbf{A}_{1,2}^T, q)$. Denote $\mathbf{A}_1 = (\mathbf{A}_{1,1}\|\mathbf{A}_{1,2})$.

For $j = 1, \ldots, N$, define $\bar{\mathbf{A}}_j = (\mathbf{A}_1\|\mathbf{A}_{2,1} + j\mathbf{A}_{2,2})$, extract a random basis $\mathbf{T}_{\bar{\mathbf{A}}_j} \leftarrow \mathsf{ExtRndBasis}(\bar{\mathbf{A}}_j, \mathbf{T}_{\mathbf{A}_{1,1}}, s)$. Finally, $\mathcal{B}$ gives the group public key $\mathsf{gpk} = \{\mathbf{A}_{1,1}, \mathbf{A}_{1,2}, \mathbf{A}_{2,1}, \mathbf{A}_{2,2}, \mathbf{B}_1, \mathbf{B}_{2,1}, \mathbf{B}_{2,2}\}$, and the group member's secret key $\mathbf{gsk} = \{\mathsf{gsk}_j = \mathbf{T}_{\bar{\mathbf{A}}_j}\}_{j\in\{1,\ldots,N\}}$ to $\mathcal{A}$, and keeps $\mathbf{T}_{\mathbf{B}_{2,2}}$ secret. Note that the distribution of gpk, $\mathbf{gsk}$ is statistically close to that in Game G_3 and G_4.

If $\mathcal{A}$ submits a valid signature $\sigma = (\mathsf{svk}^*, \sigma_1, \sigma_2) \neq \sigma^*$ to the open oracle, $\mathcal{B}$ aborts as in Game G_3. Otherwise, $\mathcal{B}$ defines $\hat{\mathbf{B}}_{\mathsf{svk}} = (\mathbf{B}_1\|\mathbf{B}_{2,1} + \mathsf{FRD}(\mathsf{svk})\mathbf{B}_{2,2}) = (\mathbf{B}_1\|\mathbf{B}_1\mathbf{R} + (H(\mathsf{svk}) - \mathsf{FRD}(\mathsf{svk}^*))\mathbf{B}_{2,2})$ and extracts $\mathbf{T}_{\hat{\mathbf{B}}_{\mathsf{svk}}} = \mathsf{ExtBasisRight}(\hat{\mathbf{B}}_{\mathsf{svk}}, \mathbf{R}, \mathbf{T}_{\mathbf{B}_{2,2}}, s\sqrt{m}\cdot\omega(\sqrt{\log m}))$ (we note that $\mathbf{T}_{\mathbf{B}_{2,2}}$ is also a basis for $\Lambda_q^\perp(\mathbf{C}\mathbf{B}_{2,2})$ for any full-rank $\mathbf{C} \in \mathbb{Z}_q^{n\times n}$, since $\Lambda_q^\perp(\mathbf{C}\mathbf{B}_{2,2}) = \Lambda_q^\perp(\mathbf{B}_{2,2})$). Finally, it uses $\mathbf{T}_{\hat{\mathbf{B}}_{\mathsf{svk}}}$ as described in the Open algorithm to answer this query.

When generating the challenge signature, $\mathcal{B}$ works the same as in G_2, except that it computes $\mathbf{c} = p\begin{pmatrix}\mathbf{u}\\ \mathbf{R}^T\mathbf{u}\end{pmatrix} + \mathbf{x}_1$. We note that if $(\hat{\mathbf{B}}, \mathbf{u})$ is a LWE tuple for distribution χ_α, $\mathbf{c}$ is the same as in G_3. Otherwise, we show that $\begin{pmatrix}\mathbf{u}\\ \mathbf{R}^T\mathbf{u}\end{pmatrix}$ is statistically close to uniform over $\mathbb{Z}_q^{2m}$ in Lemma 7.3, which shows that the distribution of $\mathbf{c}$ is statistically close to that in G_4. Thus, if $\mathcal{A}$ can distinguish G_3 from G_4 with advantage ϵ, then $\mathcal{B}$ can break the LWE assumption with advantage $\epsilon - \mathsf{negl}(\kappa)$. This completes the proof. □

Lemma 7.3 *Given randomly chosen* $\mathbf{B}_1 \xleftarrow{\$} \mathbb{Z}_q^{n\times m}$ *and* $\mathbf{u} \xleftarrow{\$} \mathbb{Z}_q^m$, *the distribution of* $(\mathbf{B}_1\mathbf{R}, \mathbf{u}, \mathbf{R}^T\mathbf{u})$ *is statistically close to uniform over* $\mathbb{Z}_q^{n\times m} \times \mathbb{Z}_q^m \times \mathbb{Z}_q^m$, *where* $\mathbf{R}$ *is randomly chosen from* $\{-1, 1\}^{m\times m}$.

Proof By Lemma 2.2, the distribution $(\mathbf{B}_1\mathbf{R}, \mathbf{u}, \mathbf{R}^T\mathbf{u})$ is statistically close to $(\mathbf{C}, \mathbf{u}, \mathbf{R}^T\mathbf{u})$, where $\mathbf{C}$ is randomly chosen from $\mathbb{Z}_q^{n\times m}$. Since the hash function defined by inner-product is a family of universal hash functions we have that the statistical distance between $(\mathbf{u}, \mathbf{R}^T\mathbf{u})$ and $(\mathbf{u}, \mathbf{v})$ for uniformly chosen $\mathbf{v} \in \mathbb{Z}_q^m$ is at most $\frac{1}{2}\sqrt{2^{-m^2+m\log q}}$ by Lemma 2.4, which is negligible in n if $m \geq 2n + \log q$ (thus, the choice of $m = \Omega(n \log q)$ suffices). This shows that $(\mathbf{B}_1\mathbf{R}, \mathbf{u}, \mathbf{R}^T\mathbf{u})$ is statistically close to $(\mathbf{C}, \mathbf{u}, \mathbf{v})$. □

In Game G_5, the challenger behaves almost the same as in G_4, except that it randomly chooses $\mathbf{c}^* \xleftarrow{\$} \mathbb{Z}_q^{2m}$. We have that Games G_5 and G_4 are identical, since this change from game G_4 to G_5 is just conceptual.

Lemma 7.4 *In Game* G_5, *the probability that* $b' = b$ *is exactly* $1/2$.

Proof This lemma follows from the fact that the signature σ^* in G_5 is independent from the choice of i_b. □

In all, we have that the group signature Π_{gs} is fully anonymous in the random oracle model. □

7.4.6 Full Traceability

Theorem 7.10 (Full Traceability) *Under the SIS assumption, the group signature* Π_{gs} *is fully traceable in the random oracle model.*

Proof Assume that there is an adversary $\mathcal{A}$ that breaks the full traceability of Π_{gs}, we construct an algorithm $\mathcal{B}$ that breaks the SIS assumption. Formally, $\mathcal{B}$ is given a SIS matrix $\hat{\mathbf{A}} \in \mathbb{Z}_q^m$ and tries to find a solution $\hat{\mathbf{x}} \in \mathbb{Z}_q^m$ such that $\|\hat{\mathbf{x}}\| \leq poly(m)$ and $\hat{\mathbf{A}}\hat{\mathbf{x}} = \mathbf{0}$.

Setup. The algorithm $\mathcal{B}$ first sets $\mathbf{A}_{1,1} = \hat{\mathbf{A}}$, and computes $(\mathbf{A}_{2,2}, \mathbf{T}_{\mathbf{A}_{2,2}}) \leftarrow \mathsf{BasisGen}(1^n, 1^m, q)$. Randomly choose $\mathbf{R}_1, \mathbf{R}_2 \xleftarrow{\$} \{-1, 1\}^{m\times m}$ and $j^* \xleftarrow{\$} \{-4m^{2.5}N + 1, \ldots, 4m^{2.5}N - 1\}$, and compute $\mathbf{A}_{1,2} = \mathbf{A}_{1,1}\mathbf{R}_1$, $\mathbf{A}_{2,1} = \mathbf{A}_{1,1}\mathbf{R}_2 - j^*\mathbf{A}_{2,2}$. Let $\mathbf{A}_1 = (\mathbf{A}_{1,1}\|\mathbf{A}_{1,2}) \in \mathbb{Z}_q^{n\times 2m}$. Randomly choose $\mathbf{B}_{2,1} \xleftarrow{\$} \mathbb{Z}_q^{n\times m}$, and compute $(\mathbf{B}_1, \mathbf{T}_{\mathbf{B}_1}) \leftarrow \mathsf{SuperSamp}(\mathbf{A}_{1,1}, -\mathbf{A}_{1,2}\mathbf{B}_{2,1}^T, q)$. Then, compute $(\mathbf{B}_{2,2}, \mathbf{T}_{\mathbf{B}_{2,2}}) \leftarrow \mathsf{SuperSamp}(\mathbf{A}_{1,2}, \mathbf{0}, q)$. Give the group public key $gpk =$

$\{\mathbf{A}_{1,1}, \mathbf{A}_{1,2}, \mathbf{A}_{2,1}, \mathbf{A}_{2,2}, \mathbf{B}_1, \mathbf{B}_{2,1}, \mathbf{B}_{2,2}\}$, the group manager's secret key $gmsk = \mathbf{T}_{\mathbf{B}_1}$ to the adversary $\mathcal{A}$.

Secret Key Queries. Upon receiving the secret key query for user j from $\mathcal{A}$, $\mathcal{B}$ aborts if $j = j^*$ or $j \notin \{1, \ldots, N\}$. Otherwise, define $\bar{\mathbf{A}}_j = (\mathbf{A}_1 \| \mathbf{A}_{2,1} + j\mathbf{A}_{2,2}) = (\mathbf{A}_1 \| \mathbf{A}_{2,1}\mathbf{R}_2 + (j - j^*)\mathbf{A}_{2,2})$, return the secret key $\mathbf{T}_{\bar{\mathbf{A}}_j} \leftarrow \mathsf{ExtBasisRight}(\bar{\mathbf{A}}_j, \mathbf{R}_2, \mathbf{T}_{\mathbf{A}_{2,2}}, s)$ to $\mathcal{A}$.

Sign Queries. Upon receiving a signing query for message M under user j from $\mathcal{A}$, $\mathcal{B}$ returns $\perp$ if $j \notin \{1, \ldots, N\}$. Else if $j = j^*$, $\mathcal{B}$ generates a signature on M using the NIZKPoK simulators for π_1 and π_2. Otherwise, it generates the signature by first extracting the corresponding key as in answering the secret key queries.

Forge. Upon receiving a forged valid signature $\sigma = (svk, \sigma_1 = (\mathbf{c}, \mathbf{x}_2, \pi_1, \pi_2), \sigma_2)$ with probability ϵ, $\mathcal{B}$ extracts the knowledge $\hat{\mathbf{e}}$, $\mathbf{x}_1$ and $\mathbf{v}_j$ with normal at most $4\eta m^2$ by programming the random oracle twice to generate two different "challenges". By the forking lemma of [8], $\mathcal{B}$ can succeed with probability at least $\epsilon(\epsilon/q_h - 2^{-t})$, where h is the maximum number of hash queries of $\mathcal{A}$. Then, $\mathcal{B}$ decrypts $\mathbf{c}$ to obtain $\hat{\mathbf{e}}'$, $\mathbf{x}_1'$, and distinguishes the following two cases:

- If $(\mathbf{x}_1', \hat{\mathbf{e}}') \neq (\mathbf{x}_1, \hat{\mathbf{e}})$, we have $\mathbf{A}_1\mathbf{c} = p\mathbf{A}_1\hat{\mathbf{e}}' + \mathbf{A}_1\mathbf{x}_1' = p\mathbf{A}_1\hat{\mathbf{e}} + \mathbf{A}_1\mathbf{x}_1$. Let $\mathbf{y} = (\mathbf{y}_1; \mathbf{y}_2) = p(\hat{\mathbf{e}} - \hat{\mathbf{e}}') + (\mathbf{x}_1 - \mathbf{x}_1')$, we have $\mathbf{0} = \mathbf{A}_1\mathbf{y} = \mathbf{A}_{1,1}\mathbf{y}_1 + \mathbf{A}_{1,2}\mathbf{y}_2 = \mathbf{A}_{1,1}(\mathbf{y}_1 + \mathbf{R}_1\mathbf{y}_2)$. Namely, $\hat{\mathbf{x}} = \mathbf{y}_1 + \mathbf{R}_1\mathbf{y}_2$ is a solution of the SIS problem, $\mathcal{B}$ returns $\hat{\mathbf{x}}$ as its own solution. Note that in this case, we have $\|\hat{\mathbf{x}}\| \leq (p+1)\eta m^{2.5} \cdot \omega(\sqrt{\log m}) = m^{8.5}\omega(\log^{3.5} m)$.
- Else if $(\mathbf{x}_1', \hat{\mathbf{e}}') = (\mathbf{x}_1, \hat{\mathbf{e}})$, we have $\mathbf{A}_1\mathbf{x}_1 + \mathbf{A}_{2,1}\mathbf{x}_2 + j\mathbf{A}_{2,2}\mathbf{x}_2 = \mathbf{0}$ according to equation (7.2) of π_2, where $j = \sum_{i=0}^{\ell-1} v_i\eta^i$ and $\mathbf{v}_j = (v_0, \ldots, v_{\ell-1})$. A simple calculation indicates that $|j| < 4m^{2.5}N < q$. Since $\mathbf{A}_{2,2}\mathbf{x}_2 \neq \mathbf{0}$ and q is a prime, the open algorithm will always output j, namely, it will never output $\perp$. In addition, if $j \neq j^*$, $\mathcal{B}$ aborts. Otherwise, $\mathcal{B}$ returns $\hat{\mathbf{x}} = \mathbf{x}_{1,1} + \mathbf{R}_1\mathbf{x}_{1,2} + \mathbf{R}_2\mathbf{x}_2$ as its own solution, where $\mathbf{x}_1 = (\mathbf{x}_{1,1}; \mathbf{x}_{1,2})$.

Since j^* is randomly chosen from $\{-4m^{2.5}N + 1, \ldots, 4m^{2.5}N - 1\}$, we have the probability that $j^* = j$ is at least $\frac{1}{8m^{2.5}N}$. Conditioned on $j^* = j$, we have $\mathbf{A}_{1,1}\mathbf{x}_{1,1} + \mathbf{A}_{1,2}\mathbf{x}_{1,2} + \mathbf{A}_{2,1}\mathbf{x}_2 + j\mathbf{A}_{2,2}\mathbf{x}_2 = \mathbf{A}_{1,1}(\mathbf{x}_{1,1} + \mathbf{R}_1\mathbf{x}_{1,2} + \mathbf{R}_2\mathbf{x}_2) = \mathbf{0}$, which shows that $\hat{\mathbf{x}} = \mathbf{x}_{1,1} + \mathbf{R}_1\mathbf{x}_{1,2} + \mathbf{R}_2\mathbf{x}_2$ is a solution of the SIS problem. In particular, we have that $\|\hat{\mathbf{x}}\| \leq \eta m^{2.5} \cdot \omega(\sqrt{\log m}) = m^{4.5} \cdot \omega(\log^2 m)$ by [2, Lem. 5].

In all, the probability that $\mathcal{B}$ solves the SIS problem is at least $\frac{\epsilon(\epsilon/q_h - 2^{-t})}{8m^{2.5}N}$, which is non-negligible if ϵ is non-negligible. Moreover, the norm of $\hat{\mathbf{x}}$ is polynomial in m. □

7.5 Background and Further Reading

It is well-known that digital signature schemes [44] can be constructed from general assumptions, such as one-way functions. Nevertheless, these generic signature schemes suffer from either large signatures or large verification keys, thus a main open problem is to reduce the signature size as well as the verification key size. The first direct constructions of lattice-based signature schemes were given in [34, 53]. Later, many works (e.g., [6, 30, 52]) significantly improved the efficiency of lattice-based signature schemes in the random oracle model. In comparison, the progress in constructing efficient lattice-based signature schemes in the standard model was relatively slow. At Eurocrypt 2010, Cash et al. [26] proposed a signature scheme with a linear number of vectors in the signatures. The first standard model short signature scheme with signatures consisting of a single lattice vector was due to Boyen [16], which was later improved by Micciancio and Peikert [55]. However, the verification keys of both schemes in [16, 55] consist of a linear number of matrices.

In 2013, Böhl et al. [11] constructed a lattice-based signature scheme with constant verification keys by introducing the confined guessing proof technique. Later, Ducas and Micciancio [31] adapted the confined guessing proof technique to ideal lattices, and proposed a short signature scheme with logarithmic verification keys. Recently, Alperin-Sheriff [4] constructed a short signature with constant verification keys based on a stronger hardness assumption by using the idea of homomorphic trapdoor functions [36]. Due to the use of the confined guessing technique, the above three signature schemes [4, 11, 31] shared two undesired byproducts. First, the security can only be directly proven to be existentially unforgeable against non-adaptive chosen message attacks (EUF-naCMA). Even if an EUF-naCMA secure scheme can be transformed into an EUF-CMA secure one by using known techniques such as chameleon hash functions [45], in the lattice setting [31] this usually introduces an additional tag to each signature and roughly increases the signature size by twice. Second, a reduction loss about $(Q^2/\epsilon)^c$ for some parameter $c > 1$ seems unavoidable, where Q is the number of signing queries of the forger $\mathcal{F}$, and ϵ is the success probability of $\mathcal{F}$. Therefore, it is desirable to directly construct an EUF-CMA secure scheme that has short signatures, short verification keys, as well as a relatively tight security proof.

Since their introduction by Chaum and van Heyst [27], group signatures have attracted much attention from the research community. Bellare, Micciancio and Warinschi (BMW) [7] formalized the security of group signatures for static groups (where the group members are fixed in the system setup phase) in two main notions, i.e., *full anonymity* and *full traceability*. Informally, *full anonymity* requires that an adversary without the group manager secret key should not be able to determine the signer's identity from a signature, even if it can access an open oracle that returns the identity of any other (valid) signature. And *full traceability* implies that no collusion of group members can create a valid signature which cannot be traced back to one of them (by the group manager using the group manager secret key). Bellare et al. [7] also gave a theoretical construction based on the existence of trapdoor permutations. In a weak variant of the BMW model where the adversary against anonymity is not given access to the open oracle (i.e., CPA-anonymity), Boneh et al. [13] constructed a short group signature scheme based on the strong Diffie-Hellman (SDH) [12] and decision linear (DLIN) [13] assumptions in the random oracle model [9]. Besides, many papers focused on designing various group signatures based on different assumptions [1, 5, 17, 18, 23, 38, 39, 48, 49].

In 2010, Gordon et al. [37] made the first step in constructing secure group signatures from lattices. They elegantly combined several powerful lattice-based tools [34, 56, 59] to build a group signature scheme where the sizes of both the group public key and signature was linear in the maximum number N of group members. Later, Camenisch el al. [24] proposed a variant of [37] with improvements both in efficiency (i.e., shorter group public key) and security (i.e., stronger adversary against anonymity), but the signature size of their scheme was still linear in N. Recently, two papers [46, 47] have significantly decreased the signature size. By first representing the identity of group members as a bit-string [16], and then applying the "encrypt-and-prove" paradigm of [7, 37], Laguillaumie et al. [46] constructed an efficient lattice-based group signature where both the sizes of the group public key and the signature are proportional to $\log N$ (i.e., with bit-length slightly greater than $O(n^2 \log^2 n \log N)$ and $O(n \log^3 n \log N)$, respectively). Using similar identity representations together with the non-interactive zero-knowledge (NIZK) proof in [50], Langlois et al. [47] proposed a nice scheme without encryption, which achieves almost the same asymptotical efficiency as that of [46], and provides an additional property called verifier-local revocation [15]. Another interesting group signature is due to Benhamouda et al. [10], for which privacy holds under a lattice-based assumption but the security is discrete-logarithm-based, i.e., it is not a pure lattice-based group signature.

References

1. Abe, M., Fuchsbauer, G., Groth, J., Haralambiev, K., Ohkubo, M.: Structure-preserving signatures and commitments to group elements. In: Rabin, T. (ed.) Advances in Cryptology - CRYPTO 2010. Lecture Notes in Computer Science, vol. 6223, pp. 209–236. Springer, Heidelberg (2010)
2. Agrawal, S., Boneh, D., Boyen, X.: Efficient lattice (H)IBE in the standard model. In: Gilbert, H. (ed.) Advances in Cryptology - EUROCRYPT 2010. Lecture Notes in Computer Science, vol. 6110, pp. 553–572. Springer, Heidelberg (2010)
3. Agrawal, S., Freeman, D., Vaikuntanathan, V.: Functional encryption for inner product predicates from learning with errors. In: Lee, D., Wang, X. (eds.) Advances in Cryptology - ASIACRYPT 2011. Lecture Notes in Computer Science, vol. 7073, pp. 21–40. Springer, Heidelberg (2011)
4. Alperin-Sheriff, J.: Short signatures with short public keys from homomorphic trapdoor functions. In: Katz, J. (ed.) Public-Key Cryptography - PKC 2015. Lecture Notes in Computer Science, vol. 9020, pp. 236–255. Springer, Heidelberg (2015)
5. Ateniese, G., Camenisch, J., Joye, M., Tsudik, G.: A practical and provably secure coalition-resistant group signature scheme. In: Bellare, M. (ed.) Advances in Cryptology - CRYPTO 2000. Lecture Notes in Computer Science, vol. 1880, pp. 255–270. Springer, Heidelberg (2000)
6. Bai, S., Galbraith, S.: An improved compression technique for signatures based on learning with errors. In: Benaloh, J. (ed.) Topics in Cryptology – CT-RSA 2014. Lecture Notes in Computer Science, vol. 8366, pp. 28–47. Springer International Publishing (2014)
7. Bellare, M., Micciancio, D., Warinschi, B.: Foundations of group signatures: formal definitions, simplified requirements, and a construction based on general assumptions. In: Biham, E. (ed.) Advances in Cryptology - EUROCRYPT 2003. Lecture Notes in Computer Science, vol. 2656, pp. 614–629. Springer, Heidelberg (2003)

8. Bellare, M., Neven, G.: Multi-signatures in the plain public-key model and a general forking lemma. In: Proceedings of the 13th ACM Conference on Computer and Communications Security, CCS '06, pp. 390–399. ACM (2006)
9. Bellare, M., Rogaway, P.: Random oracles are practical: a paradigm for designing efficient protocols. In: Proceedings of the 1st ACM Conference on Computer and Communications Security, CCS '93, pp. 62–73. ACM (1993). https://doi.org/10.1145/168588.168596, http://doi.acm.org/10.1145/168588.168596
10. Benhamouda, F., Camenisch, J., Krenn, S., Lyubashevsky, V., Neven, G.: Better zero-knowledge proofs for lattice encryption and their application to group signatures. In: Sarkar, P., Iwata, T. (eds.) Advances in Cryptology - ASIACRYPT 2014. Lecture Notes in Computer Science, vol. 8873, pp. 551–572. Springer, Heidelberg (2014)
11. Böhl, F., Hofheinz, D., Jager, T., Koch, J., Seo, J., Striecks, C.: Practical signatures from standard assumptions. In: Johansson, T., Nguyen, P. (eds.) Advances in Cryptology - EUROCRYPT 2013. Lecture Notes in Computer Science, vol. 7881, pp. 461–485. Springer, Heidelberg (2013)
12. Boneh, D., Boyen, X.: Short signatures without random oracles. In: Cachin, C., Camenisch, J. (eds.) Advances in Cryptology - EUROCRYPT 2004. Lecture Notes in Computer Science, vol. 3027, pp. 56–73. Springer, Heidelberg (2004)
13. Boneh, D., Boyen, X., Shacham, H.: Short group signatures. In: Franklin, M. (ed.) Advances in Cryptology - CRYPTO 2004. Lecture Notes in Computer Science, vol. 3152, pp. 41–55. Springer, Heidelberg (2004)
14. Boneh, D., Nikolaenko, V., Segev, G.: Attribute-based encryption for arithmetic circuits. Cryptology ePrint Archive, Report 2013/669 (2013)
15. Boneh, D., Shacham, H.: Group signatures with verifier-local revocation. In: Proceedings of the 11th ACM Conference on Computer and Communications Security, CCS '04, pp. 168–177. ACM (2004)
16. Boyen, X.: Lattice mixing and vanishing trapdoors: a framework for fully secure short signatures and more. In: Nguyen, P., Pointcheval, D. (eds.) Public Key Cryptography - PKC 2010. Lecture Notes in Computer Science, vol. 6056, pp. 499–517. Springer, Heidelberg (2010)
17. Boyen, X., Waters, B.: Compact group signatures without random oracles. In: Vaudenay, S. (ed.) Advances in Cryptology - EUROCRYPT 2006. Lecture Notes in Computer Science, vol. 4004, pp. 427–444. Springer, Heidelberg (2006)
18. Boyen, X., Waters, B.: Full-domain subgroup hiding and constant-size group signatures. In: Okamoto, T., Wang, X. (eds.) Public Key Cryptography - PKC 2007. Lecture Notes in Computer Science, vol. 4450, pp. 1–15. Springer, Heidelberg (2007)
19. Brakerski, Z., Gentry, C., Vaikuntanathan, V.: Fully homomorphic encryption without bootstrapping. Innovations in Theoretical Computer Science, ITCS, pp. 309–325 (2012)
20. Brakerski, Z., Vaikuntanathan, V.: Efficient fully homomorphic encryption from (standard) LWE. In: 2011 IEEE 52nd Annual Symposium on Foundations of Computer Science (FOCS), pp. 97 –106 (2011). https://doi.org/10.1109/FOCS.2011.12
21. Brakerski, Z., Vaikuntanathan, V.: Lattice-based FHE as secure as PKE. In: Proceedings of the 5th Conference on Innovations in Theoretical Computer Science, ITCS '14, pp. 1–12. ACM (2014)
22. Brickell, E., Camenisch, J., Chen, L.: Direct anonymous attestation. In: Proceedings of the 11th ACM Conference on Computer and Communications Security, CCS '04, pp. 132–145. ACM (2004). https://doi.org/10.1145/1030083.1030103, http://doi.acm.org/10.1145/1030083.1030103
23. Camenisch, J., Lysyanskaya, A.: Signature schemes and anonymous credentials from bilinear maps. In: Franklin, M. (ed.) Advances in Cryptology - CRYPTO 2004. Lecture Notes in Computer Science, vol. 3152, pp. 56–72. Springer, Heidelberg (2004)
24. Camenisch, J., Neven, G., Rückert, M.: Fully anonymous attribute tokens from lattices. In: Visconti, I., Prisco, R. (eds.) Security and Cryptography for Networks. Lecture Notes in Computer Science, vol. 7485, pp. 57–75. Springer, Heidelberg (2012)
25. Canetti, R., Halevi, S., Katz, J.: Chosen-ciphertext security from identity-based encryption. In: Cachin, C., Camenisch, J. (eds.) Advances in Cryptology - EUROCRYPT 2004. Lecture Notes in Computer Science, vol. 3027, pp. 207–222. Springer, Berlin/Heidelberg (2004)

26. Cash, D., Hofheinz, D., Kiltz, E., Peikert, C.: Bonsai trees, or how to delegate a lattice basis. In: Gilbert, H. (ed.) Advances in Cryptology - EUROCRYPT 2010. Lecture Notes in Computer Science, vol. 6110, pp. 523–552. Springer, Berlin/Heidelberg (2010)
27. Chaum, D., van Heyst, E.: Group signatures. In: Davies, D. (ed.) Advances in Cryptology - EUROCRYPT '91. Lecture Notes in Computer Science, vol. 547, pp. 257–265. Springer, Heidelberg (1991)
28. Chen, L., Li, J.: Flexible and scalable digital signatures in TPM 2.0. In: Proceedings of the 2013 ACM SIGSAC Conference on Computer and Communications Security, CCS '13, pp. 37–48. ACM (2013)
29. Diffie, W., Hellman, M.: New directions in cryptography. IEEE Trans. Inf. Theory **22**(6), 644–654 (1976)
30. Ducas, L., Durmus, A., Lepoint, T., Lyubashevsky, V.: Lattice signatures and bimodal gaussians. In: Canetti, R., Garay, J. (eds.) Advances in Cryptology - CRYPTO 2013. Lecture Notes in Computer Science, vol. 8042, pp. 40–56. Springer, Heidelberg (2013)
31. Ducas, L., Micciancio, D.: Improved short lattice signatures in the standard model. In: Garay, J., Gennaro, R. (eds.) Advances in Cryptology - CRYPTO 2014. Lecture Notes in Computer Science, vol. 8616, pp. 335–352. Springer, Heidelberg (2014)
32. Fiat, A., Shamir, A.: How to prove yourself: practical solutions to identification and signature problems. In: Odlyzko, A. (ed.) Advances in Cryptology - CRYPTO '86. Lecture Notes in Computer Science, vol. 263, pp. 186–194. Springer, Heidelberg (1987)
33. Gentry, C., Gorbunov, S., Halevi, S., Vaikuntanathan, V., Vinayagamurthy, D.: How to compress (reusable) garbled circuits. Cryptology ePrint Archive, Report 2013/687 (2013)
34. Gentry, C., Peikert, C., Vaikuntanathan, V.: Trapdoors for hard lattices and new cryptographic constructions. In: Proceedings of the 40th annual ACM symposium on Theory of computing, STOC '08, pp. 197–206. ACM (2008)
35. Goldwasser, S., Micali, S., Rivest, R.L.: A digital signature scheme secure against adaptive chosen-message attacks. SIAM J. Comput. **17**(2), 281–308 (1988)
36. Gorbunov, S., Vaikuntanathan, V., Wichs, D.: Leveled fully homomorphic signatures from standard lattices. In: STOC 2015, pp. 469–477. ACM (2015). https://doi.org/10.1145/2746539.2746576
37. Gordon, S., Katz, J., Vaikuntanathan, V.: A group signature scheme from lattice assumptions. In: Abe, M. (ed.) Advances in Cryptology - ASIACRYPT 2010. Lecture Notes in Computer Science, vol. 6477, pp. 395–412. Springer, Heidelberg (2010)
38. Groth, J.: Simulation-sound NIZK proofs for a practical language and constant size group signatures. In: Lai, X., Chen, K. (eds.) Advances in Cryptology - ASIACRYPT 2006. Lecture Notes in Computer Science, vol. 4284, pp. 444–459. Springer, Heidelberg (2006)
39. Groth, J.: Fully anonymous group signatures without random oracles. In: Kurosawa, K. (ed.) Advances in Cryptology - ASIACRYPT 2007. Lecture Notes in Computer Science, vol. 4833, pp. 164–180. Springer, Heidelberg (2007)
40. Group, T.C.: TCG TPM specification 1.2. http://www.trustedcomputinggroup.org (2003)
41. Group, T.C.: TCG TPM specification 2.0. http://www.trustedcomputinggroup.org/resources/tpm_library_specification (2013)
42. Group, VSC Project, I.P.W.: Dedicated short range communications (DSRC) (2003)
43. Hofheinz, D., Kiltz, E.: Programmable hash functions and their applications. In: Wagner, D. (ed.) Advances in Cryptology - CRYPTO 2008. Lecture Notes in Computer Science, vol. 5157, pp. 21–38. Springer, Heidelberg (2008)
44. Katz, J.: Digital Signatures. Springer (2010)
45. Krawczyk, H., Rabin, T.: Chameleon signatures. In: Proceedings of NDSS 2000. Internet Society (2000)
46. Laguillaumie, F., Langlois, A., Libert, B., Stehlé, D.: Lattice-based group signatures with logarithmic signature size. In: Sako, K., Sarkar, P. (eds.) Advances in Cryptology - ASIACRYPT 2013. Lecture Notes in Computer Science, vol. 8270, pp. 41–61. Springer, Heidelberg (2013)
47. Langlois, A., Ling, S., Nguyen, K., Wang, H.: Lattice-based group signature scheme with verifier-local revocation. In: Krawczyk, H. (ed.) Public-Key Cryptography - PKC 2014. Lecture Notes in Computer Science, vol. 8383, pp. 345–361. Springer, Heidelberg (2014)

48. Libert, B., Peters, T., Yung, M.: Group signatures with almost-for-free revocation. In: Safavi-Naini, R., Canetti, R. (eds.) Advances in Cryptology - CRYPTO 2012. Lecture Notes in Computer Science, vol. 7417, pp. 571–589. Springer, Heidelberg (2012)
49. Libert, B., Peters, T., Yung, M.: Scalable group signatures with revocation. In: Pointcheval, D., Johansson, T. (eds.) Advances in Cryptology - EUROCRYPT 2012. Lecture Notes in Computer Science, vol. 7237, pp. 609–627. Springer, Heidelberg (2012)
50. Ling, S., Nguyen, K., Stehlé, D., Wang, H.: Improved zero-knowledge proofs of knowledge for the ISIS problem, and applications. In: Kurosawa, K., Hanaoka, G. (eds.) Public-Key Cryptography - PKC 2013. Lecture Notes in Computer Science, vol. 7778, pp. 107–124. Springer, Heidelberg (2013)
51. Lyubashevsky, V.: Lattice-based identification schemes secure under active attacks. In: Cramer, R. (ed.) Public Key Cryptography - PKC 2008. Lecture Notes in Computer Science, vol. 4939, pp. 162–179. Springer, Heidelberg (2008)
52. Lyubashevsky, V.: Lattice signatures without trapdoors. In: Pointcheval, D., Johansson, T. (eds.) Advances in Cryptology - EUROCRYPT 2012. Lecture Notes in Computer Science, vol. 7237, pp. 738–755. Springer, Heidelberg (2012)
53. Lyubashevsky, V., Micciancio, D.: Asymptotically efficient lattice-based digital signatures. In: Canetti, R. (ed.) Theory of Cryptography. Lecture Notes in Computer Science, vol. 4948, pp. 37–54. Springer, Heidelberg (2008)
54. Micciancio, D., Mol, P.: Pseudorandom knapsacks and the sample complexity of LWE search-to-decision reductions. In: Rogaway, P. (ed.) Advances in Cryptology - CRYPTO 2011. Lecture Notes in Computer Science, vol. 6841, pp. 465–484. Springer, Heidelberg (2011)
55. Micciancio, D., Peikert, C.: Trapdoors for lattices: simpler, tighter, faster, smaller. In: Pointcheval, D., Johansson, T. (eds.) Advances in Cryptology - EUROCRYPT 2012. Lecture Notes in Computer Science, vol. 7237, pp. 700–718. Springer, Heidelberg (2012)
56. Micciancio, D., Vadhan, S.: Statistical zero-knowledge proofs with efficient provers: lattice problems and more. In: Boneh, D. (ed.) Advances in Cryptology - CRYPTO 2003. Lecture Notes in Computer Science, vol. 2729, pp. 282–298. Springer, Heidelberg (2003)
57. Nguyen, P., Zhang, J., Zhang, Z.: Simpler efficient group signatures from lattices. In: Katz, J. (ed.) Public-Key Cryptography – PKC 2015, pp. 401–426. Springer, Heidelberg (2015). https://doi.org/10.1007/978-3-662-46447-2_18
58. Peikert, C., Rosen, A.: Efficient collision-resistant hashing from worst-case assumptions on cyclic lattices. In: Halevi, S., Rabin, T. (eds.) Theory of Cryptography. Lecture Notes in Computer Science, vol. 3876, pp. 145–166. Springer, Berlin/Heidelberg (2006)
59. Regev, O.: On lattices, learning with errors, random linear codes, and cryptography. In: Proceedings of the Thirty-seventh Annual ACM Symposium on Theory of Computing, STOC '05, pp. 84–93. ACM (2005)
60. Rivest, R.L., Shamir, A., Adleman, L.: A method for obtaining digital signatures and public-key cryptosystems. Commun. ACM **21**(2), 120–126 (1978)
61. Zhang, J., Chen, Y., Zhang, Z.: Programmable hash functions from lattices: short signatures and IBEs with small key sizes. In: Robshaw, M., Katz, J. (eds.) Advances in Cryptology - CRYPTO 2016, pp. 303–332. Springer, Heidelberg (2016)

Printed in the United States
by Baker & Taylor Publisher Services